Cyclope(

of

Engineering

A General Reference Work on

STEAM BOILERS, PUMPS, ENGINES, AND TURBINES, GAS AND OIL ENGINES, AUTOMOBILES, MARINE AND LOCOMOTIVE WORK, HEATING AND VENTILATING, COMPRESSED AIR, REFRIGERATION, DYNAMOS, MOTORS, ELECTRIC WIRING, ELECTRIC LIGHTING, ELEVATORS, ETC.

Editor-in-Chief

LOUIS DERR, M. A., S. B.

PROFESSOR OF PHYSICS, MASSACHUSETTS INSTITUTE OF TECHNOLOGY

Assisted by a Staff of

CONSULTING ENGINEERS, TECHNICAL EXPERTS, AND DESIGNERS OF THE HIGHEST PROFESSIONAL STANDING

Illustrated with over Two Thousand Engravings

SEVEN VOLUMES

CHICAGO

AMERICAN TECHNICAL SOCIETY

1910

A LARGE REFRIGERATING MACHINE.
Frick Company.

Editor-in-Chief

LOUIS DERR, M. A., S. B.

Professor of Physics, Massachusetts Institute of Technology

Authors and Collaborators

LIONEL S. MARKS, S. B., M. M. E.

Assistant Professor of Mechanical Engineering, Harvard University
American Society of Mechanical Engineers

LLEWELLYN V. LUDY, M. E.

Professor of Mechanical Engineering, Purdue University
American Society of Mechanical Engineers

LUCIUS I. WIGHTMAN, E. E.

Electrical and Mechanical Engineer, Ingersoll-Rand Co., New York

FRANCIS B. CROCKER, E. M., Ph. D.

Head of Department of Electrical Engineering, Columbia University, New York
Past President, American Institute of Electrical Engineers

CHARLES L. GRIFFIN, S. B.

Assistant Engineer, the Solvay-Process Co.
American Society of Mechanical Engineers

VICTOR C. ALDERSON, D. Sc.

President, Colorado School of Mines
Formerly Dean, Armour Institute of Technology

WALTER S. LELAND, S. B.

Assistant Professor of Naval Architecture, Massachusetts Institute of Technology
American Society of Naval Architects and Marine Engineers

Authors and Collaborators—Continued

CHARLES L. HUBBARD, S. B., M. E.
Consulting Engineer on Heating, Ventilating, Lighting, and Power

ARTHUR L. RICE, M. M. E.
Editor, *The Practical Engineer*

WALTER B. SNOW, S. B.
Formerly Mechanical Engineer, B. F. Sturtevant Co.
American Society of Mechanical Engineers

HUGO DIEMER, M. E.
Professor of Mechanical Engineering, Pennsylvania State College
American Society of Mechanical Engineers

SAMUEL S. WYER, M. E.
American Society of Mechanical Engineers
Author of "Gas-Producers and Producer Gas"

WILLIAM G. SNOW, S. B.
Steam Heating Specialist
American Society of Mechanical Engineers

GLENN M. HOBBS, Ph. D.
Secretary, American School of Correspondence
Formerly Instructor in Physics, University of Chicago
American Physical Society

LOUIS DERR, M. A., S. B.
Professor of Physics, Massachusetts Institute of Technology

JOHN H. JALLINGS
Mechanical Engineer and Elevator Expert

HOWARD MONROE RAYMOND, B. S.
Dean of Engineering, and Professor of Physics, Armour Institute of Technology

Authors and Collaborators—Continued

WILLIAM T. McCLEMENT, A. M., D. Sc.

Head of Department of Botany, Queen's University, Kingston, Canada

GEORGE C. SHAAD, E. E.

Head of Department of Electrical Engineering, University of Kansas

GEORGE L. FOWLER, A. B., M. E.

Consulting Engineer
American Society of Mechanical Engineers

RALPH H. SWEETSER, S. B.

Superintendent, Columbus Iron & Steel Co.
American Institute of Mining Engineers

CHARLES E. KNOX, E. E.

Consulting Electrical Engineer
American Institute of Electrical Engineers

MILTON W. ARROWOOD

Graduate, United States Naval Academy
Refrigerating and Mechanical Engineer, with the Triumph Ice Machine Company

R. F. SCHUCHARDT, B. S.

Testing Engineer, Commonwealth Edison Co., Chicago

WILLIAM S. NEWELL, S. B.

With Bath Iron Works
Formerly Instructor, Massachusetts Institute of Technology

GEORGE F. GEBHARDT, M. E., M. A.

Professor of Mechanical Engineering, Armour Institute of Technology

HARRIS C. TROW, S. B., *Managing Editor*

Editor-in-Chief, Textbook Department, American School of Correspondence
American Institute of Electrical Engineers

Authorities Consulted

THE editors have freely consulted the standard technical literature of Europe and America in the preparation of these volumes. They desire to express their indebtedness, particularly to the following eminent authorities, whose well-known treatises should be in the library of every engineer.

Grateful acknowledgment is here made also for the invaluable co-operation of the foremost engineering firms, in making these volumes thoroughly representative of the best and latest practice in the design and construction of steam and electrical machines; also for the valuable drawings and data, suggestions, criticisms, and other courtesies.

JAMES AMBROSE MOYER, S. B., A. M.

Member of The American Society of Mechanical Engineers; American Institute of Electrical Engineers, etc.; Engineer, Westinghouse, Church, Kerr & Co.
Author of "The Steam Turbine," etc.

E. G. CONSTANTINE

Member of the Institution of Mechanical Engineers; Associate Member of the Institution of Civil Engineers.
Author of "Marine Engineers."

C. W. MACCORD, A. M.

Professor of Mechanical Drawing, Stevens Institute of Technology.
Author of "Movement of Slide Valves by Eccentrics."

CECIL H. PEABODY, S. B.

Professor of Marine Engineering and Naval Architecture, Massachusetts Institute of Technology.
Author of "Thermodynamics of the Steam Engine," "Tables of the Properties of Saturated Steam," "Valve Gears to Steam Engines," etc.

FRANCIS BACON CROCKER, M. E., Ph. D.

Head of Department of Electrical Engineering, Columbia University; Past President American Institute of Electrical Engineers.
Author of "Electric Lighting," "Practical Management of Dynamos and Motors."

SAMUEL S. WYER

Mechanical Engineer; American Society of Mechanical Engineers.
Author of "Treatise on Producer Gas and Gas-Producers," "Catechism on Producer Gas."

E. W. ROBERTS, M. E.

Member, American Society of Mechanical Engineers.
Author of "Gas-Engine Handbook," "Gas Engines and Their Troubles," "The Automobile Pocket-book," etc

Authorities Consulted—Continued

GARDNER D. HISCOX, M. E.

Author of "Compressed Air," "Gas, Gasoline, and Oil-Engines," "Mechanical Movements," "Horseless Vehicles, Automobiles, and Motor-Cycles," "Hydraulic Engineering," "Modern Steam Engineering," etc.

EDWARD F. MILLER

Professor of Steam Engineering, Massachusetts Institute of Technology.
Author of "Steam Boilers."

ROBERT M. NEILSON

Associate Member, Institution of Mechanical Engineers; Member of Cleveland Institution of Engineers; Chief of the Technical Department of Richardsons, Westgarth and Co., Ltd.
Author of "The Steam Turbine."

ROBERT WILSON

Author of "Treatise on Steam Boilers," "Boiler and Factory Chimneys," etc.

CHARLES PROTEUS STEINMETZ

Consulting Engineer, with the General Electric Co.; Professor of Electrical Engineering, Union College.
Author of "The Theory and Calculation of Alternating-Current Phenomena," "Theoretical Elements of Electrical Engineering," etc.

JAMES J. LAWLER

Author of "Modern Plumbing, Steam and Hot-Water Heating."

WILLIAM F. DURAND, Ph. D.

Professor of Marine Engineering, Cornell University.
Author of "Resistance and Propulsion of Ships," "Practical Marine Engineering."

HORATIO A. FOSTER

Member, American Institute of Electrical Engineers; American Society of Mechanical Engineers; Consulting Engineer.
Author of "Electrical Engineer's Pocket-book."

ROBERT GRIMSHAW, M. E.

Author of "Steam Engine Catechism," "Boiler Catechism," "Locomotive Catechism," "Engine Runners' Catechism," "Shop Kinks," etc.

SCHUYLER S. WHEELER, D. Sc.

Electrical Expert of the Board of Electrical Control, New York City; Member American Societies of Civil and Mechanical Engineers.
Author of "Practical Management of Dynamos and Motors."

J. A. EWING, C. B., LL. D., F. R. S.

Member, Institute of Civil Engineers; formerly Professor of Mechanism and Applied Mechanics in the University of Cambridge; Director of Naval Education.
Author of "The Mechanical Production of Cold," "The Steam Engine and Other Heat Engines."

LESTER G. FRENCH, S. B.

Mechanical Engineer.
Author of "Steam Turbines."

ROLLA C. CARPENTER, M. S., C. E., M. M. E.

Professor of Experimental Engineering, Cornell University; Member of American Society Heating and Ventilating Engineers; Member American Society Mechanical Engineers.
Author of "Heating and Ventilating Buildings."

J. E. SIEBEL

Director, Zymotechnic Institute, Chicago.
Author of "Compend of Mechanical Refrigeration."

WILLIAM KENT, M. E.

Consulting Engineer; Member of American Society of Mechanical Engineers, etc.
Author of "Strength of Materials," "Mechanical Engineer's Pocket-book," etc.

WILLIAM M. BARR

Member American Society of Mechanical Engineers.
Author of "Boilers and Furnaces," "Pumping Machinery," "Chimneys of Brick and Metal," etc.

WILLIAM RIPPER

Professor of Mechanical Engineering in the Sheffield Technical School; Member of the Institute of Mechanical Engineers.
Author of "Machine Drawing and Design," "Practical Chemistry," "Steam," etc.

J. FISHER-HINNEN

Late Chief of the Drawing Department at the Oerlikon Works.
Author of "Continuous Current Dynamos."

SYLVANUS P. THOMPSON, D. Sc., B. A., F. R. S., F. R. A. S.

Principal and Professor of Physics in the City and Guilds of London Technical College.
Author of "Electricity and Magnetism," "Dynamo-Electric Machinery," etc.

ROBERT H. THURSTON, C. E., Ph. B., A. M., LL. D.

Director of Sibley College, Cornell University.
Author of "Manual of the Steam Engine," "Manual of Steam Boilers," "History of the Steam Engine," etc.

Authorities Consulted—Continued

JOSEPH G. BRANCH, B. S., M. E.

Chief of the Department of Inspection, Boilers and Elevators; Member of the Board of Examining Engineers for the City of St. Louis.
Author of "Stationary Engineering," "Heat and Light from Municipal and Other Waste," etc.

JOSHUA ROSE, M. E.

Author of "Mechanical Drawing Self Taught," "Modern Steam Engineering," "Steam Boilers," "The Slide Valve," "Pattern Maker's Assistant," "Complete Machinist," etc.

CHARLES H. INNES, M. A.

Lecturer on Engineering at Rutherford College.
Author of "Air Compressors and Blowing Engines," "Problems in Machine Design," "Centrifugal Pumps, Turbines, and Water Motors," etc.

GEORGE C. V. HOLMES

Whitworth Scholar; Secretary of the Institute of Naval Architects, etc.
Author of "The Steam Engine."

FREDERIC REMSEN HUTTON, E. M., Ph. D.

Emeritus Professor of Medical Engineering in Columbia University; Past Secretary and President of American Society of Mechanical Engineers.
Author of "The Gas Engine," "Mechanical Engineering of Power Plants," etc.

MAURICE A. OUDIN, M. S.

Member of American Institute of Electrical Engineers.
Author of "Standard Polyphase Apparatus and Systems."

WILLIAM JOHN MACQUORN RANKINE, LL. D., F. R. S. S.

Civil Engineer; Late Regius Professor of Civil Engineering in University of Glasgow.
Author of "Applied Mechanics," "The Steam Engine," "Civil Engineering," "Useful Rules and Tables," "Machinery and Mill Work," "A Mechanical Textbook."

DUGALD C. JACKSON, C. E.

Head of Department of Electrical Engineering, Massachusetts Institute of Technology; Member of American Institute of Electrical Engineers.
Author of "A Textbook on Electro-Magnetism and the Construction of Dynamos," "Alternating Currents and Alternating-Current Machinery."

A. E. SEATON

Author of "A Manual of Marine Engineering."

WILLIAM C. UNWIN, F. R. S., M. Inst. C. E.

Professor of Civil and Mechanical Engineering, Central Technical College, City and Guilds of London Institute, etc.
Author of "Machine Design," "The Development and Transmission of Power," etc.

AN INGERSOLL ELECTRIC-AIR DRILL BRUSHING AN ENTRY IN A COAL MINE
A Small Motor and Air Compressor are Direct-Connected and Mounted on Trucks. In This Way, the Tool is Driven from a Compressor Nearby, and the Disadvantages of Long Distance Transmission of Air are Overcome.

Foreword

THE rapid advances made in recent years in all lines of engineering, as seen in the evolution of improved types of machinery, new mechanical processes and methods, and even new materials of workmanship, have created a distinct necessity for an authoritative work of general reference embodying the accumulated results of modern experience and the latest approved practice. The Cyclopedia of Engineering is designed to fill this acknowledged need.

¶ The aim of the publishers has been to create a work which, while adequate to meet all demands of the technically trained expert, will appeal equally to the self-taught practical man, who may have been denied the advantages of training at a resident technical school. The Cyclopedia not only covers the fundamentals that underlie all engineering, but places the reader in direct contact with the experience of teachers fresh from practical work, thus putting him abreast of the latest progress and furnishing him that adjustment to advanced modern needs and conditions which is a necessity even to the technical graduate.

¶ The Cyclopedia of Engineering is based upon the method which the American School of Correspondence has developed and successfully used for many years in teaching the principles and practice of Engineering in its different branches.

¶ The success which the American School of Correspondence has attained as a factor in the machinery of modern technical and scientific education is in itself the best possible guarantee for the present work. Therefore, while these volumes are a marked innovation in technical literature—representing, as they do, the best ideas and methods of a large number of different authors, each an acknowledged authority in his work—they are by no means an experiment, but are, in fact, based on what has proved itself to be the most successful method yet devised for the education of the busy man. The formulæ of the higher mathematics have been avoided as far as possible, and every care exercised to elucidate the text by abundant and appropriate illustrations.

¶ Numerous examples for practice are inserted at intervals; these, with the text questions, help the reader to fix in mind the essential points, thus combining the advantages of a text-book with those of a reference work.

¶ The Cyclopedia has been compiled with the idea of making it a work thoroughly technical yet easily comprehended by the man who has but little time in which to acquaint himself with the fundamental branches of practical engineering. If, therefore, it should benefit any of the large number of workers who need, yet lack, technical training, the publishers will feel that its mission has been accomplished.

¶ Grateful acknowledgment is due the corps of authors and collaborators—engineers and designers of wide practical experience, and teachers of well-recognized ability—without whose co-operation this work would have been impossible.

Table of Contents

VOLUME VI

---

* For page numbers, see foot of pages.

† For professional standing of authors, see list of Authors and Collaborators at front of volume.

350-TON REFRIGERATING MACHINE
Duplex Double-Acting Compressors. Driven by Cross-Compound Corliss Engine.
The Vilter Mfg. Co., Milwaukee, Wis.

COMPRESSED AIR

PART I

Engineering practice recognizes three primary subdivisions of any power system. The first of these is the *power producing* or *generating* element, which is always a converter of energy from some present form into another form more convenient for transmission and application to some specific purpose in hand. The second subdivision covers the *power transmitting* or *distributing* element, by means of which the newly converted energy is delivered to its point of application. The third division is the *power applying* or *utilizing* element, which is also a power converter, the delivered power being transformed into some form of useful work.

These logical subdivisions will be used in the present treatment of the subject of Compressed Air, which will be considered under the three general sections of *Production*, *Transmission*, and *Application*.

In the first element of a compressed-air power system, which consists of the compressor, the power of steam, electricity, or other energy is converted into the energy of compressed air. In the transmission system, this compressed air is carried to a point where it is to be used. In the third division, the compressed air is applied in mechanical devices, by means of which its power is converted into such work as driving machinery, drilling rock, channeling stone, chipping metal, driving rivets, pumping, operating switch and signal systems, and the other well-known applications of compressed air.

PRODUCTION OF COMPRESSED AIR

General Definitions. *Free air* is air at ordinary atmospheric pressure and temperature, whatever these may be. It shows no pressure as registered by the gauge, but has an absolute pressure due to its own weight. Thus free air is in reality compressed air unconfined within any limits. Throughout this treatise, "free air," unless otherwise specified, will be considered as air at an absolute pressure of 14.7 lbs. and at a temperature of 32° F.

Atmospheric pressure is the pressure in the free air due to its own weight, or the weight of all the air above it over a unit-area. Atmospheric pressure is variable, depending upon altitude or height above sea-level and upon temperature. At sea-level and at a temperature of 32° F., atmospheric pressure is 14.6963 lbs. per square inch, or, as usually written, 14.7 lbs. Atmospheric pressure grows less as altitudes increase, and increases as depths below sea-level increase.

An *atmosphere* is an arbitrary unit of pressure representing 14.7 lbs. by the gauge. Thus "a pressure of two atmospheres" means a pressure of 29.4 lbs., gauge. Free air always has one atmosphere of absolute pressure, which must always be taken into account in computations involving absolute pressure. "Two atmospheres" gauge pressure is thus 29.4 lbs. gauge, or 44.1 lbs. absolute pressure.

Absolute pressure is the pressure registered by the gauge, *plus* the atmospheric pressure. Absolute pressure is therefore just as variable a quantity as atmospheric pressure, and is affected by the same conditions. For all practical purposes, however, it is usually based on the pressure of air at sea-level and at 32° F., or 14.7 lbs. Thus any absolute pressure is found by adding 14.7 lbs. to the gauge pressure.

Absolute temperature is the temperature as indicated by the Fahrenheit scale + 461°. Thus, at 80° F., the absolute temperature is 80 + 461, or 541° absolute; and at −20° F., the absolute temperature is −20 + 461, or 441° absolute. On the Centigrade scale, the value 461 is to be replaced by 273. Thus 40° C. represents an absolute temperature of 40 + 273, or 313° C. absolute. The Fahrenheit scale, however, will be used throughout the present discussion.

Composition of Air. Air is a mechanical mixture composed principally of two gases—*oxygen* and *nitrogen.* The proportions of the two by weight are 23 parts of oxygen to 77 parts of nitrogen. The proportions by volume are 20.7 parts of oxygen to 79.3 parts of nitrogen. By *mechanical mixture* is meant that these two gases are not chemically united in air, as are the oxygen and hydrogen which combine in water with the symbol H_2O. Air has no chemical symbol.

Air is never found absolutely pure. It may contain other gaseous substances in varying and minute quantities, among which the most common are carbonic acid gas and vapor of water. The relative proportions of oxygen and nitrogen are also known to vary within narrow limits under different conditions. The proportion of vapor of water contained in air varies with the temperature and with the pressure or density of the air.

Some Physical Properties of Air. A *perfect gas* is one which cannot be liquefied. Repeated experiments have practically demonstrated the fact that all gases can be reduced to liquid form under suitable conditions of pressure and temperature, and it may therefore be said that there is no perfect gas; but since nearly all gases are liquefied only with great difficulty and under the greatest extremes of temperature and pressure, most gases may for ordinary purposes be considered as perfect gases.

The gases composing air are in the latter class; and air itself, therefore, for all practical purposes, may be looked upon as a perfect gas. In fact, the behavior of air is found in practice to conform so closely with the laws of perfect gases that the discrepancies may be ignored in ordinary calculations such as those involved in the present treatment of the subject.

Table I, from Richards' "Compressed Air," lists the weight of air per cubic foot at various temperatures Fahrenheit; also the volume of one pound of air at the same temperatures.

The weight of a cubic foot of dry air at any atmospheric pressure and at any temperature Fahrenheit, may be found by dividing the constant 39.819 by the absolute temperature (temperature Fahrenheit + 461.) The volume of one pound of dry air at atmospheric pressure and any temperature, is found by dividing the absolute temperature by the constant 39.819. Thus,

Let T = Temperature, Fahrenheit.
V = Volume, in cubic feet.
W = Weight, in pounds.

Then,

$$W = 39.819 \div (T + 461);$$
$$V = (T + 461) \div 39.819.$$

Where the temperature and pressure of the air are both variable quantities, the weight per cubic foot is found by multiplying the con-

TABLE I

Weights and Volumes of Dry Air at Atmospheric Pressure and at Various Temperatures

(Richards)

Temperature, Degrees Fahrenheit	Weight of One Cubic Foot in Pounds	Volume of One Pound in Cubic Feet
0	.0863	11.582
10	.0845	11.834
20	.0827	12.084
30	.0811	12.336
32	.0807	12.387
40	.0794	12.587
50	.0779	12.838
60	.0764	13.089
70	.0750	13.340
80	.0736	13.592
90	.0722	13.843
100	.0710	14.094
110	.0697	14.345
120	.0685	14.596
130	.0674	14.847
140	.0662	15.098
150	.0651	15.350
160	.0641	15.601
170	.0631	15.852
180	.0621	16.103
190	.0612	16.354
200	.0602	16.605
210	.0593	16.856
220	.0591	16.907

stant 2.7093 by the absolute pressure in pounds, and dividing by the absolute temperature. Under these conditions, the volume per pound is found by dividing the absolute temperature by the product of the absolute pressure and the constant 2.7093. Thus, taking the same symbols as above, with the addition of P = pressure in pounds, gauge

$$W = 2.7093 \times (P + \text{atmosphere}) \div (T + 461);$$
$$V = (T + 461) \div \left\{ (P + \text{atmosphere}) \times 2.7093 \right\}$$

Boyle's Law. The First Law of Gases is generally known as *Boyle's Law,* and expresses the relation between pressure and volume in a perfect gas. This law states that, *at a constant temperature, the volume of a gas varies inversely as its pressure*; or, *at a constant temperature, the product of the pressure and volume of a gas is a constant.*

Let P = Pounds pressure per square inch, absolute;
V = Volume in cubic feet.

Then,

$$PV = C.$$

This is the algebraic expression of Boyle's law. The value of the constant C is not the same for all gases, and indeed varies for the same gas under different temperatures.

One pound of air at 32° F. and atmospheric pressure, occupies 12.387 cubic feet. In this case, P = 14.7 lbs. absolute; and V = 12.387 cubic feet. The constant C, therefore, for air at 32° F. and atmospheric pressure, is 14.7 × 12.387, or 182.08.

Isothermal Compression. Fig. 1 is a graphic representation of the relations expressed by Boyle's law. Pressures are laid off on the vertical line OY, and volumes on the horizontal line OX. Any vertical line representing pressure will, with a horizontal line representing volume, enclose a rectangle, the area of which will be $P \times V$. According to Boyle's law, this product, and therefore this area, must always be a constant, and with air at 32° F. must be 182.08. If a series of such rectangles be laid out, a line joining their upper right-hand corners becomes a curve mathematically known as a *rectangular hyperbola.* In thermodynamics, this curve is known as the *Isothermal Curve,* or curve of compression at constant temperatures.

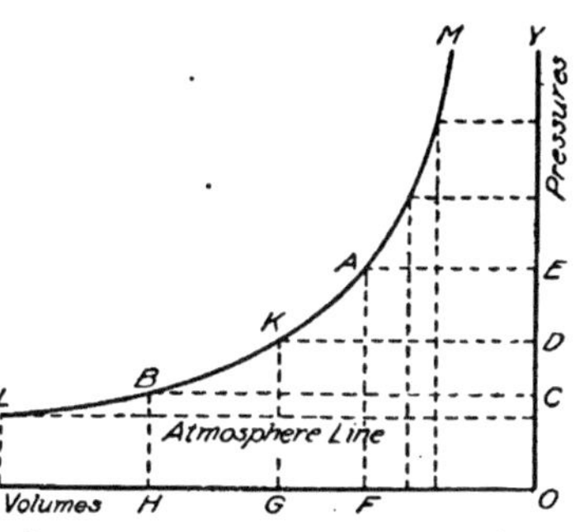

Fig. 1. Isothermal Curve Illustrating Boyle's Law of Pressures and Volumes.

Isothermal compression of air is compression without any change in the temperature of the air volume during compression.

Charles's Law. The Second Law of Gases, known as *Charles's Law,* defines the relations between pressure, volume, and temperature of a gas. This law states that, *at a constant pressure, the volume of a gas is proportional to its absolute temperature*; or, *at a constant volume, the pressure of a gas is proportional to its absolute temperature.* In other words, the product of pressure and volume in a gas is proportional to its absolute temperature.

Let P = Pounds pressure per square inch, absolute;
V = Volume in cubic feet;
T = Absolute temperature.

Then,

$$PV = KT; \text{ or, } PV \div T = K.$$

The value of the constant K, for air, is found as follows: In the discussion of Boyle's law, it was shown that $PV = C = 182.08$ for 32° F. In this case the absolute temperature is $32 + 461 = 493°$ absolute. Substituting these figures in the above equation, it becomes $182.08 \div 493 = K = .3693$.

Phenomena of Compression. *Thermodynamics* is the science of the relation between heat and energy. It is based on two fundamental laws, only the first of which has a bearing on the discussion of compressed air.

The First Law of Thermodynamics states that *where heat is converted into mechanical energy, or mechanical energy is converted into heat, the quantity of heat is exactly equivalent to the amount of mechanical energy.*

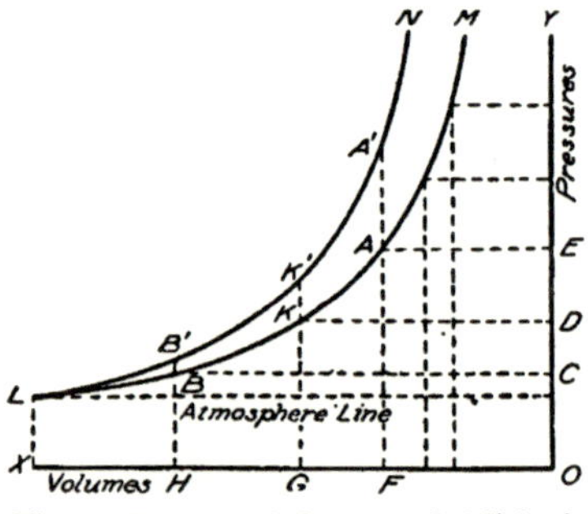

Fig. 2. Isothermal Curve and Adiabatic Curve Illustrating Charles's Law of Pressures, Volumes, and Temperatures.

This law demands that when work is done upon a volume of air in compressing it from a lower to a higher pressure, a quantity of heat must be developed exactly equivalent to the energy expended in compression.

In other words, when a volume of air is compressed to a higher pressure, all the work done upon that air volume is converted into heat; and that heat acts to increase the temperature of the air volume, whether the process of compression be slow or fast.

All modern air-compressors consist fundamentally of a cylinder with a moving piston; and the air is compressed by crowding it into a constantly diminishing space in the cylinder in advance of the moving piston. According to Boyle's law, as soon as the volume is diminished, the pressure must increase.

It has just been stated that the compression of air produces an increase in its temperature; and according to Charles's law, this increased temperature, acting on the volume of air in the compressing cylinder (where the volume at any instant is constant), produces an increase of pressure.

In the ordinary process of air-compression, therefore, two elements are at work toward the production of a higher pressure—*first,* the reduction of volume by the advancing piston; *second*, the increasing temperature due to the increasing pressure corresponding to the reduced volume.

Adiabatic Compression. This is compression without the removal or escape of any of the heat produced by the compression process.

It has just been shown that the pressure of an air volume during compression in a cylinder is at any moment the sum of two different pressures. The first pressure is that due to the reduction of volume; and since temperature does not enter into this increment of pressure, the line showing the course of this increasing pressure will be an isothermal curve. The second pressure is due to the increasing temperature, and is to be added, at any given moment or position of the piston, to the first pressure.

In Fig. 2, the curve $A\ K\ B$ is a curve of isothermal compression. The curve $A'\ K'\ B'$ is a curve of adiabatic compression, and is found by adding to the lines $F\ A$, $G\ K$, and $H\ B$, representing pressures due to decreased volume, the lines AA', KK', and BB', representing pressures due to the increase in temperature corresponding to the pressures FA, GK, and HB.

The formula for this curve is:

$$P_1V_1^N = P_2V_2^N = P_3V_3^N \text{, etc.,}$$

the exponent N providing for the heat of compression. The method by which the value of N is found has no place in this paper. It varies for different gases, and for air it is usually taken at 1.41. The equation of the adiabatic curve for air, therefore, is:

$$P_1V_1^{1.41} = P_2V_2^{1.41} = P_3V_3^{1.41} \text{, etc.}$$

P_1, P_2, P_3, etc., represent absolute pressures; V_1, V_2, V_3, etc., represent the corresponding volumes in adiabatic compression. This is simply another form of expressing Charles's law in its relation to air.

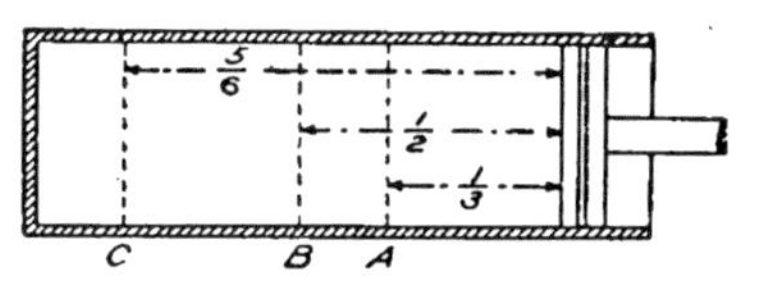

Fig. 3. Diagram Showing Effect of Heat of Compression

Pressures and Volumes in Compression. Assume a compressing cylinder (Fig. 3) with a cross-section of 1 square foot, and a piston travel or stroke of 3 feet. Let this be filled with air at atmospheric pressure of 14.7 lbs. absolute, and 60° F. temperature. If the piston be advanced to the position A, covering one-third of its stroke, the volume of air will be reduced one-third, and there will be three volumes crowded into the space of two, or the volume will be decreased in the ratio of 2 to 3; and since pressure varies inversely as the volume, according to Boyle's law, the pressure in the cylinder *due to the reduced volume* will be the original pressure increased in the ratio of 3 to 2, or $\frac{3}{2} \times 14.7 = 22.05$ lbs. absolute = $1\frac{1}{2}$ atmospheres = 7.35 lbs. gauge.

But a pressure gauge on the cylinder would at this time actually show a higher pressure than 7.35 lbs.; and the difference between this high pressure and the theoretically correct pressure of 7.35 lbs. is *due to the increase in temperature* produced in crowding three air volumes into the space of two.

Similarly, when the piston has advanced to B, or one-half stroke, the pressure should be 29.4 lbs. absolute, or 14.7 lbs. gauge; and at C, or $\frac{5}{6}$ stroke, the pressure should be 88.2 lbs. absolute, or 6 atmospheres, or 73.5 lbs. gauge. But it is actually found to be more than these figures in all cases, owing to the added pressure produced by the increased temperature.

If the cylinder were made of a non-heat-conducting material so that not a particle of the heat of compression could escape, the relations of pressures, volumes, and temperatures at any point of the stroke would be expressed by the equation for Charles's law, $PV = KT$. On the other hand, if it were possible to withdraw the heat of compression as fast as it was produced, the indicated gauge pressures would have been exactly as they theoretically should be, for compression would then follow the expression of Boyle's law, $PV = C$, the pressures varying exactly in inverse ratio with the volumes.

As a matter of fact, neither of these conditions is attained in practice. There is bound to be some escape of heat by conduction and radiation from the materials comprising the cylinder, piston, etc.; but it is not possible to remove all of this heat in practice. In actual air-compression work, means are adopted for removing the heat of compression as far as possible—which will be discussed later.

Necessity for Cooling during Compression. In the case just cited, the terminal pressure at $\frac{5}{6}$-stroke was found to be more than 73.5 lbs. gauge. If the piston is allowed to remain at this point until the temperature of the compressed air volume has dropped to 60° F., which was the temperature it had at the beginning of compression, and if it is assumed that there is no leakage, the final pressure will be found to be 73.5 lbs. gauge. If the discharge valves are then opened and the compressed air is discharged to a pipe-line, it will have a useful and stable pressure of 73.5 lbs. gauge, if the temperature of its surroundings is 60° F.

But to secure this useful and stable working pressure of 73.5 lbs., enough power had to be applied to produce the higher terminal pressure before cooling. In other words, there was a waste of power equivalent to that required to produce the difference in pressure between 73.5 lbs. and the higher but unstable terminal pressure. This waste of power, therefore, was due entirely to the heat generated in the air volume during compression.

Economical air-compression fundamentally demands, therefore, the removal of the heat of compression as fast as it is produced, so far as this is practically possible.

Ideal air-compression is compression along isothermal lines, following Boyle's law of pressures and volumes. The most uneconomical air-compression is compression along adiabatic lines, following Charles's law of pressures, volumes, and temperatures.

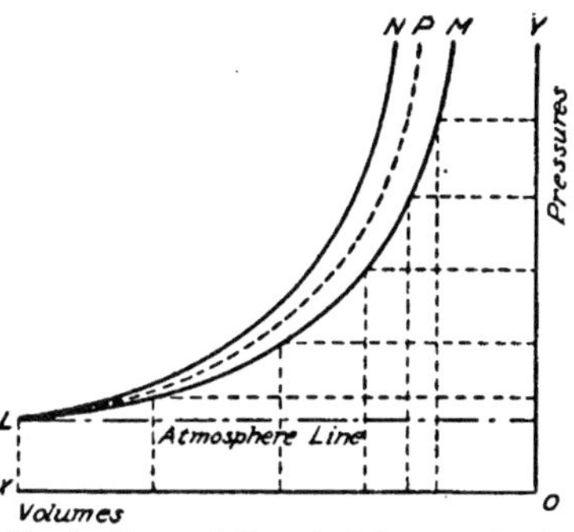

Fig. 4. Curve of Practical Compression in Relation to Isothermal and Adiabatic Curves.

It is possible in practice to avoid the latter extreme, but it is entirely impossible to realize the ideal expressed in the former extreme. The best air-compressor practice of to-day approaches that shown by the curve *LP* in Fig. 4, which is the mean between the adiabatic curve *LN* and the isothermal curve *LM*. Compressors of the highest refinement of design and of the highest economy, may show a curve following this intermediate line; but the simpler types of compressors, particularly in small sizes, will give a compression curve running closer to the adiabatic.

The question of compression economy is thus seen to be primarily one of adequate cooling during compression; and the accepted methods in use to-day will be discussed later in the proper place.

TABLE II

Final Temperatures Due to Adiabatic Compression to Various Gauge Pressures, from Initial Pressure of One Atmosphere, and from Various Initial Temperatures

(Richards)

Final Pressure, Lbs. Gauge	Initial Temperature 0°F.	Initial Temperature 32°F.	Initial Temperature 60°F.	Initial Temperature 100°F.
1	8	41	70	111
2	16	50	79	121
3	25	59	88	132
4	33	67	97	140
5	41	75	106	150
10	74	113	144	191
15	104	144	177	226
20	130	171	207	258
25	153	196	233	287
30	175	219	258	313
35	195	240	280	337
40	213	260	301	360
45	231	279	321	381
50	247	296	339	401
55	262	316	357	420
60	277	328	373	437
65	291	343	389	454
70	304	358	404	471
75	317	371	419	486
80	330	384	433	501
85	342	397	446	516
90	353	410	459	530
95	364	422	472	543
100	375	435	484	556
125	425	486	540	617
150	468	532	588	669
175	507	574	633	717
200	542	612	672	781

Heating Effect of Compression. While repeated reference has been made to the heat produced by compression, no definite values have as yet been assigned in these pages. Table II is accordingly presented, being copied from the book "Compressed Air," by courtesy of the author, Mr. Frank Richards. It shows the temperature corresponding to the various pressures in adiabatic compression of free air, from initial temperatures of 0°, 32°, 60°, and 100° F. Of course it need never be expected to find in practice these terminal temperatures after compressing to the pressures given. This table assumes that none of the heat of compression escapes. As a matter of fact, however, there is sure to be some loss of heat by conduction and radiation from the materials of which the compressor is built.

TABLE III

Loss of Work Due to Heat in Compressing Air from Atmospheric Pressure to Various Gauge Pressures by Simple and Compound Compression

(E. F. Schaefer)

Air in each cylinder at initial temperature of 60° F.

Gauge Pressure	One-Stage		Two-Stage		Three-Stage		Four-Stage	
	Percentage of Work Lost, in Terms of							
	Isothermal Compression	Adiabatic Compression	Isothermal Compression	Adiabatic Compression	Isothermal Compression	Adiabatic Compression	Isothermal Compression	Adiabatic Compression
60	29.9	23.0	13.4	11.8	8.6	7.9	4.7	4.5
70	30.6	23.4	14.1	12.4	8.7	8.0	6.1	5.7
80	32.7	24.6	14.7	12.8	9.7	8.9	6.4	6.0
90	34.7	25.8	16.1	13.8	10.5	9.5	7.3	6.8
100	36.7	26.8	16.9	14.5	10.9	9.8	7.8	7.3
125	41.1	29.2	18.5	15.6	11.6	10.4	8.8	8.1
150	44.8	30.9	20.1	16.7	12.3	10.9	9.1	8.4
200	51.2	33.9	22.2	18.1	14.0	12.3	10.5	9.5
300	61.2	37.9	25.7	20.5	16.6	14.2	12.0	10.7
400	68.7	40.7	28.9	22.4	18.2	15.4	13.1	11.5
500	70.6	41.4	31.2	23.8	19.3	16.2	14.1	12.3
600	80.4	44.5	32.8	24.7	20.4	16.9	14.9	13.0
700	85.0	46.0	34.6	25.7	21.3	17.6	16.1	13.8
800	89.5	47.2	35.7	26.3	22.0	18.1	16.2	13.9
900	93.0	48.2	37.1	27.0	22.6	18.5	16.6	14.4
1,000	96.1	49.0	37.9	27.5	23.2	18.8	16.9	14.5
1,200		50.7	40.3	28.8	24.8	19.9	17.7	15.0
1,400		52.0	41.5	29.3	25.9	20.5	18.6	15.7
1,600		53.1	43.5	30.3	26.5	20.9	19.2	16.1
1,800		54.0	44.8	31.0	27.3	21.2	19.6	16.4
2,000		55.0	45.8	31.4	27.5	21.5	19.9	16.5

Loss of Work Due to Heat of Compression. Table III, computed by E. F. Schaefer, and used here by courtesy of the Ingersoll-Rand Company, shows the magnitude of the losses in air-compression due to the heat of compression, the figures at all pressures being given for both single-, two-, three-, and four-stage compression. The values assume an initial temperature in all cylinders of 60° F.; and since this is a condition which would probably never be fully realized in practice, the actual losses may be assumed to be even larger than those here given.

Compressed-Air Indicator Card. The general theory of the indicator card in its relation to the steam engine may here be assumed to be known by the reader; but the importance of the subject of the compressed-air indicator card justifies a somewhat extended special treatment.

To avoid confusion on the part of those familiar only with the steam engine indicator card, it is to be remembered that in the air-compressor indicator card all the operations are exactly reversed. because the former is a record of expansions and the latter a record of compressions.

To be exact, the admission line of the steam card corresponds to the delivery or discharge line of the air card; the expansion line of the former, to the compression line of the latter; the exhaust or back-pressure line of the former, to the admission or intake line of the latter; and the compression line of the former, to the re-expansion line of the latter.

At the outset, let it be understood that an air-compressor card, to be a correct record of the machine's performance, must be taken only after the compressor has been run long enough to get thoroughly *warmed up*, so that all temperature conditions affecting compression have attained their normal running maximum—which is usually not until the machine has been running an hour or more.

An ideal indicator card from an air-compressor is set forth in Fig. 5. In this figure, GH is the line of zero pressure, drawn at the proper distance below the atmospheric line EF, according to the scale. DA is the admission or intake line; AB, the compression line; BC, the delivery or discharge line; and CD, the re-expansion line. The admission line DA is here drawn at an exaggerated distance below the atmospheric line EF, simply for the sake of clearness. In the best practice, these two lines so nearly coincide as to be hardly distinguishable. The line CO having been dropped vertically from the point C, the distance EO represents the stroke of the piston. The line CO (or PR) represents the maximum gauge pressure.

The rectangle $E\ N\ C\ O$ represents the actual volume displaced by the piston in moving from E to O. This, however, is not the total capacity of the cylinder, for, at the end of the stroke, there remains a small volume, represented by the rectangle $O\ C\ K\ J$, in the clearance space of the cylinder. This small volume is, of course, compressed to maximum pressure. On the return stroke, it expands to atmosphere along the curve CD, its expansion occurring so quickly as to be practically adiabatic. Thus the total volume of air acted upon by the piston is represented by the rectangle $E\ N\ K\ J$.

The distance OJ (or CK), representing the clearance percentage, is found as follows: Assume that the scale of the spring used in taking the card is 30, and the terminal pressure is 70 lbs. gauge. The volume of air in the clearance space, after expansion, may be found from column 5 of Table VI (page 21). Here the volume of air ex-

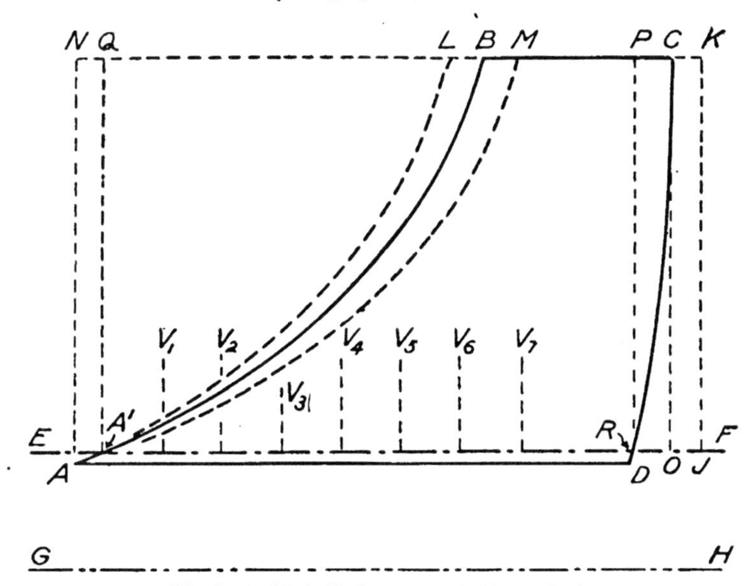

Fig. 5. An Ideal Air-Compressor Indicator Card.

panding without cooling, or adiabatically, from 70 lbs. to atmosphere, is found to be 0.288, as compared with the unit-volume. Let X = the distance OJ. Then,

$$X : DO + X = 0.288 : 1.00.$$

The length DO can be measured from the diagram, and may be here assumed as 0.25 inch. Then,

$$X : 0.25 + X = 0.288 : 1.00;$$

whence,

$$X = 0.101 \text{ inch (say 0.1 inch).}$$

It will be noted that, since compression began a little below atmosphere, the compression line AB intersects the atmosphere line EF at A', which, being the point at which compression from atmosphere begins, may be taken as the initial point from which the entire cycle of operations represented in the diagram starts.

Erecting the perpendicular $A'Q$, the rectangle $A'QKJ$ represents the total volume of air *at atmospheric pressure* upon which power is exerted for compression.

With the piston at the end of its stroke, there still remains the air at terminal or discharge pressure in the clearance space represented by $OCKJ$. On the return stroke, this air expands to atmosphere, the re-expansion line CD intersecting the atmosphere line at R. This volume, while not delivered or discharged, and having required an expenditure of power to compress it, gives back this power by its re-expansion, helping the piston on its return stroke.

Clearance in an air-compressor cylinder thus represents, not a loss of power, but a loss of capacity.

With this understanding of the quantities represented on the indicator card, an investigation of its possibilities as a source of information on the performance of a machine may be undertaken.

In computing the work done in the cylinder, and the mean effective pressure (M.E.P.) throughout the stroke, the entire area enclosed by the diagram $ABCD$ must be used. This area is to be measured by the means usually employed in such cases—that is, by a planimeter or by the use of ordinates. This area, of course, represents only the work done in one stroke, representing only one end of a double-acting compression cylinder. It is much more accurate to have deductions based upon the mean of cards from both ends of the cylinder; but the present card will serve as a guide for all cases.

The area $AA'RD$, lying below the atmosphere line, represents the resistance on the return stroke; but it may be assumed that the cards from both ends of the cylinder are similar, and this area may therefore be safely considered in connection with the area above the atmosphere line for the single stroke.

Computing the I. H. P. of the Air Cylinder. The *indicated horse-power* (I. H. P.) of a compressor cylinder is figured in the same manner as for a steam cylinder. Assume that the M.E.P. has been found to be 30 lbs., and that the card under examination was taken from a compressor having a 24-inch cylinder and a 30-inch stroke, running at 75 revolutions per minute (R.P.M.). Then

$P = 30$ lbs.; $L = 2 \times 30 \div 12 = 5$ ft.; $A = 24^2 \times .7854 = 452.4$ sq. in.; $N = 75$ R.P.M. Therefore,

$$\text{I.H.P.} = P \times L \times A \times N \div 33{,}000 = 154.23 \text{ H.P}$$

This, it will be noted, is simply the horse-power representing the work done in compressing the air. It takes no account of friction or other losses.

Computing the Mechanical Efficiency of a Compressor. The *mechanical efficiency* of an air-compressor is the ratio of the power developed in the air cylinder to the power applied in the steam cylinder. In other words, it is the ratio of the I.H.P. of the air cylinder to the I.H.P. of the steam cylinder, if the compressor is steam-driven. The difference between these two represents the power required to keep the machine going, overcome friction, etc. In a power-driven compressor, the mechanical efficiency is the ratio of the I.H.P. of the air cylinder to the *brake horse-power* (B.H.P.) applied to the driving wheel or shaft.

A comparison of the steam and air cards taken simultaneously from a steam-driven compressor is thus seen to show exactly what percentage of the power applied is used in overcoming the friction of the machine, and what percentage is used in actually compressing the air. These proportions can also be determined exactly in the case of a power-driven compressor where the applied power can be actually measured, as by a brake or other dynamometer. There is probably no other machine in everyday use in which the mechanical efficiency can be determined so accurately as in an air-compressor.

Determining the Volumetric Efficiency of a Compressor. The *volumetric efficiency* of an air-compressor is the ratio of the actual volume of air compressed and discharged, to the volume theoretically displaced by the piston in a given time. This is shown with great accuracy by the air-compressor indicator card.

The point A' (Fig. 5), where the compression line crosses the atmosphere line, indicates that the cylinder is full of air at atmospheric pressure at this point in the stroke. The distance EA' represents the portion of the stroke required to compress to atmospheric pressure the rarified air indicated by the intake line. This distance is therefore to be deducted from the full stroke EO as non-effective in the compression and delivery of atmospheric air under pressure. This discussion emphasizes the importance of bringing the intake line as near as possible to the atmosphere line, as the distance EA' evidently increases as the intake line drops below atmosphere.

At the end of the stroke, the clearance air remains at terminal pressure, and on the return stroke re-expands to atmosphere, its re-expansion line crossing the atmospheric line at the point R. Evidently atmospheric air cannot enter through the inlet valves of a compressor until this air in the clearance space has expanded to atmospheric pressure. The distance RO, therefore, represents the proportion of the stroke occupied in compressing and re-expanding the clearance air. It is non-effective in the admission of atmospheric air, and it also must be deducted from the full stroke EO.

In the diagram (Fig. 5), EO measures approximately 4 inches; RO is about $\frac{1}{4}$ inch, and EA' about $\frac{3}{16}$ inch. The ratio of the effective stroke to actual stroke is thus seen to be $3\frac{9}{16}:4$, or 0.89. This ratio is evidently also the ratio of volumes, and therefore the volumetric efficiency represented by this card is 89 per cent. This is lower than is found in the best practice; but in this assumed card many distances are exaggerated for the sake of greater clearness.

It would at first seem that this method would give exactly the volumetric efficiency of the cylinder under examination; but one serious shortcoming of the indicator card as a record of performance must here be noted. While the method just described gives accurately the ratio of the air volume compressed and discharged to the cylinder displacement, it does not take into consideration the temperature of the air. It is not usually safe to assume that the air admitted to the cylinder is actually of the density of the surrounding free air, for it may be heated, and expanded accordingly, in entering the cylinder through heated ports and valves. An example will illustrate this. If the atmospheric temperature is 70° F., and the air in entering the cylinder is heated to 120 ° F., the ratio of densities in the two cases, from Table I (page 4), will be 0.0685:0.075, or 0.913. This means that only 0.913 of the air volume apparently compressed is actually compressed. More than this, it means that the full power as shown by the steam card is being used to compress this reduced volume of air.

There is no way of determining exactly the initial temperature *in the cylinder*; and the best that can be done, the conditions being understood, is to arrange as far as possible to avoid the heating of

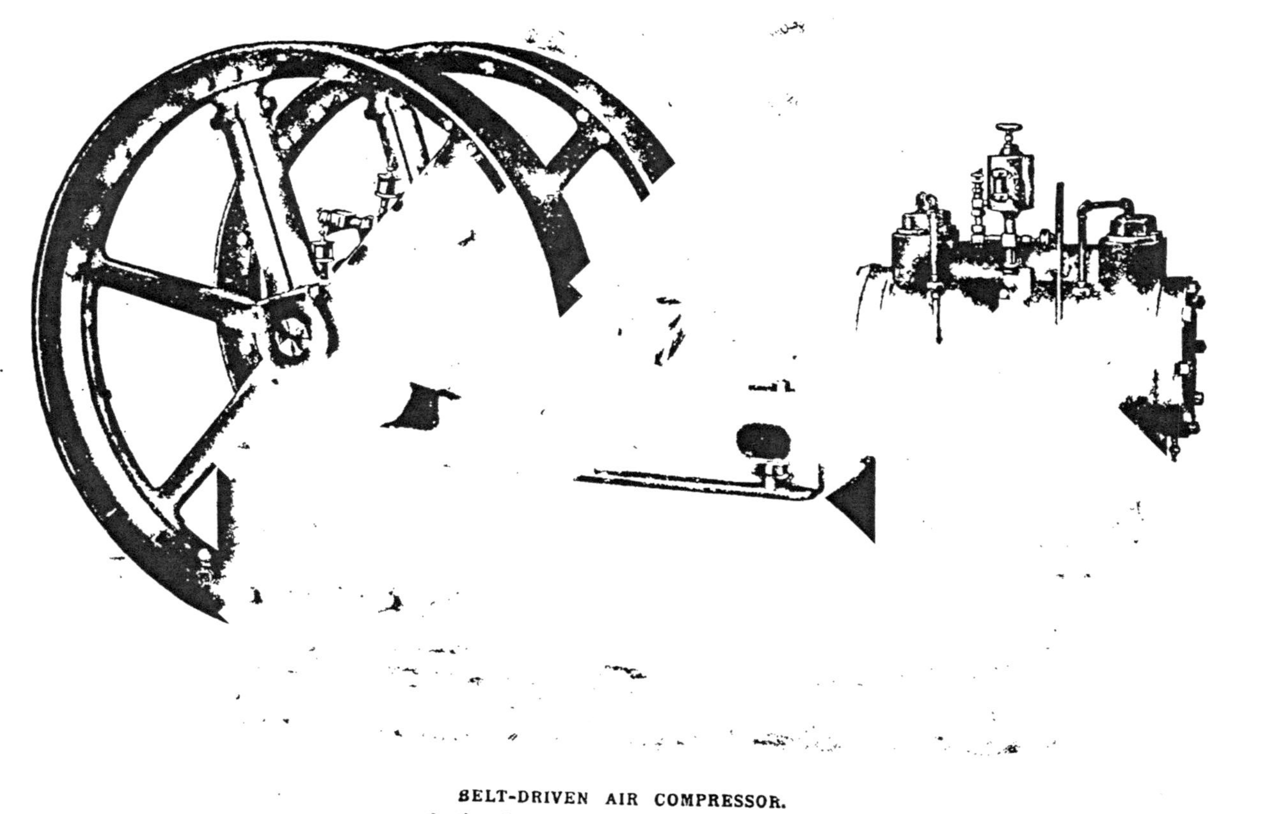

BELT-DRIVEN AIR COMPRESSOR.
Sectionalized for Mule-back Transportation.
Sullivan Machinery Company.

the intake air. This matter will receive more attention in a later section where cooling is discussed.

Much can be done, however, by seeing that the air is delivered *to the compressor intake* as cool as possible. The air of the average engine room is heated, and should not be admitted to the cylinder. It is well worth while to lay an inlet duct from the compressor intake to where the coolest possible air can be had—for example, outside the engine room, from the north side of a building, etc. Table IV shows the importance of this point. An examination of its figures reveals the fact that for every 10 degrees reduction in intake temperature, there is a gain of practically 2 per cent in capacity and volumetric efficiency. This is a small economy often overlooked, though its only cost is that of laying a suitable intake duct or tube to the right place, making it, of course, large enough so that the air flows freely and no *suction load* is imposed upon the piston.

TABLE IV

Effect of Various Initial or Intake Temperatures of Air on the Capacity of the Compressor and the Efficiency of Compression

Initial Temperature		Relative Capacity and Efficiency	Initial Temperature		Relative Capacity and Efficiency
Fahrenheit	Absolute		Fahrenheit	Absolute	
−20	441	1.180	70	531	.980
−10	451	1.155	80	541	.961
0	461	1.130	90	551	.944
10	471	1.104	100	561	.928
20	481	1.083	110	571	.912
30	491	1.061	120	581	.896
32	493	1.058	130	591	.880
40	501	1.040	140	601	.866
50	511	1.020	150	611	.852
60	521	1.000	160	621	.838

Effect of Clearance Space. There must of necessity be some clearance between the piston and cylinder heads, merely on the grounds of safety. In addition, there are usually spaces left after the inlet and discharge valves are closed—large or small according to the mechanical design. These spaces which, in the aggregate, constitute the clearance space, are filled with air at terminal pressure at the end of the stroke; and when the return stroke begins, this air re-expands to atmosphere or intake pressure. This air is compressed and re-expanded at each end of the cylinder, once each revolution or double stroke. Power is spent to compress it; and on expansion, it gives back practically all this power as work applied in helping the piston on the return stroke. In this respect, therefore, clearance does not represent a loss of power.

But, as has been stated, no atmospheric air can enter the cylinder until the clearance air is expanded to atmosphere. Thus a portion of the admission stroke is really no admission at all, and a full cylinder of air is never admitted in practice. Clearance, therefore, does mean a loss of capacity, and should be kept down to the lowest possible percentage. Furthermore, since air-compressors are rated by piston displacement, and since the efficiencies are based on this rating rather than on the actual delivered volume (which can seldom be measured), the loss of capacity due to clearance is really in effect a proportionate loss of power, if the theoretical power is figured according to piston displacement.

Determining the Compression Efficiency. The *compression efficiency* of an air-compressor is the ratio of the minimum power theoretically required to produce a given pressure to the power actually expended in the air cylinder in producing that pressure. Since the minimum power is that of isothermal compression, the compression efficiency is the ratio of the isothermal horse-power to the indicated horse-power as registered by the indicator card.

The power developed or expended in the air cylinder is proportional to the M.E.P., since, in the horse-power formula $HP = P\,L\,A\,N \div 33{,}000$, all factors except P may be considered as constant throughout the stroke. Accordingly the ratio of powers in computing compression efficiency is the same as the ratio of M.E.P.'s for isothermal conditions and for the actual conditions as the indicator card reveals them.

The simplest way, therefore, to figure the compression efficiency of a compressor, is to compare the M.E.P. of its indicator card with the M.E.P. corresponding to the given pressure, as given in column 6 of Table VI (page 21). It is to be borne in mind, however, that Table VI is correct only for sea-level conditions.

Laying Out the Isothermal and Adiabatic Curves. The method just described for determining the compression efficiency gives all the information ordinarily required; but it is sometimes useful, as a graphic demonstration, to lay out the isothermal and adiabatic lines on the indicator card in their proper relations to the lines of the diagram. The method is as follows:

In isothermal compression, $PV = C$, a constant; or $P_1 V_1 = P_2V_2 = P_3V_3$, etc. Transposing, $P_2 = P_1 (V_1 \div V_2)$; $P_3 = P_1 (V_1 \div V_3)$, etc. Referring once more to Fig. 5, it will be remembered that the rectangle $A'QKJ$ represents the total volume of atmospheric air undergoing compression, and the division of this rectangle by vertical lines would indicate proportionate volumes at corresponding points of stroke. Let these lines V_1, V_2, V_3, etc., be drawn, dividing this rectangle into ten equal parts. Let the volume represented by $A'J$ be 1; then the volume represented by V_1J will be 0.9; by V_2J, 0.8; etc. If the initial pressure of the card is 14.7 lbs. absolute, this will be P_1. The values in the above equation, then, become:

P_2 = 14.7 (1.0/0.9) = 16.317 lbs. absolute.
P_3 = 14.7 (1.0/0.8) = 18.375 lbs. "
P_4 = 14.7 (1.0/0.7) = 20.874 lbs. "

In like manner the absolute pressures at any percentage of the stroke may be figured; and if these pressures on the proper scale are laid off above the atmospheric line on the corresponding vertical lines of the diagram, a curve joining the points thus determined will be the *isothermal curve* for that diagram.

For any other initial pressure than 14.7 lbs., the method is exactly the same, with the exception that this new value of P_1 is to be substituted in the formula. Further precaution must be taken to see that the atmospheric line is properly located in relation to the admission line.

To plot the adiabatic curve, the method is the same as for the isothermal, except that the mathematics involved demands the use of logarithms because of the fractional exponent. The formula to be used in this case is $P_2 = P_1 (V_1 / V_2)^{1.41}$; $P_3 = P_1 (V_1 / V_3)^{1.41}$; etc. This equation is solved by logarithms thus:

$$\log. P_2 = \log. P_1 + \left\{(\log. V_1 - \log. V_2) \times 1.41\right\}$$

Pressures thus obtained for the various percentages of stroke are laid off on the corresponding vertical lines, and the points thus determined are joined by a line, which is the *adiabatic curve.*

TABLE V

Multipliers for Determining the Isothermal and Adiabatic Pressures in Air-Compression from any Initial Pressure, at Various Percentages of Stroke

Percentage of Stroke	Isothermal	Adiabatic	Percentage of Stroke	Isothermal	Adiabatic
0	1.000	1.000	50	2.000	2.657
5	1.052	1.074	55	2.222	3.083
10	1.111	1.160	60	2.500	3.640
15	1.176	1.256	65	2.857	4.394
20	1.250	1.369	70	3.333	5.460
25	1.333	1.499	75	4.000	7.052
30	1.428	1.652	80	5.000	9.672
35	1.538	1.835	85	6.666	14.501
40	1.666	2.054	90	10.000	25.704
45	1.818	2.323	95	20.000	68.305

Table V gives the multipliers for determining the isothermal and the adiabatic pressures at the percentages of the stroke given, *per pound of initial pressure*. These multipliers are, for isothermal pressures, the value of V_1/V_2; and for adiabatic pressures, the value of $(V_1/V_2)^{1.41}$, etc. To use this table in plotting curves, simply multiply the atmospheric or intake pressure in pounds absolute by these constants; lay out the pressures thus secured at the proper percentages of stroke; and join the points by a curve.

Relation of Pressures, Volume, and Temperatures. Table VI (page 21) is reproduced by courtesy of Mr. Frank Richards, from his work on "Compressed Air," and was prepared by him for use in working up indicator cards and in other compressed-air computations. The basis of these figures is air at atmospheric pressure of 14.7 lbs. and a temperature of 60° F.; and the values in the table start with the air in this condition.

Column 1 lists the *gauge pressure*, which would be the ordinary working pressure of the air after compression.

Column 2 gives the corresponding *absolute pressure*, and the figures are found by adding 14.7 lbs. to the gauge pressure in column 1.

Column 3 gives the *pressure in atmospheres*, and is found by dividing the corresponding figure in column 2 by 14.7.

TABLE VI

Volumes, Mean Pressures, Temperatures, etc., in Compressing Air from One Atmosphere and 60° F.

(Richards)

1	2	3	4	5	6	7	8	9	10	11	12
Gauge Pressure	Absolute Pressure	Pressure in Atmospheres	Volume with Air at Constant Temperature	Volume with Air not Cooled	M.E.P. per Stroke, Air at Constant Temperature	M.E.P. per Stroke, Air not Cooled	M.E.P. per Stroke, Average Conditions of Practice	M.E.P. for Compression Only, Air at Constant Temperature	M.E.P. for Compression Only, Air not Cooled	M.E.P. for Compression Only, Average Conditions of Practice	Final or Terminal Temperature, Air not Cooled
0	14.7	1.000	1.0000	1.00	0.00	0.000	0.000	0.00	0.00	0.000	60
1	15.7	1.068	.9363	.95	.96	.975	.967	.43	.44	.435	71
2	16.7	1.136	.8803	.91	1.87	1.91	1.89	.95	.96	.955	80.4
3	17.7	1.204	.8305	.876	2.72	2.8	2.76	1.4	1.41	1.405	88.9
4	18.7	1.272	.7861	.84	3.53	3.67	3.60	1.84	1.86	1.85	98
5	19.7	1.34	.7462	.81	4.3	4.5	4.40	2.22	2.26	2.24	106
10	24.7	1.68	.5952	.69	7.62	8.27	7.94	4.14	4.26	4.20	145
15	29.7	2.02	.495	.606	10.33	11.51	10.92	5.77	5.99	5.88	178
20	34.7	2.36	.4237	.543	12.62	14.4	13.52	7.2	7.58	7.30	207
25	39.7	2.7	.3703	.494	14.59	17.01	15.80	8.49	9.05	8.77	234
30	44.7	3.04	.3289	.4638	16.34	19.4	17.87	9.66	10.39	10.02	255
35	49.7	3.381	.2957	.42	17.92	21.6	19.76	10.72	11.59	11.15	281
40	54.7	3.721	.2687	.393	19.32	23.66	21.49	11.7	12.8	12.25	302
45	59.7	4.061	.2462	.37	20.52	25.59	23.05	12.62	13.95	13.28	321
50	64.7	4.401	.2272	.35	21.79	27.39	24.59	13.48	15.05	14.26	339
55	69.7	4.741	.2109	.331	22.77	29.11	25.94	14.3	15.98	15.14	357
60	74.7	5.081	.1968	.3144	23.84	30.75	27.29	15.05	16.89	15.97	375
65	79.7	5.423	.1844	.301	24.77	31.69	28.23	15.76	17.88	16.82	389
70	84.7	5.762	.1735	.288	26.00	33.73	29.86	16.43	18.74	17.58	405
75	89.7	6.102	.1639	.276	26.65	35.23	30.94	17.09	19.54	18.31	420
80	94.7	6.442	.1552	.267	27.33	36.6	31.96	17.7	20.5	19.10	432
85	99.7	6.782	.1474	.2566	28.05	37.94	32.99	18.3	21.22	19.76	447
90	104.7	7.122	.1404	.248	28.78	39.18	33.98	18.87	22.0	20.43	459
95	109.7	7.462	.134	.24	29.53	40.4	34.96	19.4	22.77	21.08	472
100	114.7	7.802	.1281	.232	30.07	41.6	35.83	19.92	23.43	21.67	485
105	119.7	8.142	.1228	.2254	30.81	42.78	36.79	20.43	24.17	22.30	496
110	124.7	8.483	.1178	.2189	31.39	43.91	37.65	20.9	24.85	22.87	507
115	129.7	8.823	.1133	.2129	31.98	44.98	38.48	21.39	25.54	23.46	518
120	134.7	9.163	.1091	.2073	32.54	46.04	39.29	21.84	26.2	24.02	529
125	139.7	9.503	.1052	.202	33.07	47.06	40.06	22.26	26.81	24.53	540
130	144.7	9.843	.1015	.1969	33.57	48.1	40.83	22.69	27.42	25.05	550
135	149.7	10.183	.0981	.1922	34.05	49.1	41.57	23.08	28.05	25.56	560
140	154.7	10.523	.095	.1878	34.57	50.02	42.29	23.41	28.66	26.03	570
145	159.7	10.864	.0921	.1837	35.09	51.0	43.04	23.97	29.26	26.61	580
150	164.7	11.204	.0892	.1796	35.48	51.89	43.68	24.28	29.82	27.05	589
160	174.7	11.88	.0841	.1722	36.29	53.65	44.97	24.97	30.91	27.94	607
170	184.7	12.56	.0796	.1657	37.2	55.39	46.25	25.71	32.03	28.87	624
180	194.7	13.24	.0755	.1595	37.96	57.01	47.48	26.36	33.04	29.70	640
190	204.7	13.92	.0718	.154	38.68	58.57	48.62	27.02	34.06	30.54	657
200	214.7	14.6	.0685	.149	39.42	60.14	49.78	27.71	35.02	31.36	672

Column 4 gives the volumes (compared with initial volume of 1) after *isothermal compression* to the pressure given. Its values assume that the heat of compression has been removed as fast as produced, or that the air has been allowed to cool after compression to its initial temperature of 60° F.

Column 5 presents the volumes after *adiabatic compression* to the stated pressures, assuming that all the heat of compression is allowed to remain in the air.

Column 6 gives the M.E.P. on the air-piston during the stroke of compression, assuming that compression is *isothermal,* the air remaining at 60° F. throughout the stroke. The figures in this column represent the ideal of air-compression, and can never be attained in practice, since no cooling device is sufficiently effective. They are, however, the standard with which the results of actual practice are to be compared, in figuring compression efficiency.

Column 7 lists the M.E.P. on the air-piston during the stroke in *adiabatic* compression to the corresponding gauge pressures, and assumes that no cooling of the air whatever occurs during compression. Since there is always some loss of heat, however, these figures are not found in practice. If column 6 presents the ideal toward which practice aims, column 7 is equally important as the extreme to be avoided as far as possible.

Column 8 gives figures which are the mean or average of those in columns 6 and 7. In other words, these are the M.E.P.'s closely approximating those found in actual work with the best air-compressors dealing with these pressures. Small high-speed machines will probably not equal these values. An indicator card showing these M.E.P.'s, or very near to them, may be considered a very good card.

Column 9 gives the M.E.P. during the *compression* part of the stroke, assuming isothermal conditions. The actual process in an air cylinder may be divided into two parts—*first,* the compression of the air volume from atmosphere to terminal pressure; *second,* the delivery or discharge of this air at terminal pressure from the cylinder. Columns 6, 7, and 8 have to do with the entire process, and cover both compression and delivery. Columns 9, 10, and 11, on the contrary, have to do only with the first part of the process.

Column 10 lists the M.E.P. during the compression part of the stroke, assuming adiabatic compression.

Column 11 gives the M.E.P.'s which are the mean or average of those in columns 9 and 10, and which represent very fairly the ordinary results found in practice as derived from indicator cards of strictly high-grade compressors.

Column 12 gives the final or terminal temperatures after *adiabatic* compression to the pressures listed. The inevitable escape of heat during compression, even where no cooling devices are used, makes these figures as they appear here higher than they are ever found in actual work; but their magnitude emphasizes the importance of using some means of cooling to avoid destructive temperatures in air-compression.

Indicator Cards from Multi-Stage Compressors. The discussion thus far has dealt only with conditions involving compression to the required terminal pressure in a single cylinder; but, as will be seen later, the modern tendency in compressor design is toward compound or multi-stage compression, in which the compression process is divided between two, three, or even four cylinders. While in such cases the performance of each cylinder is exactly recorded on the indicator card from that cylinder, still it is sometimes desirable to show the *combined card*, presenting in one diagram the conditions of compression in all the air cylinders of the machine.

The general method is the same as in the case of steam-engine indicators, for making combined steam-engine diagrams, with which the reader is assumed to be familiar. The low-pressure card is left at its original size, and the cards of the other cylinders are reduced to the same scale by altering the length (representing volume) in proportion to the ratio of cylinder areas, and the height (representing pressures) in proportion to the scale of the springs used in taking the cards. The isothermal and adiabatic lines are then drawn, and the phenomena of compression in all cylinders are graphically presented in one diagram.

Some Representative Indicator Cards. Figs. 6 to 14, inclusive, are reproductions of indicator cards taken from the air ends of machines of various types and under various conditions. They are here introduced to give an idea of what may be expected from air-compressors in everyday work, as these cards are all taken from machines running under practical conditions.

Fig. 6 is a card from a single-stage straight-line compressor with a cylinder 26 inches in diameter and 30 inches stroke, running at a speed of 88 R.P.M. and compressing to 61 lbs. gauge. The atmospheric pressure is 14.7 lbs. absolute; and it will be noted that, while the intake line starts on the atmospheric line, it drops below toward the end of the admission stroke, the intake pressure at the end of the stroke being 13.5 lbs. absolute. This would indicate an insufficient inlet-valve area, or that a partial vacuum or rarification of the intake air was produced. The compression line runs rather closer than usual to the adiabatic. The waves in the discharge line are due largely to vibration in the springs of the poppet discharge valves,

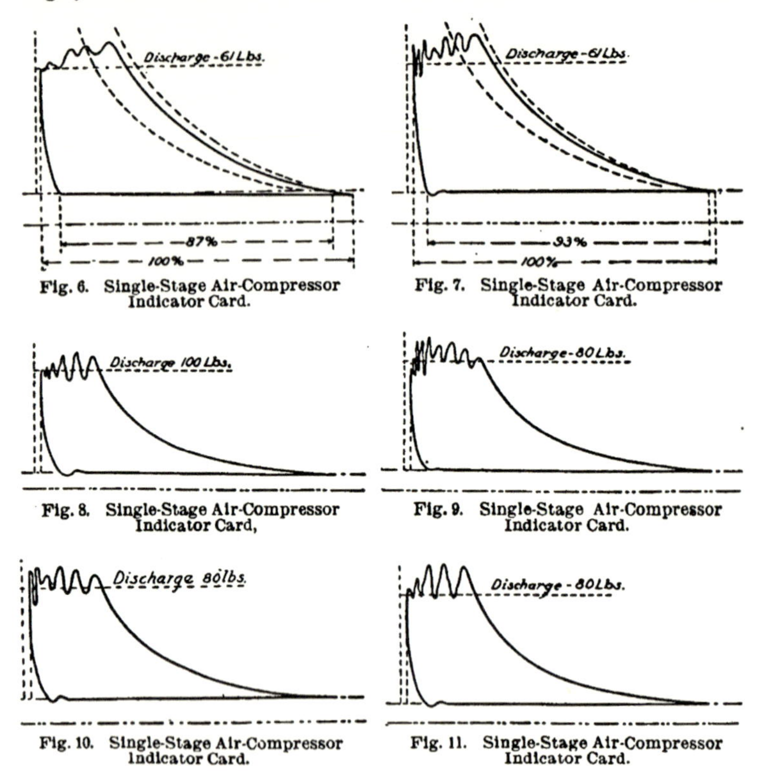

Fig. 6. Single-Stage Air-Compressor Indicator Card.

Fig. 7. Single-Stage Air-Compressor Indicator Card.

Fig. 8. Single-Stage Air-Compressor Indicator Card,

Fig. 9. Single-Stage Air-Compressor Indicator Card.

Fig. 10. Single-Stage Air-Compressor Indicator Card.

Fig. 11. Single-Stage Air-Compressor Indicator Card.

the variations being amplified by the vibration of the indicator spring. The re-expansion line is good. The volumetric efficiency shown by this card is 87 per cent. The I. H. P. of the card is 201.5; and the efficiency, compared with perfect isothermal single-stage compression, is 93 per cent.

Fig. 7 is from another 26 by 30-inch single-stage air-cylinder, the terminal pressure being again 61 lbs. gauge, and speed 88 R.P.M. In fact, this card was taken especially for comparison with the card shown in Fig. 6, to show the result of some improvements in the second machine, which was, however, in general type, identical with the first. The operating conditions were therefore made the same in the two cases. It will be noted that the intake line coincides with the atmospheric line for almost the entire stroke; the slight drop in the intake line at beginning of admission suggests the failure of the inlet valve to open fully and instantly, but is really caused by the spring of the indicator; and the slight rise in pressure immediately following this initial drop is due to the reaction of the indicator spring. In this card the volumetric efficiency is 93 per cent—an excellent result for these conditions. The I. H. P. and compression efficiency are not given.

Fig. 8 is another card from this same machine, running, in this case, at 100 R.P.M., and compressing to 100 lbs. gauge.

Figs. 9, 10, and 11 are cards from the same machine with 26 by 30-inch air cylinder at 80 lbs. gauge pressure, and at speeds of 60, 80, and 100 R.P.M. The volumetric efficiency remains practically at 93 per cent at all these speeds, evidencing a very good performance.

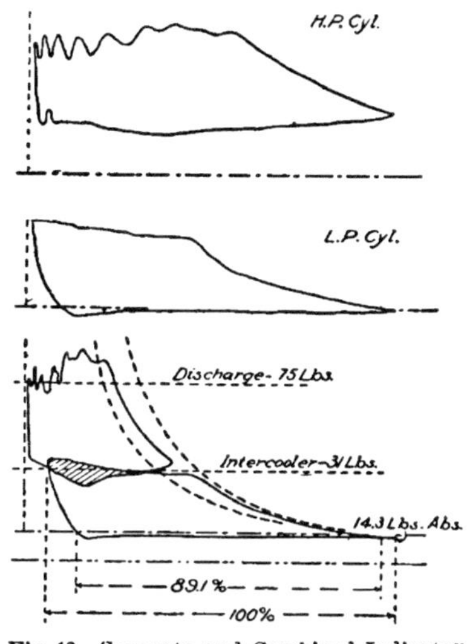

Fig. 12. Separate and Combined Indicator Cards from a Tandem Compound Two-Stage Compressor.

Fig. 12 shows the cards from the high- and low-pressure cylinders of a tandem compound compressor of 28 inches stroke and air cylinders 28 and 17½ inches in diameter. The combined card is also shown. The intercooler pressure was 31 lbs.; and the final pressure, 75 lbs. gauge. Volumetric efficiency is 89.1 per cent; I.H.P. of low-pressure cylinder, 129; I.H.P. of high-pressure cylinder, 108; total I.H.P., 237. The efficiency, compared with perfect isothermal compression in two stages, is 80.9 per cent. A considerable vacuum is shown at the beginning of the intake line, which reduces toward the end of the stroke. This would suggest that the inlet valves are not working properly. The re-expansion line indicates a large clearance space in the low-pressure cylinder. The upward slanting low-pressure discharge line suggests a restricted discharge area, or too early closing of the discharge valves. The *peak* on the high-pressure discharge line is rather higher than would ordinarily be considered desirable. The shaded area in the combined card represents work which has been done twice, once in the low-pressure cylinder and later in the high-pressure cylinder. It may be due to a restricted low-pressure discharge, an insufficient high-pressure intake, or an inadequate cross-section in the intercooler.

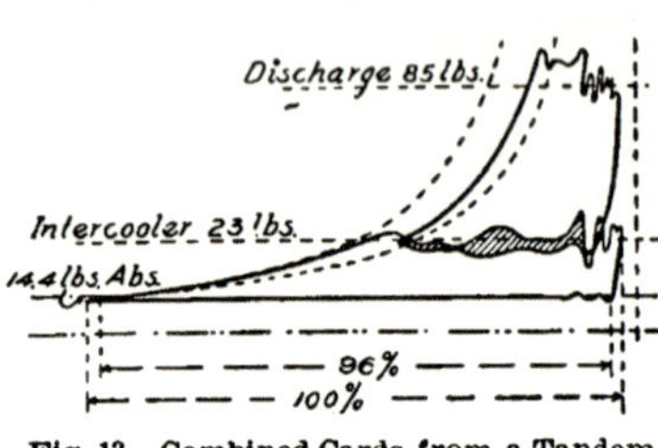

Fig. 13. Combined Cards from a Tandem Compound Two-Stage Compressor.

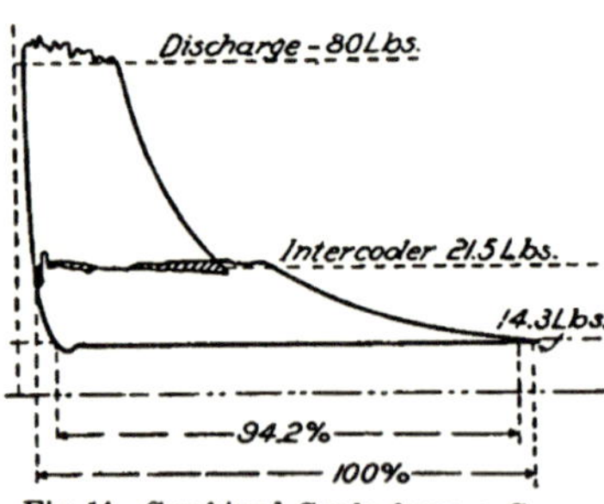

Fig. 14. Combined Cards from a Cross-Compound Two-Stage Compressor.

Fig. 13 presents the separate and combined cards from the air cylinders of another tandem compound machine of a still different type, the size being 20 inches and 13 inches by 12 inches stroke. The speed was 100 R.P.M., and the terminal pressure 85 lbs. gauge, with atmosphere at 14.7 lbs. absolute. The volumetric efficiency is 96 per cent — an exceptionally good showing. The I.H.P. of the low-pressure cylinder is 44; and of the high-pressure cylinder, 52.25; total I.H.P., 96.25. The efficiency of compression, compared with perfect isothermal two-stage, is 91.25 per cent. Taken as a whole, this is a card rather above the average from this type of machine, although the shaded area in the combined card suggests possibilities of improvement in the intercooler and in the valve action.

Fig. 14 gives separate and combined cards from a double cross-compound compressor with an air end 34 inches and 24 inches by 24 inches stroke, running at 94 R.P.M., and compressing at sea-level to 80 lbs. gauge. The total I.H.P. is 324; and the compression efficiency compared with two-stage isothermal, is 91.7 per cent. Volumetric efficiency is 94.2 per cent. Taken as a whole, this is a splendid record of performance; though some slight defects may be noted in the details of the cards.

TABLE VII

Horse-Power Developed in Compressing One Cubic Foot of Free Air from Atmospheric Pressure (14.7 Pounds) to Various Gauge Pressures

INITIAL TEMPERATURE OF THE AIR IN EACH CYLINDER TAKEN AS 60° F.
(Jacket Cooling Not Considered)

Gauge Pressure	Isothermal Compression	Adiabatic Compression			
		One-Stage	Two-Stage	Three-Stage	Four-Stage
10	.0332	.0358			
20	.0551	.0623			
30	.0713	.0842			
40	.0842	.1026			
50	.0950	.1187			
60	.1042	.1331			
70	.1122	.1465	.128	.122	.119
80	.1194	.1585	.137	.131	.127
90	.1258	.1695	.146	.139	.135
100	.1317	.1800	.154	.146	.142
125	.1443	.2036	.171	.161	.157
150	.1549	.2244	.186	.174	.169
200	.1719	.2600	.210	.196	.190
300	.1964	.3164	.247	.229	.220
400	.2141	.3613	.276	.253	.242
500	.2279	.3889	.299	.272	.260
600	.2393	.4318	.318	.288	.275
700	.2489	.4608	.335	.302	.289
800	.2573	.4873	.349	.314	.299
900	.2649	.5114	.363	.325	.310
1,000	.2720	.5337	.375	.335	.318
1,200	.2829	.5742	.397	.353	.333
1,400	.2924	.6102	.414	.368	.347
1,600	.3012	.6427	.432	.381	.359
1,800	.3087	.6724	.447	.393	.369
2,000	.3154	.7003	.460	.403	.379

NOTE—The above values are for sea-level conditions only. For the power developed at altitudes, see Table IX (p. 30).

Power Required to Compress Air. Table VII, compiled by F. M. Hitchcock, lists the power required or developed per cubic foot of free air in compressing from atmospheric pressure of 14.7 lbs. absolute to the gauge pressures listed in the first column. The second column gives the isothermal or theoretical minimum horse-power required, never equaled in practice. In the next four columns are given the horse-powers required for adiabatic compression in one, two, three, and four stages. It will be noted that this table does not consider stage compression for pressures below 70 lbs. gauge; and a comparison of the figures in these last four columns, for any given pressure, throws an interesting light on the way in which the possibilities of gain by compounding in various degrees grow larger as pressures increase.

It is to be noted, also, that this table is based on sea-level conditions, and assumes that, whether in single- or multi-stage compression, the initial temperature of the air in each cylinder is 60° F. In other words, in the compound machines, the intercooler is supposed to reduce the temperature of the air to the initial temperature in the low-pressure cylinder. These figures are theoretical only, and no allowance has been made for the usual losses found in actual compressor operation. Increasing these figures by 10 or 15 per cent would liberally cover such losses and serve as a safe approximation to the conditions of everyday work.

TABLE VIII

Multipliers for Determining the Volume of Free Air at Various Altitudes, which, when Compressed to Various Pressures, is Equivalent in Effect to a Given Volume of Free Air at Sea-Level

Altitude in Feet	Barometric Pressure		Multiplier				
	Inches of Mercury	Pounds per Square Inch	Gauge Pressure (Pounds)				
			60	80	100	125	150
0	30.00	14.75	1.000	1.000	1.000	1.000	1.000
1,000	28.88	14.20	1.032	1.033	1.034	1.035	1.036
2,000	27.80	13.67	1.064	1.066	1.068	1.071	1.072
3,000	26.76	13.16	1.097	1.102	1.105	1.107	1.109
4,000	25.76	12.67	1.132	1.139	1.142	1.147	1.149
5,000	24.79	12.20	1.168	1.178	1.182	1.187	1.190
6,000	23.86	11.73	1.206	1.218	1.224	1.231	1.234
7,000	22.97	11.30	1.245	1.258	1.267	1.274	1.278
8,000	22.11	10.87	1.287	1.300	1.310	1.319	1.326
9,000	21.29	10.46	1.329	1.346	1.356	1.366	1.374
10,000	20.49	10.07	1.373	1.394	1.404	1.416	1.424

Air-Compression at Altitudes. As altitude above sea-level increases, the atmospheric pressure diminishes, and the density of the air is correspondingly reduced. The first three columns of Table VIII, compiled by F. M. Hitchcock, show the relation between altitude and atmospheric pressure. This phenomenon of progressive rarification of the air as heights increase, has an important bearing upon air-compression at altitudes.

Air-compressors are rated in cubic feet of free air per minute, but no atmospheric pressure is usually specified. This capacity is ordinarily given as the actual piston displacement of the machine; but this is not correct, since there are always certain losses reducing the volumetric efficiency below unity. When, however, a given amount of work is to be done by the compressor, the atmospheric pressure is an important factor in the problem. For instance, a 2-inch rock-drill is usually rated at an air consumption of 50 cubic feet of free air per minute, compressed to 60 lbs. gauge. If it requires this volume of free air, and the equivalent volume of compressed air at 60 lbs. pressure, at sea-level, when operating at (say) 9,000 feet altitude and supplied with 50 cubic feet of free air per minute compressed to 60 lbs. pressure, it will fall far short of its full rated performance. This is because the equivalent volume at 60 lbs. pressure of 50 cubic feet of free air at sea-level, or 14.7 lbs. absolute pressure, is considerably more than the equivalent volume at 60 lbs. gauge of 50 cubic feet of free air at 9,000 feet altitude, where the atmospheric pressure is only 10.46 lbs. absolute. To provide for this, therefore, a larger volume of free air must be compressed at altitudes than at sea-level, to get a given amount of work out of a machine having a certain rating in cubic feet of compressed air at sea-level.

Columns 4 to 8 of Table VIII give the multipliers for determining what volume of free air at various altitudes and atmospheric pressures it will be necessary to compress in order to produce the equivalent in air under pressure of a given volume of sea-level free air. For example, ten 3¼-inch rock-drills, each rated at 100 cubic feet of free sea-level air at 60 lbs. gauge pressure, are to be operated simultaneously at 6,000 feet altitude; what volume of free air must be supplied them to give rated performance at this altitude? At sea-level the ten drills would require 1,000 cubic feet of free air per minute. The multiplier, from the table, for 6,000 feet altitude and 60 lbs. pressure, is 1.206. The required volume of free air at the higher altitude is therefore 1,000 times 1.206, or 1,206 cubic feet of free air per minute compressed to 60 lbs. gauge. It is evident from this example and from the accompanying discussion, that an increase in altitude is equivalent to a reduction in the volumetric efficiency of the compressor.

Horse-Power for Compression at Altitudes. The power required to compress a given volume of free air to a given pressure, diminishes as the altitude increases. This will be apparent upon a little consideration of the compressed-air indicator card. It will be remembered that such a card is made up of two areas representing two distinct stages of work—*first,* compression to the stated pressure; *second,* discharge at that pressure. As the density of the free air (or the atmospheric pressure) diminishes with increased altitude or otherwise, it is evident that the piston will have to travel further into the cylinder before the required terminal pressure is reached and discharge begins. This will increase the area of compression work and reduce the area showing discharge work. But the latter decreases more rapidly than the former increases; hence the total power represented by the combined areas will be less.

TABLE IX

Horse-Power Developed, with Allowance for Usual Losses, in Compressing One Cubic Foot of Free Air at Various Altitudes from Atmospheric to Various Pressures

Initial Temperature of the Air in Each Cylinder Taken as 60° F.
(Jacket Cooling Not Considered)

Altitude in Feet	Horse-Power Developed							
	Simple Compression			Two-Stage Compression				
	Gauge Pressure (Pounds)			Gauge Pressure (Pounds)				
	60	80	100	60	80	100	125	150
0	.1533	.1824	.2075	.1354	.1580	.1765	.1964	.2138
1,000	.1511	.1795	.2040	.1332	.1553	.1734	.1926	.2093
2,000	.1489	.1766	.2006	.1310	.1524	.1700	.1887	.2048
3,000	.1469	.1739	.1971	.1286	.1493	.1666	.1848	.2003
4,000	.1448	.1712	.1939	.1263	.1464	.1635	.1810	.1963
5,000	.1425	.1685	.1906	.1241	.1438	.1600	.1772	.1921
6,000	.1402	.1656	.1872	.1218	.1409	.1566	.1737	.1879
7,000	.1379	.1628	.1839	.1197	.1383	.1536	.1700	.1838
8,000	.1358	.1600	.1807	.1173	.1358	.1504	.1662	.1797
9,000	.1337	.1572	.1774	.1151	.1329	.1473	.1627	.1758
10,000	.1316	.1547	.1743	.1132	.1303	.1442	.1592	.1717

Table IX, also compiled by F. M. Hitchcock, gives the horse-power required to compress a cubic foot of free air at various altitudes to various pressures, by one- and two-stage compression. In the figures given, a fair allowance has been made for the ordinary losses, so that these values may be taken as representing average conditions of practice. This table, in connection with Table VIII, affords the solution of a very common problem. For instance, what power will be required to secure by two-stage compression at 9,000 feet altitude, the equivalent of 1,000 cubic feet of free air at sea-level at 100 lbs. gauge pressure? From Table VIII, the multiplier for determining the volume is found to be 1.356; from Table IX, the horse-power required per cubic foot of free air compressed at this altitude to this pressure is found to be .1473. The required result, therefore, is 1,000 × 1.356 × .1473, or 199.73 horse-power.

Tables VII, VIII, and IX are here used by permission of the Ingersoll-Rand Company.

TABLE X

Relation of Volumetric Efficiencies and Horse-Power Required in Air-Compression at Various Altitudes

Altitude in Feet	Atmospheric Pressure, Lbs. Abs.	Volumetric Efficiency	Loss of Volumetric Efficiency	Per Cent Power Required	Per Cent Reduction in Power Required
Sea-Level	14.75	1.00	.00	1.000	.000
1,000	14.20	.97	.03	.982	.018
2,000	13.67	.93	.07	.965	.035
3,000	13.16	.90	.10	.948	.052
4,000	12.67	.87	.13	.931	.069
5,000	12.20	.84	.16	.915	.085
6,000	11.73	.81	.19	.899	.101
7,000	11.30	.78	.22	.884	.116
8,000	10.87	.76	.24	.869	.131
9,000	10.46	.73	.27	.854	.146
10,000	10.07	.70	.30	.839	.161
11,000	9.70	.68	.32	.824	.176
12,000	9.34	.65	.35	.809	.191
13,000	8.98	.63	.37	.794	.206
14,000	8.65	.60	.40	.779	.221
15,000	8.32	.58	.42	.765	.235

Table X shows the relation of volumetric efficiencies and horse-power required in air-compression at various altitudes. It will be noted that the volumetric efficiency decreases more rapidly than the required power.

Unavoidable Losses in Air-Compression. From the preceding discussion of the general theory and phenomena of air-compression, the following concrete deductions may be made as to the losses which are unavoidable in practical air-compression work—losses which can be kept down to the minimum by correct design and construction, but which can never be entirely eliminated.

The *first loss* is that due to friction in the mechanical structure which transmits the power from the driving to the compressing element.

The *second loss* is that arising from the heating of the air during admission to the cylinder, resulting in a reduction in its density and a diminished volumetric efficiency of the cylinder.

The *third loss* arises from the heat produced by compression, and its effect upon the air during compression, calling for more power.

The *fourth loss* comes from the reduced volumetric efficiency of the cylinder due to the fact that the intake air is seldom at atmospheric pressure, owing to friction in the inlet valves and passages.

The *fifth loss* is also one of volumetric efficiency, due to the effect of the clearance air in reducing the effective length of the admission stroke.

It is not possible to give even average percentages for any of these losses, which will occur in different combinations depending on the mechanical details, the general type, the speed, and the capacity of the compressor. After all these losses have been deducted, the *net efficiency* of the machine remains; and of this, Mr. Frank Richards, in his work on "Compressed Air," says: "It is safe to say that the ultimate efficiency never goes as high as 80 per cent, while it often goes below 60 per cent."

CLASSIFICATION OF COMPRESSOR TYPES

Any air-compressor may be considered as made up of three primary parts—the *driving element;* the *compressing element;* and the mechanical structure uniting these two, which may be designated as the *power-transmitting element.* It is not always possible to draw the line clearly between these several elements, but this subdivision can always be made with more or less exactness.

The first classification of compressor types will be based upon the motive power or method of drive—the driving element. This division affords two basic types—*Steam-Driven* and *Power-Driven* compressors The former type includes all those compressors in which a steam cylinder (or cylinders) is an integral part of the ma-

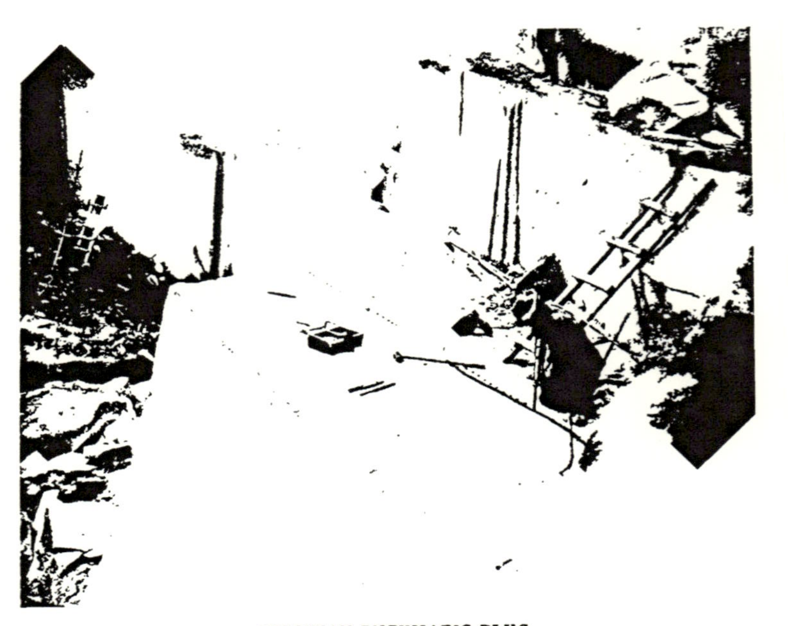

SULLIVAN PNEUMATIC PLUG
Hammer Drills in Vermont Quarry

chine. The latter includes all those compressors where power is applied from a source outside of the machine, such as a driving shaft, an electric motor, or a water wheel.

Power-driven compressors may be still further classified by the means employed for applying the power, into *Belt-Driven*, *Rope-Driven*, *Chain-Driven*, *Gear-Driven*, and *Direct Shaft Connected* (with motor or water wheel).

Another classification is based upon the direction in which the reciprocating parts of the machine work; and its divisions include *Vertical Compressors* and *Horizontal Compressors.*

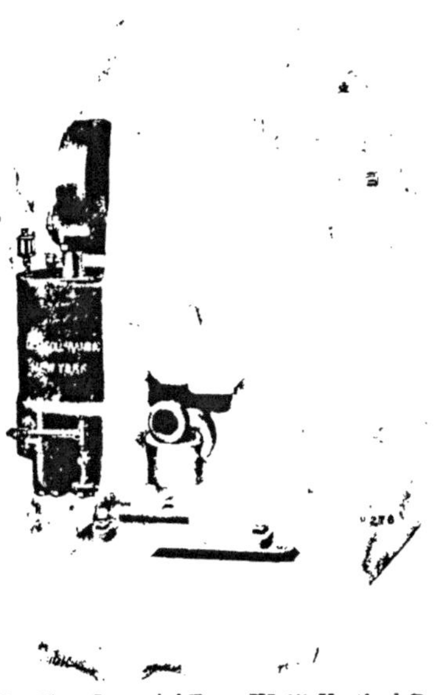

Fig. 15. "Imperial Type XI-1" Vertical Duplex Power-Driven Air-Compressor. Ingersoll-Rand Co., New York City.

The third general classification refers to the relative position of the driving and compressing elements. When a steam-driven compressor has all its steam and air cylinders arranged on an axial line with their pistons on one long rod, the type is designated as a *Straight-Line Compressor.* When the compressor consists of two distinct elements, each comprising a driving and a compressing element, and these two elements are arranged side by side with a common shaft, the type is known as a *Duplex Compressor.* These two latter types, of course, originated in the steam-driven machine; but the general design has been carried on into the power-driven class, so that there are straight-line and duplex power-driven compressors as well.

The fourth general classification has to do entirely with the compressing end. A compressor in which all the process of compression is carried on in one cylinder, is known as a *Single-Stage Compressor.* A duplex compressor having two air-cylinders, in each of which compression is carried from initial pressure to terminal or discharge pressure, is called a *Duplex Single-Stage Compressor.* When the

Fig. 16. Horizontal Straight-Line Simple Steam Single-Stage Compressor, Class BS. Bury Compressor Company, Erie, Pa.

Fig. 17. Class NT-32 Double Cross-Compound Steam Driven Air-Compressor. J. Geo. Leyner Engineering Works Company, Littleton, Col.

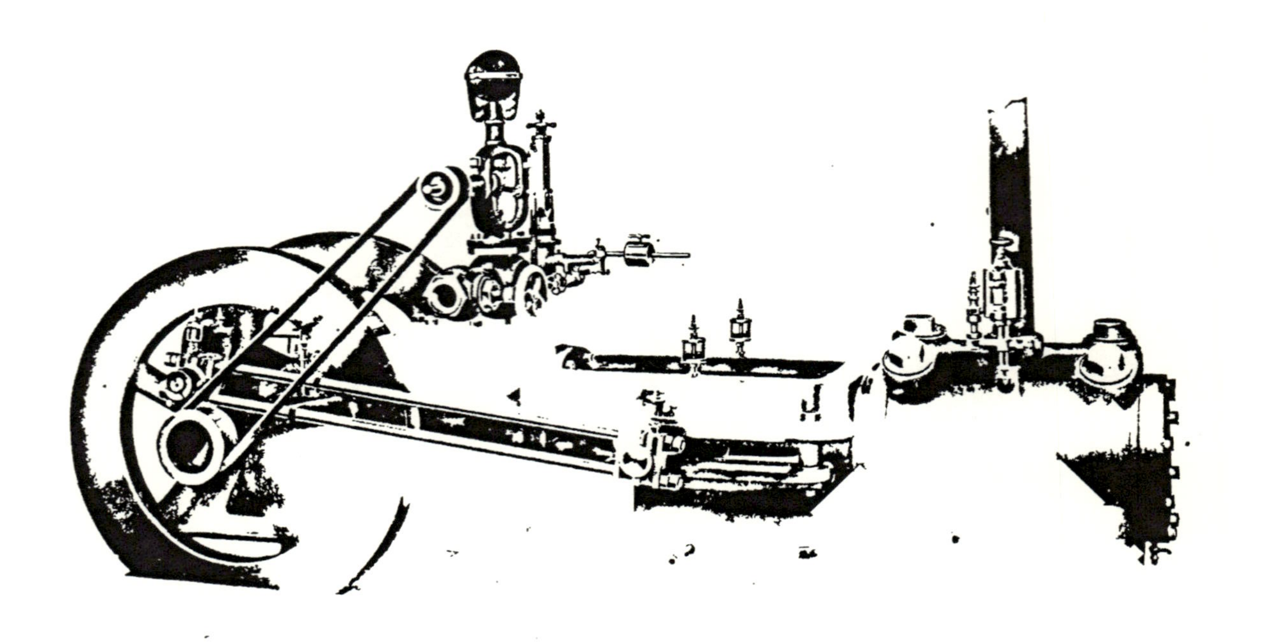

Fig. 18. Class WA-3 Straight Line Simple Steam Single-Stage Air Compressor. Sullivan Machinery Co., Chicago, Ill.

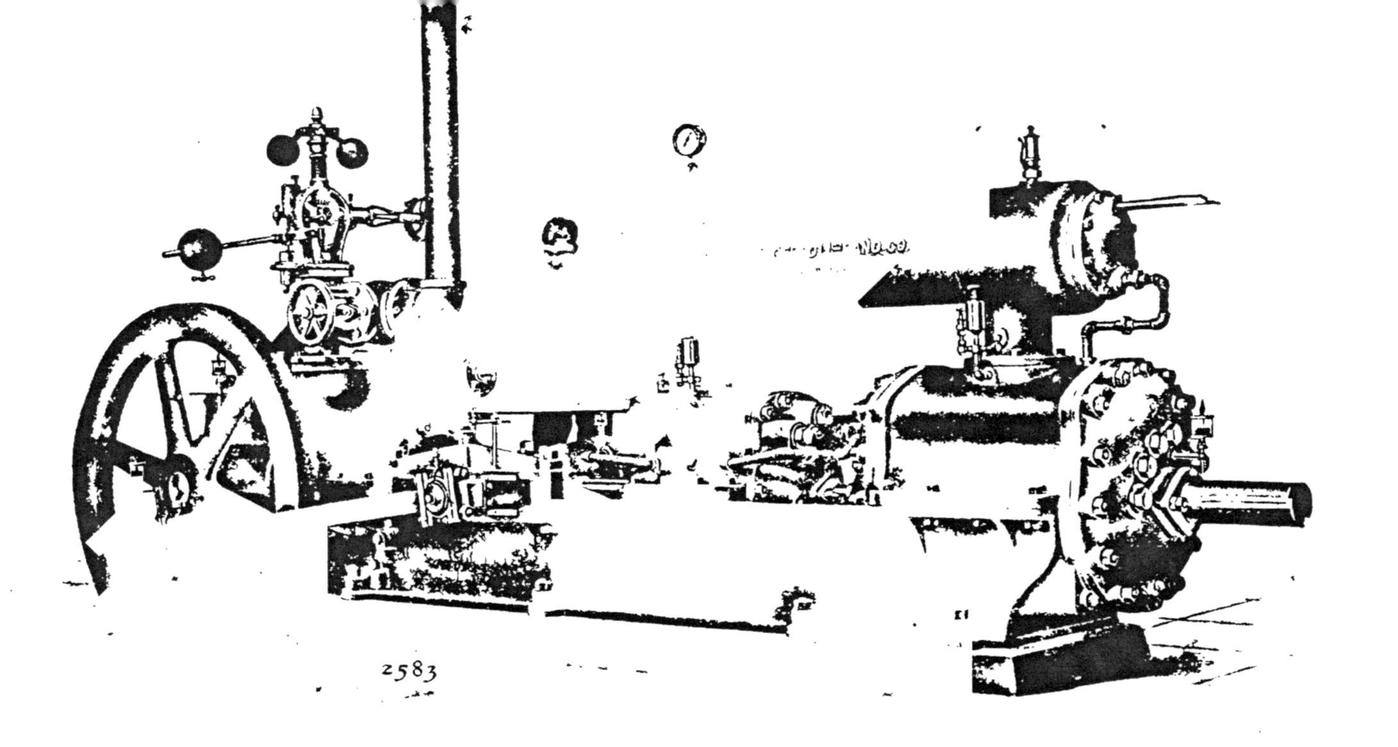

Fig. 19. Class A-2 Simple Steam Tandem Compound Straight-Line Air-Compressor.
A typical 18-inch stroke machine with tandem compound air cylinders and horizontal overhead intercooler.
Ingersoll-Rand Co., New York City.

Fig. 20. Class WB-2 Straight-Line Simple Steam Tandem Compound Air Compressor. Sullivan Machinery Co., Chicago, Ill

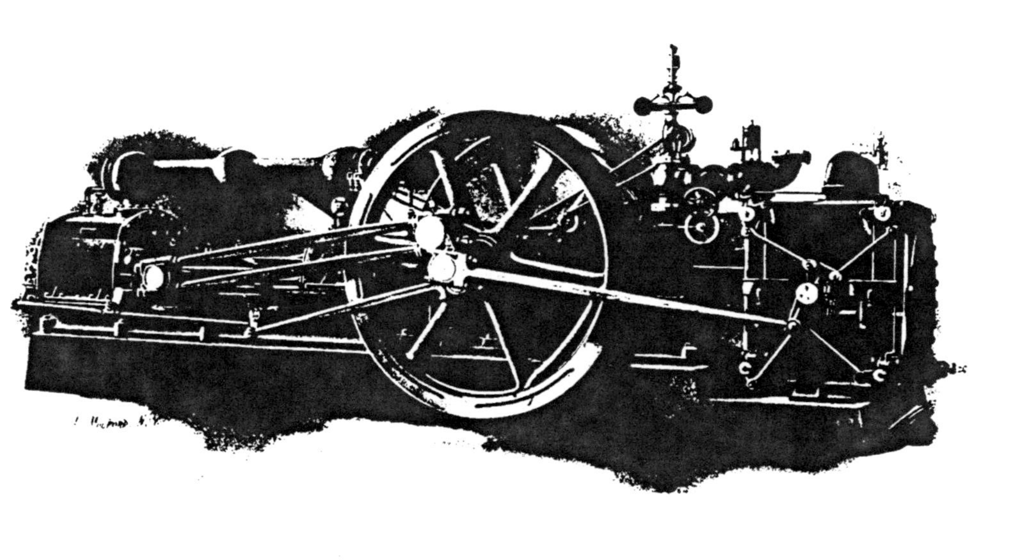

Fig. 21. Tandem Double-Compound Straight-Line Air-Compressor.
Norwalk Iron Works Company, South Norwalk, Conn.

Fig. 22. Franklin Corliss Double Cross-Compound Steam-Driven Air-Compressor. Compound steam cylinders and two-stage air cylinders. (Separate intercooler not shown.) Chicago Pneumatic Tool Company, Chicago, Ill.

Fig. 23. Class WX Corliss Double Cross-Compound Steam-Driven Air Compressor. Sullivan Machinery Co., Chicago, Ill.

compression process is divided between two cylinders, in the first of which a certain pressure is reached, and in the second of which the pressure is carried from the terminal pressure of the first cylinder to final discharge pressure, the machine is called a *Two-Stage Compressor.* Where three degrees of compression in three cylinders are used, the machine is a *Three-Stage Compressor*; and with four degrees or stages of compression in four cylinders, a *Four-Stage Compressor.* The term *compound,* as applied to air cylinders, usually implies a two-stage machine, while *multi-stage* is usually applied to three and four stages.

A straight-line compressor with simple steam cylinder and a single-stage air cylinder, is spoken of as a *Simple Steam Single-Stage Straight-Line* or a *Simple Tandem Straight-Line.* When a two-stage or three-stage air end is used in a straight-line machine in connection with a simple air cylinder, the machine is designated as a *Simple Steam Tandem Compound* or a *Simple Steam Tandem Three-Stage.* When the steam cylinders are also compounded, the machine becomes a *Tandem Double-Compound* or a *Tandem Double-Compound Three-Stage.*

A duplex compressor with two simple steam cylinders and two single-stage air-cylinders is a *Simple Duplex*; with compound steam cylinders and two single-stage air cylinders, a *Cross-Compound Duplex Air*; with compound steam cylinders and two-stage air cylinders, a *Double Cross-Compound*; with compound steam and multi-stage air end, a *Double Cross-Compound Three-* or *Four-Stage*; with simple steam and compounded or multi-stage air cylinders, a *Duplex Steam Cross-Compound* or a *Duplex Steam Cross-Compound Three-* or *Four-Stage.*

While different methods of designation are found to hold among different manufacturers, those just given are based on rational grounds, and will be adopted in this treatise for the sake of uniformity.

There are, of course, compressor types which do not conform exactly to any of the classifications here given; but an analysis of their designs, with proper allowance for certain features more or less special in character, will place them in some of the classes here defined. There is a growing tendency toward a standardization of types among all the large builders; and in the present work on Compressed Air, only the best modern practice will be considered.

STEAM-DRIVEN COMPRESSORS

The ordinary steam-driven compressor is simply a steam engine with one or more air cylinders directly coupled on the extended piston-rod. With this understanding, it will be unnecessary here to enter into any detailed discussion of the steam end of the air-compressor.

It will be enough to state that the smaller steam-driven compressors generally use a plain steam valve—either of D or piston type—with a fixed point of cut-off adjusted for the best steam economy under average conditions. Following this first and simplest class of steam ends, comes the class in which adjustable cut-off steam valves are used, either of the flat or piston type, balanced or unbalanced as the case may be. A third general class is that in which Corliss steam valves, more or less modified, are used; and these are usually found on the larger sizes of compressors, where the very best steam economy is sought. These Corliss valves are either of the "drop-release" pattern, or of the fixed type with no release. Corliss steam valves, to realize their best economy, should be used with a condensing apparatus.

Distinction between Steam-Engine and Compressor Practice. There is one vital point of difference between steam-engine and air-compressor practice which is to be borne in mind throughout all the discussion that follows. While the steam engine for ordinary power service is designed to maintain a constant speed irrespective of load, the steam-driven air-compressor must run at varying speed, since its output of air is proportional to its piston displacement, and the latter in turn is proportional to the speed. This means that while the steam engine in ordinary power service has a variable load per stroke, in air-compressor practice the steam cylinder has a *constant load per stroke*, except for the small discrepancy due to varying friction and compressive efficiencies at varying speeds.

Compound Steam Cylinders on Air-Compressors. It is clearly understood that the fundamental object of compounding steam cylinders is to secure a higher economy by the more complete expansion of the steam—that is, to make more complete the abstraction of the heat energy of the steam by increasing the difference between the initial and terminal temperatures.

Another advantage of compounded steam cylinders, which may

be referred to later, is that the maximum pressure in compound cylinders—and therefore the maximum load on valves, valve gears, etc. —is less than in an equivalent simple steam cylinder. This means lower friction losses, lower terminal pressures on bearings, etc., and a general improvement in the operating conditions.

The first essential of successful steam compounding is in securing the proper ratio of volumes in high- and low-pressure cylinders, so that when a given quantity of steam (determined by the cut-off on the high-pressure cylinder) is admitted and expanded through the two cylinders, its terminal or exhaust pressure shall be such that there shall be no loss by condensation due to over-expansion. This means that for a given steam pressure, for a given exhaust or back pressure, and for a given cylinder ratio, there is one and only one proper point of cut-off for both cylinders, to give the best economy. In the relation just expressed, the first and third elements must be considered as constant, leaving the latter element as the variable to be adjusted to meet varying conditions.

Now, where a constant speed must be maintained regardless of load, this change of speed must evidently be produced by a change of cut-off; and this immediately destroys the proper adjustment, and the best economy is at once sacrificed. Constant-speed practice with compound engines must at the best be a compromise based on varying load conditions.

But in the air-compressor there is a constant load per stroke (ignoring friction for the present) so that the cut-off remains practically constant at all speeds, and the correct ratio of expansion is maintained over the full load range. The word *practically* is used above with deliberate intent, since there is bound to be a very slight departure from perfect conditions to produce the requisite change of speed under changing load, as will be seen in a subsequent section on Regulation. But enough has been said here to demonstrate that the air-compressor offers the ideal opportunity for securing in highest degree the advantages of steam compounding.

Steam Pressures for Compounding. Within practical limits, the higher the initial steam pressure, the greater are the advantages of compounded steam cylinders. Compressors have been built with triple- and quadruple-expansion cylinders using steam pressures as high as 250 lbs. These machines, however, are so rare that they need not enter into a discussion of general compressor practice; and the most that the average practical man may expect to encounter is a compound or double-expansion compressor. The peculiar operating conditions of the air-compressor, just explained, justify the adoption of compounded steam ends on pressures lower than would be advisable for compounding in ordinary steam-engine practice. The

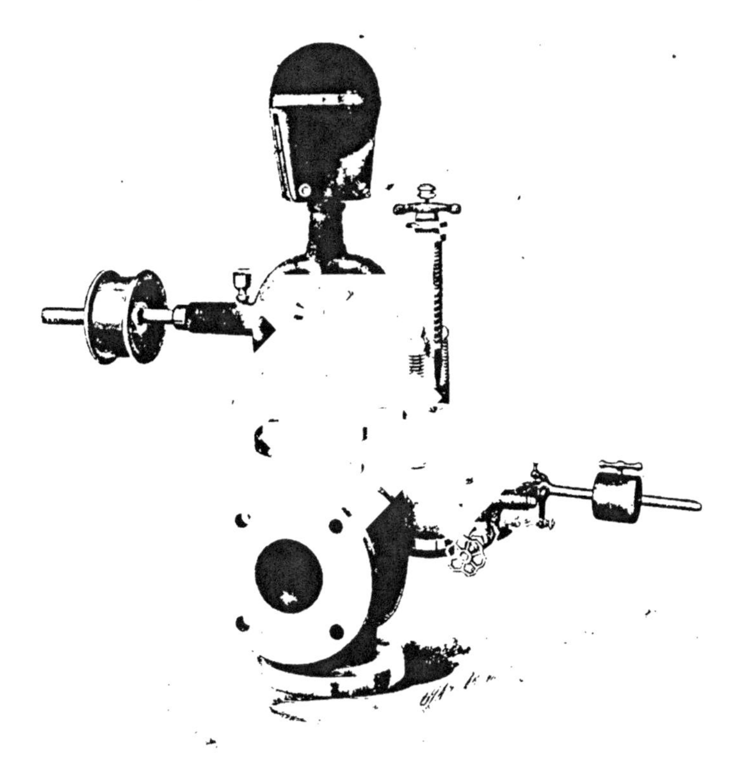

Fig. 24. Combined Speed and Pressure Regulator for Steam-Driven Compressors. Sullivan Machinery Co., Chicago, Ill.

largest builders of air and gas compressors to-day are recommending compounded steam ends for 80 lbs. boiler pressure where the machine is to be run condensing, and for 90 lbs. boiler pressure running non-condensing. Above these minima the pressures may run up to the limit of everyday practice, which may be as high as 125 or even 150 lbs. gauge.

The three fundamental advantages of compounded steam cylinders are as follows:

Lower fuel consumption and fuel cost, owing to the smaller quantity of water to be evaporated.

Lower cost of boiler feed-water, as a result of the greater power secured per pound of water by the higher expansion.

Lower cost of boiler plant and accessories, because of the reduced amount of water required to be evaporated.

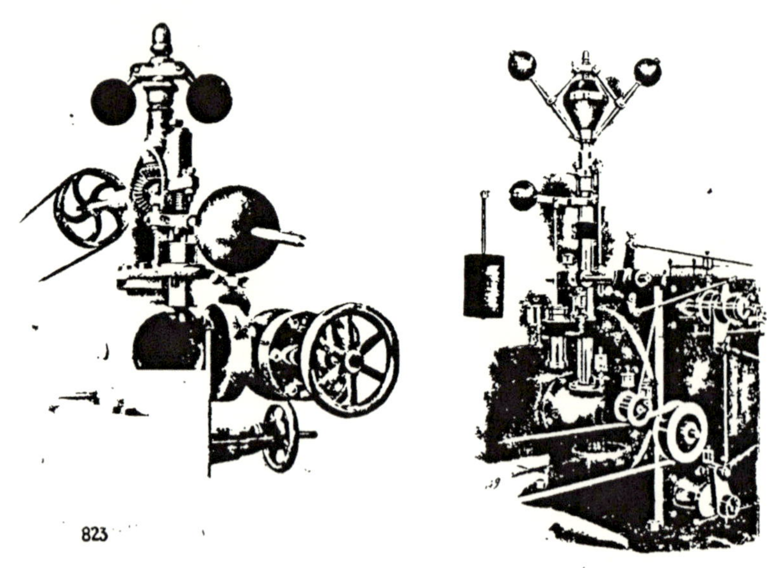

Fig. 25. Combination Air Ball Governor and Pressure Regulator for Steam-Driven Compressors. Ingersoll-Rand Co., New York City.

Fig. 26. Class A-15 Combined Speed and Pressure Regulator for Corliss Compressors. Ingersoll-Rand Co., New York City.

Other incidental advantages of steam compounding will be developed later in another connection.

Regulation of Steam-Driven Compressors. While the regulating devices or *governors* used on air-compressors vary in detail among the different builders, all may be divided, so far as the principle of operation is concerned, into two classes.

The first class is that applied to compressors with plain or adjustable cut-off valves of flat or piston type. It operates by throttling the steam supply as load diminishes. Devices in this class consist fundamentally of a valve in the steam pipe, which is opened or closed by the action of a piston in a cylinder, this piston being actuated by air pressure from the receiver. The movement of this piston is opposed by weights on a lever or by a spring; and the spring tension or the weights may be adjusted so that the governor valve is full open at any desired normal air pressure. But when pressure exceeds this limit, the tension or weight is overcome, and the valve in the steam supply is closed in a degree corresponding with the amount of excess pressure. This slows down the machine and reduces the volume of air discharged until such time as the

Fig. 27. Sergeant A-14 Steam Regulator.
Automatically regulates speed under varying load by throttling steam supply.
Ingersoll-Rand Co., New York City.

normal pressure is reached, when the weights or spring tension again open the governor valve, and full speed is restored. Evidently this partial throttling of the steam supply will result in some wire-drawing of the steam, which is about offset by the resulting superheating of the steam.

The second class includes governors applied to machines with Corliss steam valves. The mechanism consists of a pressure cylinder and piston with the opposing weights or springs as described in the preceding paragraph; but in this case the movement of the governor, instead of throttling the steam, changes the cut-off of the steam valves, reducing speed under partial load, and restoring it as load increases. The resistance per stroke being the same through-

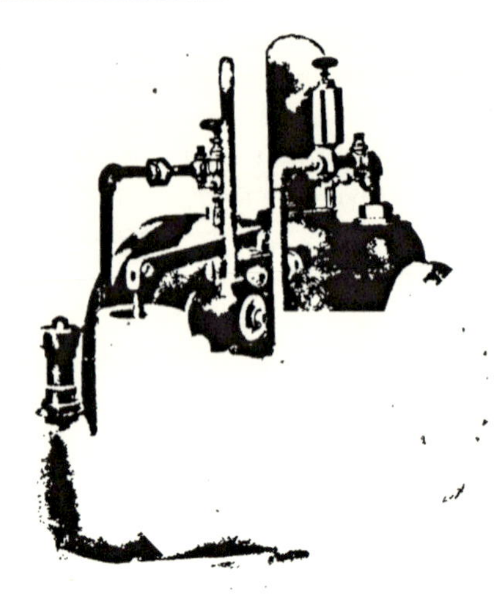

Fig. 28. SergeantA-28 Air Unloader Used with Steam Regulator Shown in Fig. 27.

out, a very slight change of cut-off, hardly affecting the economy, produces all the speed change necessary.

Both of these classes of governors include also a speed-limiting device, the more common form being the familiar fly-ball arrangement, which throttles the steam supply or greatly shortens the cut-off when speed exceeds a certain limit, as it might in case an air-pipe should break or other accident occur.

There are occasional instances in which governors of one or other of the above classes are used in connection with an *unloader* on the air end, so arranged that, as speed falls off, the load is partially taken from the air end. These are the specialties of individual builders, and as such cannot be elaborated upon here. The two general classes defined cover the general requirements of this paper.

Fig. 29. Belt-Driven Single-Stage Straight-Line Compressor, Class A. Blaisdell Machinery Co., Bradford, Pa.

SULLIVAN AIR COMPRESSOR, DRIVEN BY 30-H. P. GASOLINE ENGINE

POWER-DRIVEN COMPRESSORS

Power-driven compressors are an evolution from steam-driven types, and partake of the general characteristics of the latter. They differ, however, in one vital essential—namely, while steam-driven compressors are variable-speed, variable-load machines, power-driven

Fig. 31. Class D-1 Rope-Driven Simple Duplex Air-Compressor. Ingersoll-Rand Co., New York City.

compressors are in the vast majority of cases constant-speed variable-load machines. This arises from the fact that the latter are usually driven from some constant-speed prime mover; and it makes the problem of regulation quite distinct in the two cases. Before taking up this point in detail, however, the general subject of power-driven compressors will be examined.

Belt drive is always applicable except where a very short

belt center, or distance between centers of driving and driven wheels, is necessary. In the latter case, so high a belt tension is required to avoid slipping of the belt, that an undue pressure on the shaft bearings is produced, resulting in excessive friction, low mechanical efficiency, and rapid wear of the bearing boxes. Belt centers should always be long enough to give the requisite belt con-

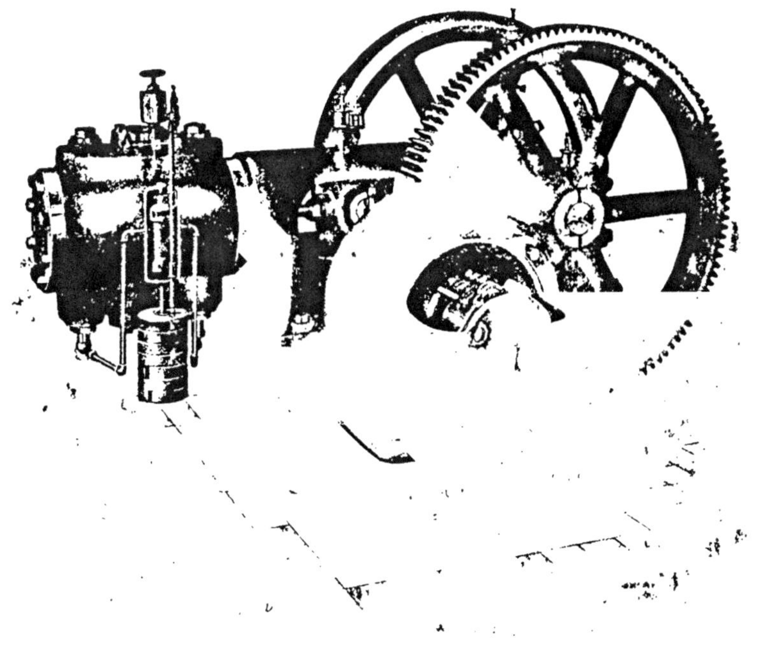

Fig. 32. Gear-Driven Straight-Line Single-Stage Compressor, Class BE. Bury Compressor Company, Erie, Pa.

tact with a comparatively slack belt; 20 to 25 feet is a good average, but 15 feet is perhaps the minimum which should be used. A good rule is to make the belt center three to four times the diameter of the larger pulley—preferably four times. The direction of belt motion should be from the top of the driving wheel to the top of the compressor wheel, this arrangement placing the slack belt on top and increasing the arc of contact.

Rope drive is always applicable, though seldom used where the power to be transmitted is small. It is particularly advantageous where a short belt center only can be had, as the grip of the rope in properly shaped grooves on the wheels gives all the pulling power necessary without undue tension and without excessive bearing friction. The durability and efficiency of the rope drive depend upon a correct shape to the rope grooves, and upon the use of a suitable number of ropes so that excessive strain on any one rope will be avoided.

Gear drive and *silent chain drive* are never desirable, and

Fig. 33. Class EE-1 Chain-Driven Single-Stage Electrical Air-Compressor. Ingersoll-Rand Co., New York City.

should be used only where conditions imperatively demand them, as where a very compact compressor unit is necessary. They are ordinarily used on compressors driven by an electric motor; and as the efficiency of either method of drive depends very largely upon maintaining proper alignment, the motor in such cases is usually mounted rigidly on an extension of the compressor frame, or else a sole-plate should unite motor and compressor bases. The best practice places the limiting power for gear drive with a rawhide pinion, at 40 H.P.; for gear drive with a steel pinion, at 60 H.P.; and for silent chain drive, at 50 H.P.

The rules for proportioning belts, ropes, gears, and chains for compressor drive, are the same as those used in ordinary mechanical work.

A very modern development is the direct-connected electric-driven compressor, with the armature or rotor of a slow-speed motor mounted directly on the compressor shaft. This direct-shaft drive has also been used where the conditions of head and volume of a water power were such as to permit mounting an impulse wheel directly on the compressor shaft. These constructions are applicable only in units of comparatively large capacity, since the rotative speed is necessarily limited, and it is only in the large powers that motors

1481

Fig. 34. Class J-2 Cross-Compound Air-Compressor Direct-Connected to Pelton Water Wheel.
Ingersoll-Rand Co., New York City.

can be used giving the required slow speed without excessive cost. The best machines of this class have very heavy mechanical structures, unusually large valve and port areas, and very effective cooling devices, so that a comparatively high speed (as compared with ordinary compressor practice) can be used without sacrifice of mechanical and volumetric efficiency. The direct-connected arrangement is the ideal one for electric or water-power drive, as the mechanical losses between motor and compressor are practically eliminated, the friction being little if any more than that of the compressor alone.

Regulation of Power-Driven Compressors. Since the power-driven compressor is almost always a constant-speed machine, the methods of regulation and governing described for variable-speed

Fig. 35. Class PE-2 Direct-Connected Cross-Compound Electrical Air-Compressor. Ingersoll-Rand Co., New York City.

steam-driven machines evidently cannot here be applied. Constant speed means constant piston displacement; and the problem of delivering a variable volume of air with constant piston displacement, becomes one of making a portion of that displacement non-effective in the compression and delivery of air. Only the fundamental principles of several methods of accomplishing this will here be discussed.

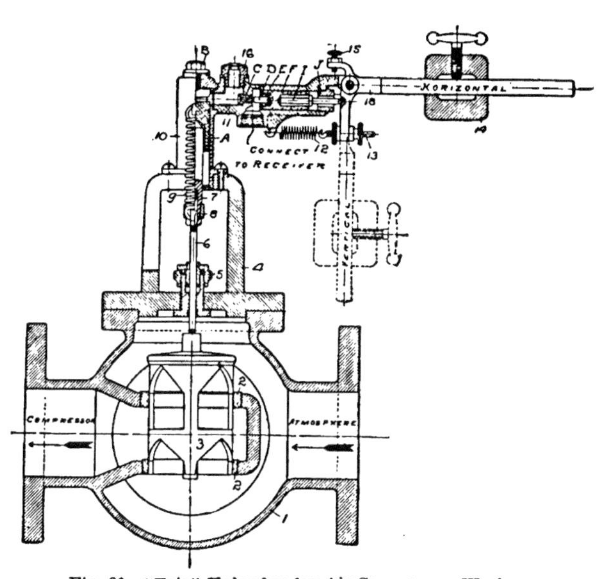

Fig. 36. "Erie" Unloader for Air-Compressor Work.
Maintains a constant pressure in receiver by governing inlet of compressor to let in only exact amount of air used, and to load compressor in exact proportion to work required.
Jarecki Manufacturing Company, Erie, Pa.

The first method is really one of *unloading*, rather than of regulating. A pressure-controlled mechanism is arranged so that when pressure exceeds normal, due to excess of delivered volume over demand, a communication is opened between the two sides of the compressing piston. Usually this is accomplished by opening and holding open one or several of the discharge valves at both ends of the cylinder, the air then simply sweeping back and forth from one side of the piston to the other through the open valves and the air-discharge passage. When normal pressure is restored, the valves are automatically closed, and compression and delivery are resumed. Evidently this is practically a total unloading of the machine for a longer or shorter period—a sudden release from load and a sudden resumption of load. Moreover, the air which is swept back and forth by the piston in its travel is air under full pressure; so that when the discharge valves suddenly close, the piston at once encounters a full cylinder of air at maximum pressure. These

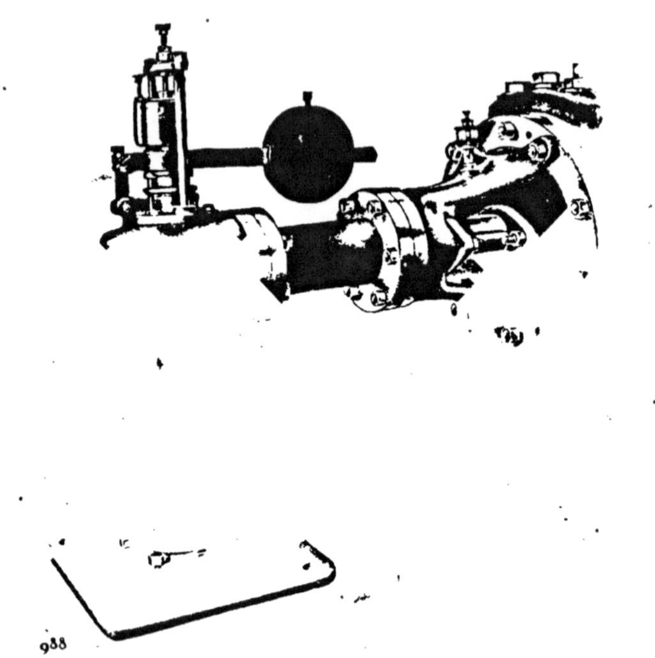

Fig. 37. Class A-18 Choking Controller for Power-Driven Compressors. Automatically proportions power to load by throttling the air intake as pressure rises above the fixed normal.
Ingersoll-Rand Co., New York City.

facts limit regulators of this class to machines of comparatively small capacity.

Another method provides, by means of a pressure-operated device, for the partial or total closing of the compressor intake under reduced load. To avoid the dangers attendant upon such an operation acting suddenly, these devices are provided with some damping mechanism so that they are compelled to operate slowly, making the release or resumption of the load gradual. The cutting down of the air intake results in a rarification of the air entering the cylinder, and a greater range between initial and discharge pressures, with a corresponding increase in the range of temperatures. This method of regulation, therefore, is not suitable for very great load variations; nor is it recommended for such conditions by the builders responsible for it.

The third method is very similar to the first, except that here the inlet valves, instead of the discharge valves, are held open when the machine is unloaded, the piston thus simply drawing in and forcing out air at atmospheric pressure. It is open to the same criticism (though in somewhat less degree) as the first method—namely, undue shock and strain on release and resumption of load.

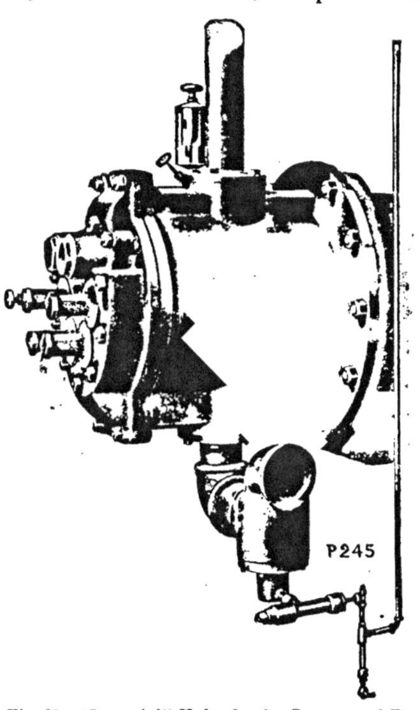

Fig. 38. "Imperial" Unloader for Steam- and Power-Driven Compressors. Ingersoll-Rand Co., New York City.

The fourth method uses a pressure-controlled valve on the compressor discharge of single-stage machines, combining also the functions of a check-valve to limit the escape of air from the receiver or air line. Excess pressure blows the discharge to atmosphere, instead of into the line. This arrangement is also used on two-stage machines by placing it on the low-pressure discharge to the intercooler. Then, when the governor valve is opened by excess pressure, the low-pressure cylinder discharges to atmosphere, and the high-pressure cylinder acts simply as a low-pressure cylinder with intake at atmospheric pressure. This device is more of a relief-valve than an unloader, for the piston must continue to compress to a pressure which will open the discharge valves; and this volume of compressed air, with its power equivalent, is wasted.

On power-driven compressors with Corliss intake valves, several different methods of unloading or regulating are used.

By one method, the Corliss valve is held open for the full admission stroke, and also for a part of the compression stroke, this latter

Fig. 39. Unloading Device for Steam and Power-Driven Air-Compressors. Sullivan Machinery Co., Chicago, Ill.

portion being determined by the degree of unloading called for. Evidently this is practically equivalent to a shortening of the stroke of the compressor.

By another method the Corliss intake valve is opened full at beginning of admission, but closes later in the admission stroke. The air admitted to that point is expanded or rarified for the remainder of the compression stroke, and then compressed, the volume of compressed air delivered being of course reduced. This arrangement is productive of an excessive temperature range in the cylinder.

Still a third method opens and holds open the intake valves at one end of the cylinder, or at opposite ends in duplex machines.

The effect of this is to make ineffective one out of every two strokes. If still further unloading is necessary, the intake valves at the other end of the cylinder or cylinders are opened and held open.

The three arrangements just outlined all operate by a pressure-controlled mechanism which actuates some form of trip or release on the Corliss air-valve gear, somewhat similar to the release mechanism of the Corliss steam valve for varying the cut-off.

Three things are to be avoided in the successful unloader or regulator for power-driven compressors: *first,* a sudden release or resumption of load, throwing heavy strains on the machine; *second,* undue rarification of the intake air, resulting in a wide range of cylinder pressures and temperatures; *third,* the blowing-off of compressed air to the atmosphere with a waste of power.

In the regulation of the power-driven compressor, less reliance must be placed upon the automatic regulation of the individual machine than upon the intelligent subdivision of the load between two or more machines and the careful management of the resulting plant. In designing a plant of these machines, maximum capacity must be cared for in the normal full output of the machines, while partial loads are provided for by starting or stopping one or more machines, the remainder running at or very near full load. This question will be taken up later in the discussion of air-power plant design.

Starting Unloaders for Power-Driven Machines. It is usually desirable to start a power-driven compressor with no load, throwing on the load gradually after normal speed has been reached. This is in fact essential in machines driven by electric motors, for the heavy inrush of current in starting under load is dangerous, particularly where power is taken from a transmission circuit supplying other motors.

Evidently almost any of the unloading devices noted in the previous section can be used for this purpose if properly arranged for manipulation. The usual form, however, is simply a by-pass valve to atmosphere on the line close to the compressor, protected by a check-valve between it and the receiver to prevent the return of air from the line when the starting unloader valve is open. This check-valve is essential where several compressors serve one line, permitting cutting in or out any machine without unloading the others. This by-pass valve is opened on starting, when the compressor simply compresses to a pressure sufficient to open its discharge valves, this air escaping then to atmosphere. When normal speed is reached, the by-pass or unloading valve is gradually closed and load resumed. On two-stage machines, an unloader valve should be provided on the low-pressure discharge to the intercooler, as well as on the high-pressure discharge to the line. In the latter case, both cylinders operate momentarily as low-pressure cylinders.

VERTICAL AND HORIZONTAL COMPRESSORS

The vertical air-compressor has never attained to any degree of popularity, except for comparatively small capacities and for special work where the smallest possible floor space is demanded, as in the high-pressure machines for torpedo service on warships. About the only advantage the vertical type has, is in its compactness; and that has never been as important a feature in compressor as in engine practice. When large capacities are reached, the horizontal type offers a decided advantage in its ready accessibility, whereas vertical compressors of equivalent capacity would tower above the floor-line out of reach for ready inspection and close attention. Vertical compressors are almost exclusively limited to the power-driven class; and in the smaller sizes, there are excellent machines of this type on the market. The horizontal type vastly predominates, however, even among small-sized units.

STRAIGHT-LINE AIR-COMPRESSORS

The straight-line air-compressor is probably the original type among machines for this purpose; for, in the beginning of pneumatic development, the first logical step would have been to attach an air cylinder to a steam engine, with its piston on an extension of the steam piston-rod. In essentials this is all that a straight-line compressor is—one or more steam cylinders and one or more air cylinders arranged in a *straight line*, with their pistons on one continuous piston-rod. An example of this type in its simplest form, is the *air-pump* on a locomotive, compressing air for the air-brakes. Add a pair of fly-wheels to equalize operations throughout the stroke, and a typical straight-line compressor is the result.

The distinguishing characteristic of a straight-line compressor is the direct application of power to load, with a minimum intermediate loss. The best machines of this type are marked by the greatest simplicity of construction, and by a freedom from auxiliary and accessory mechanisms which absorb power in friction, require frequent attention, and offer possibilities of derangement and breakdown. In fact, the simplicity of the straight-line type is its chief claim for consideration; and when attempts are made to add complicated and questionable refinements to it, the excellence of the original type is likely to be sacrificed, with no material gain.

Fig. 40. Straight-Line Steam-Driven Tandem Compound Air-Compressor. Norwalk Iron Works Company, South Norwalk, Conn.

The compactness of the straight-line compressor, and the fact that it is usually a self-contained unit entirely supported by its bedplate or frame, are other strong arguments in its favor, adapting it for use with a very simple foundation (or no foundation at all, in emergency), making its proper installation the simplest matter, and affording the greatest ease of transportation as a complete unit. This probably explains the fact that the straight-line type has found its greatest application for mining and contract work, where the location is difficult of access and where the plant is more or less temporary in character.

There is no more sturdy or dependable compressor type than the best modern straight-line within its proper sphere, if the fundamental principles which distinguish it are clearly understood and closely adhered to. It is pre-eminently the machine for everyday service in small or moderate capacities under conditions where moderate first cost, ease of installation, reliability in the hands of the average engineer, and a good day-after-day economy are the main objects sought. For other conditions—particularly where the lowest operating economy is important—preference should be given to the duplex type, which will be discussed later.

There are some points about the straight-line compressor which are not beyond criticism; but the well-known advantages of th type usually outweigh these disadvantages in the class of service for which this machine is fundamentally intended. A plain, simple

Fig. 41. Class F-1 Steam-Driven Straight-Line Air-Compressor. Ingersoll-Rand Co., New York City.

steam, single-stage, straight-line compressor is almost above reproach in its proper place. The addition of another air cylinder for two-stage compression, means a longer piston-rod, with more expansion and contraction under changes of temperature to be provided for in the cylinder clearance. If a second steam cylinder for steam compounding is added, this trouble is still further magnified. In the latter case the very long connecting rods required, also offer possibilities of trouble. But the simple steam tandem compound and the double tandem compound compressors are not to be altogether condemned, even on these grounds. There are first-class machines built in both these types.

The double tandem compound straight-line cannot be considered wholly a success under the average conditions for which a straight-line machine would probably be selected. The inherent characteristic of the type—the direct opposition of power to resistance—produces conditions offering serious objections to this construction

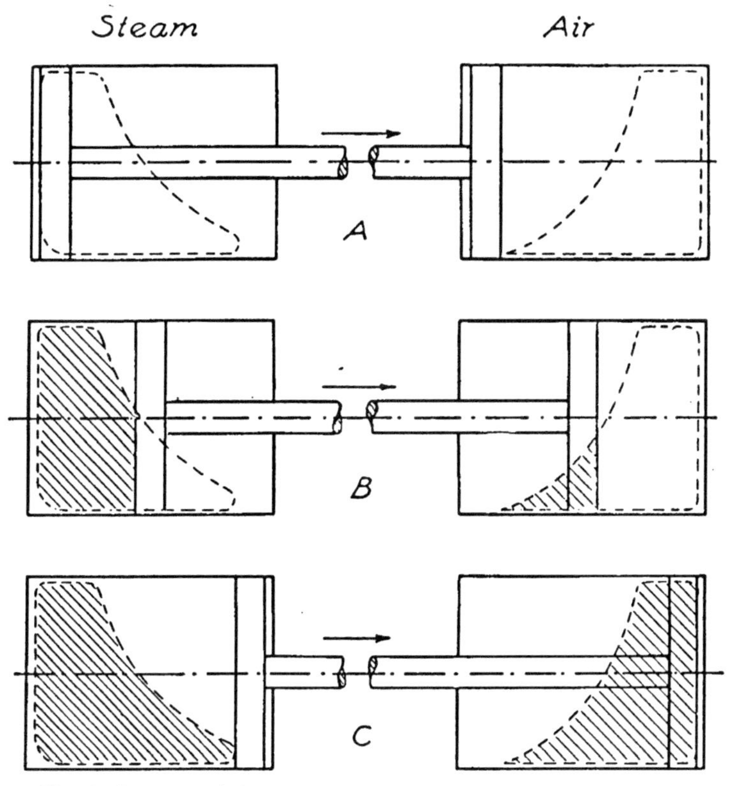

Fig. 42. Pressure Relations in the Simple Steam Single-Stage Straight-Line Air-Compressor.

under widely variable loads. This will be more clearly understood after an investigation of the pressure relations in the steam and air cylinders of the straight-line air-compressor.

Pressure Relations in the Straight-Line Type. Fig. 42 indicates diagrammatically the principle of the straight-line compressor with simple steam and air cylinders. The indicator cards of the two ends are only roughly drawn in, with no attempt at accuracy. In *A*, the

Fig. 43. Class A-1 Simple Steam Single-Stage Straight-Line Air-Compressor. Ingersoll-Rand Co., New York City.

AN ELECTRIC-COMPRESSED-AIR DRILL IN THE "SEVEN MILE TUNNEL," DUMONT, COLORADO
Ingersoll Sargant Drill Co.

stroke is just beginning, and boiler pressure is acting on the steam piston with maximum power, which will be maintained until the point of cut-off; but in the air cylinder the resistance due to compression is nothing at the beginning of the stroke; and even when cut-off has been reached in the steam end, the load on the air piston will still be very small. In *B*, the conditions at mid-stroke are shown, and steam pressure still exceeds air pressure. In *C*, the conditions at the completion of the stroke are shown; and from the moment

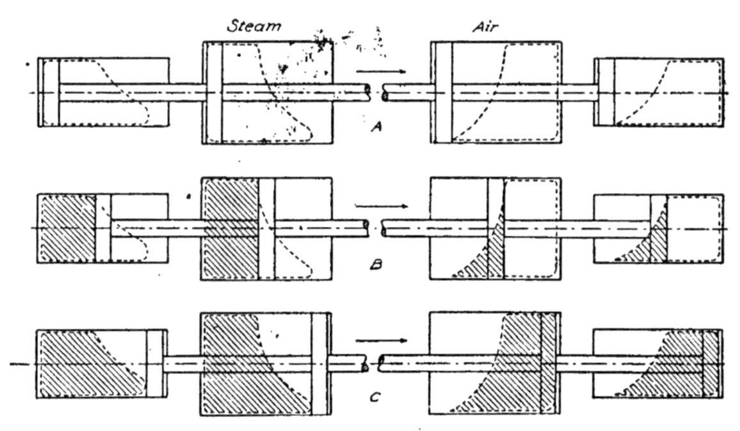

Fig. 44. Pressure Relations in the Double Tandem Compound Straight-Line Compressor.

the discharge point in the air cylinder was passed, a rapidly diminishing steam pressure has been applied to maximum air pressure.

Evidently there was surplus power during the first half of the stroke, and excess resistance during the last half. The condition is only rendered worse as cut-off in the steam cylinder is shortened to give better steam economy. Some method of equalizing the distribution of power and load must be used; and this is the function of the fly-wheels. They absorb the surplus energy during the first half-stroke, and deliver it against excess resistance during the last half-stroke; but their capacity for absorbing and giving out work depends upon their speed of rotation; and at slow speeds and short cut-off, even with very heavy fly-wheels, the straight-line compressor is apt to come to a dead stop. To be capable of regulation under wide load variation, therefore, the cut-off in the steam cylinder of the simple steam, single-stage air, straight-line

Fig. 45. Class WC Corliss Double Tandem Compound Air-Compressor. Sullivan Machinery Co., Chicago, Ill.

compressor must be late enough to permit the machine to run at slow speed, in which case there is a waste of steam at more nearly full load and at full load.

The pressure relations in the double tandem compound are sketched in Fig. 44, along the same lines as in the previous figure. The secret of compound steam economy lies in properly proportioned cylinders and in the correct adjustment of cut-off to give exactly the right expansion in both cylinders. If the cut-off is changed under varying speed to meet the conditions already described, it is evident

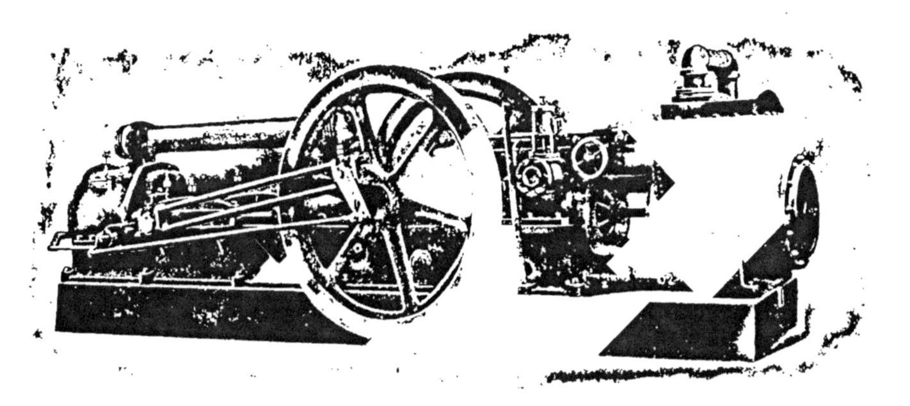

Fig. 46. Straight-Line Double Tandem Compound Air-Compressor. Norwalk Iron Works Company, South Norwalk, Conn.

that the ratio of expansions is deranged and the economy accordingly sacrificed. These facts practically forbid the successful compounding of the steam end of a straight-line compressor, except for operation under fairly steady loads. Even the plain simple steam single-stage straight-line is not economically self-regulating at less than 40 or 45 per cent of full load. Nevertheless the straight-line compressor, even in its more refined types, remains among the most popular machines for all-around service under fairly constant load, or in conditions where the mechanical superiorities of the type are of more importance than the very highest steam economy.

DUPLEX AIR-COMPRESSOR

The duplex compressor may be best described as two straight-line machines side by side, with a common shaft for the two and a fly-wheel mounted between them. With this understanding,

Fig. 47. Class C-4 Corliss Cross-Compound Duplex Air-Compressor.
Ingersoll-Rand Co., New York City

it may be stated at the outset that the duplex compressor has practically all the good features of the straight-line, with some additional advantages distinctly its own. Of course the compactness which characterizes the straight-line machine is sacrificed in the duplex compressor, as well as the portable features. Usually there are one steam cylinder and one air cylinder in each half of a duplex, either or both being cross-compounded. For three- or four-stage compression, there may be two air cylinders on each side; and sometimes duplex two-stage compressors are built with a high- and low-pressure air cylinder on each side arranged in tandem, with each side delivering one-half the total capacity in two-stage compression.

The duplex compressor is built with quartered cranks; that is, the crank-pin of one side is set one-quarter of a circle, or 90 degrees,

Fig. 48. "Cincinnati" Duplex Steam Cross-Compound Air-Compressor. Laidlaw-Dunn-Gordon Company, Cincinnati, Ohio.

in advance of the other crank-pin. The reason for this will be obvious upon examination of the four diagrams, Figs. 49 to 52, which show the pressure relations in a duplex compressor at four equal points in a revolution. For the sake of clearness, pressures on only one side of the pistons are here considered, for the back-pressure on the steam pistons, and the suction or intake pressure on the air pistons, do not materially affect the relations of power and resistance as here shown.

Pressure Relations in the Duplex Compressor. In Fig. 49, the pistons in the two lower cylinders are at the beginning of their stroke, and those in the upper cylinders are at mid-stroke. In the lower cylinders, there is no power and no load; in the upper cylinders the power greatly exceeds load, as indicated by shaded areas in the diagrams. Even if there were no fly-wheel, there would still be power enough to spare in the upper section to carry the lower section

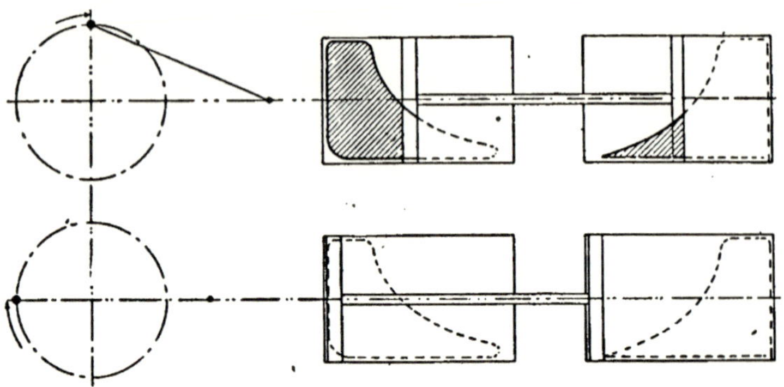

Fig. 49. Pressure Relations in Duplex Compressor, First Quarter.

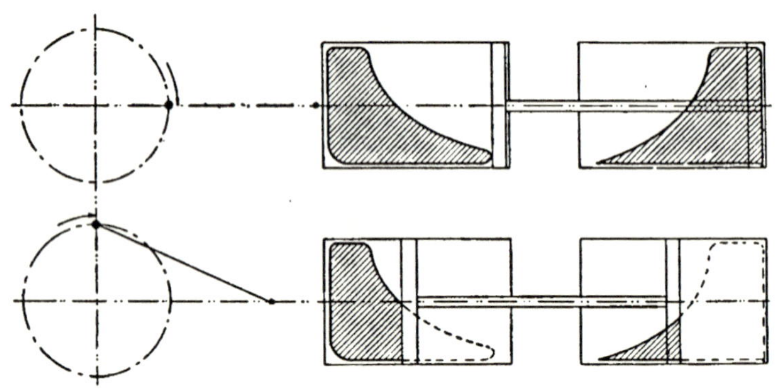

Fig. 50. Pressure Relations in Duplex Compressor, Second Quarter.

of the machine past its dead center, until steam was admitted to the lower steam cylinder.

In Fig. 50, the cranks have advanced a quarter of a revolution; and in the upper section, just completing its stroke, power has fallen to the minimum, while load has risen to the maximum, the upper section being now on its dead center. In the lower section, conditions are identical with those shown in the upper section of Fig. 49, and surplus power is available to pull the upper section past its center.

In Fig. 51, after the completion of a half-revolution, the conditions are exactly reversed from those in Fig. 50; and in Fig. 52, after three-quarters of a revolution, the conditions shown in Fig. 50 are reproduced in inverse relation. On completion of the revolution, this cycle is exactly repeated.

It is evident from the foregoing, that there is no point in the com-

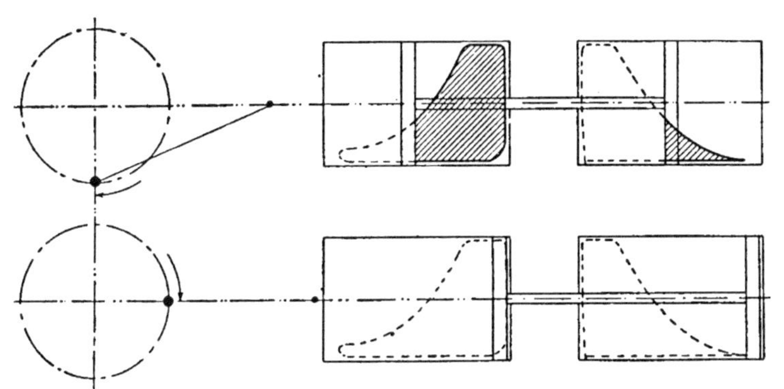

Fig. 51. Pressure Relations in Duplex Compressor, Third Quarter.

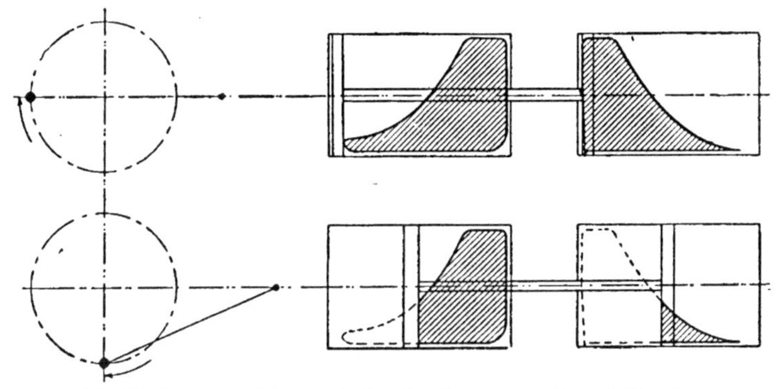

Fig. 52. Pressure Relations in Duplex Compressor, Fourth Quarter.

pression cycle of a duplex compressor at which there is not ample power applied to the load. Such a machine would probably run, even without a fly-wheel; but the latter is always used to equalize the variations in the application of power to resistance, making the machine run smoothly and uniformly, whereas, without a fly-wheel, it would probably hesitate at or near dead centers. The fly-wheel, however, need not be so heavy in proportion for a duplex machine as for a straight-line compressor.

If the cut-off in the steam cylinders of a duplex compressor is not less than one-half, there is no time during the revolution when steam at boiler pressure is not being admitted on some face of the

Fig. 53. Class WE Duplex Steam Two-Stage Air-Compressor. Sullivan Machinery Co., Chicago, Ill.

steam pistons; but so late a cut-off as this would probably not be used, except where the steam pressure was low. Even with a cut-off of three-eighths or one-quarter, the interval between steam admissions is so short that there is always ample power, taking into consideration the reserve held by the fly-wheel. It will be noted,

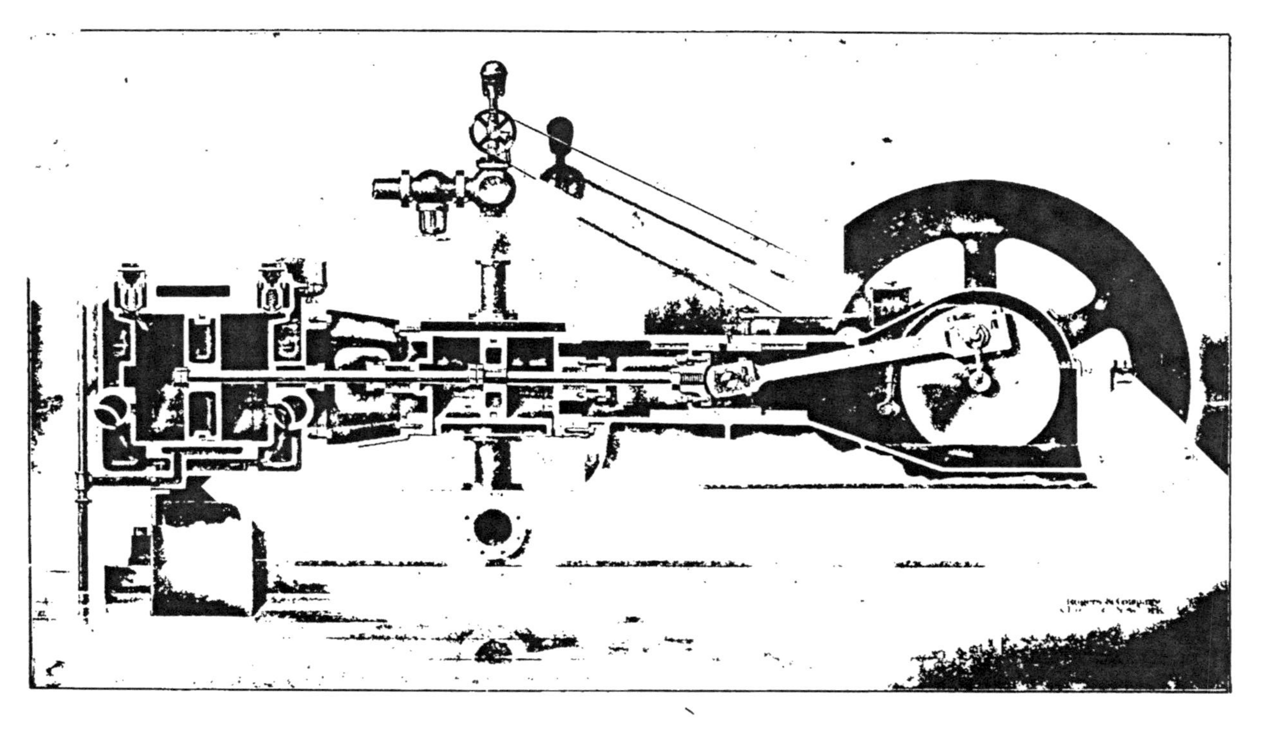

Fig. 54. Class WE Duplex Steam Two-Stage Air-Compressor, Sectional View Showing Oiling System.
Sullivan Machinery Co., Chicago, Ill.

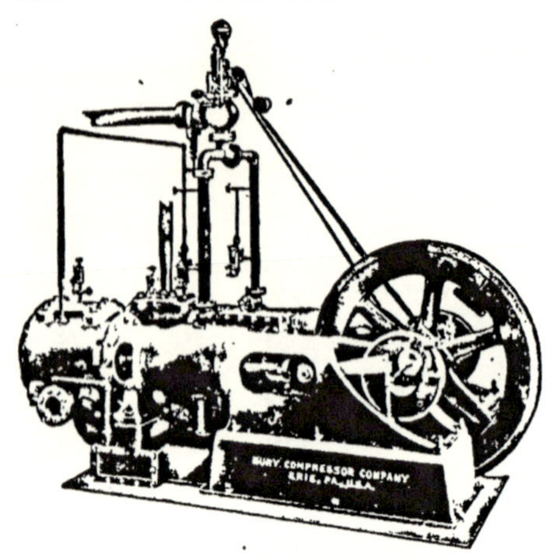

Fig. 55. Duplex Steam-Driven Compressor, with Duplex Steam and Cross-Compound Air Cylinders, Class CCS.
Bury Compressor Company, Erie, Pa.

moreover, that this is true whether the speed is slow or fast, so that it is next to impossible to "stall" a duplex, even at the lightest load and it slowest speed. In practice, in starting a duplex machine with adjustable cut-off steam valves, the cut-off is usually lengthened until normal speed and full load are reached, after which the cut-off is reduced to the most economical point; and thereafter the machine will take care of itself at from a few revolutions per minute to full speed, responding instantly and automatically by means of its governor to any change in load.

Advantages of Duplex Construction. The first advantage of the duplex compressor lies in the fact that the quartered arrangement of cranks, with a suitable fly-wheel (which need not be so heavy as on a straight-line machine of equivalent capacity), equalizes the

Fig. 56. Class SB Duplex Belt-Driven Air-Compressor.
Laidlaw-Dunn-Gordon Co., Cincinnati, Ohio.

Fig. 57. Class II-2 Steam-Driven Duplex Air-Compressor, with Duplex Steam and Cross-Compound Air Cylinders. Ingersoll-Rand Co., New York City.

application of power to resistance, resulting in a more uniform distribution of the pressures in the cylinders, and in a steadier, more uniform operation.

The second advantage lies in the improved regulation resulting from the fact that there is always power available to be applied to the resistance at any point in the stroke and at any speed.

The third advantage of the duplex is that the mechanical efficiency of a well-designed machine of this class will be higher than that of a straight-line machine of equal capacity, owing to the equalization of stresses throughout the structure by the quartered arrangement of cranks. This advantage, however, presupposes that the machine has been properly lined up and leveled on its foundation, so that the main bearings run free, without undue friction resulting from incorrect alignment. A duplex compressor, unless it is of a pattern having a continuous bed-plate or sub-base under all parts, is more dependent upon its foundation than a straight-line machine.

The fourth advantage of the duplex type lies in the fact that in it all the advantages of compounding steam and air cylinders can be secured without the addition of more cylinders, valves, and wearing parts than are required in the simple duplex. It has been seen that a straight-line compressor cannot be compounded on either steam or air end without the addition of parts and without running into difficulties peculiar to the type. As a matter of fact, it is quite probable that a double cross-compound duplex will show a higher mechanical efficiency than a simple duplex or a double tandem compound straight-line; for the terminal or maximum stresses on bearings, etc., are reduced by compounding, resulting in a lower friction load. These reduced stresses have already been mentioned in connection with compound steam cylinders (page 44), and the point will be taken up later in its relation to compounded air cylinders.

The fifth advantage of the duplex construction is that full compound steam economy is secured over the entire load range, because the 90-degree crank arrangement assures operation whatever the load, without greatly departing from the point of cut-off of maximum steam economy. A double cross-compound air-compressor will practically maintain its best economy under all loads.

The sixth advantage of the duplex comes from its balanced con-

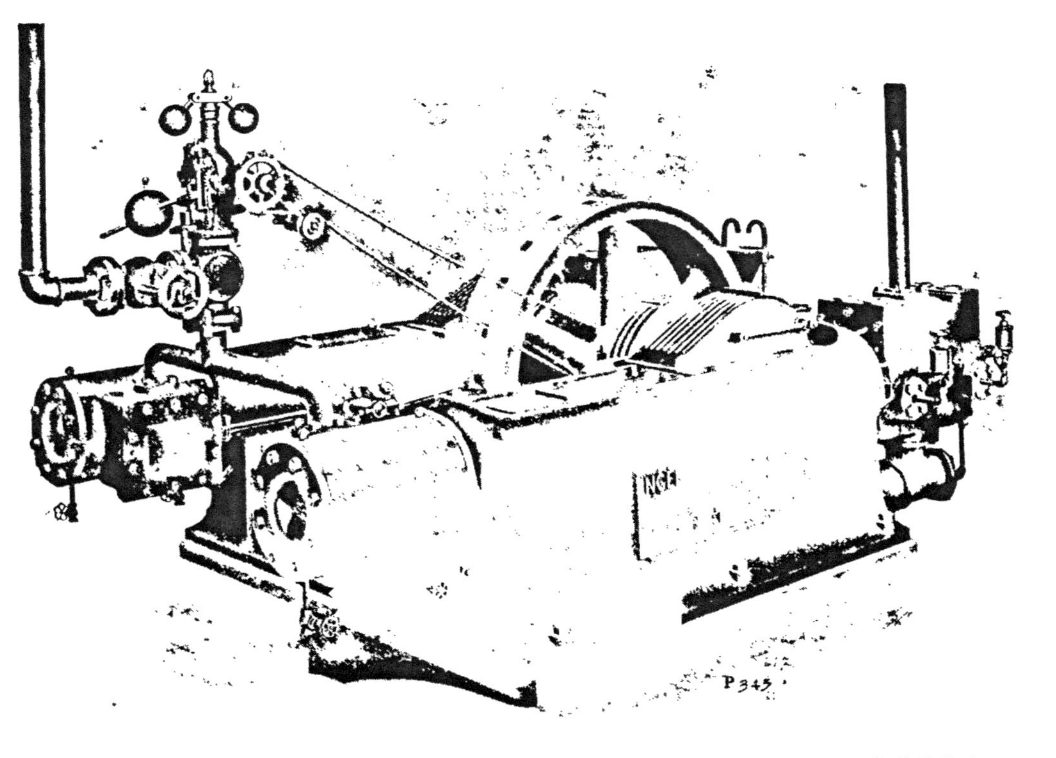

Fig. 58. "Imperial" Type X-2 Duplex Air-Compressor, with Duplex Steam and Cross-Compound Air Cylinders. Ingersoll-Rand Co., New York City.

Fig. 59. Class EE-1 Straight-Line Single-Stage Belt-Driven Air-Compressor. Ingersoll-Rand Co., New York City.

struction, in which an impulse in one direction on one side is partially or wholly equalized by an impulse in the opposite direction on the other side. The combined effect of these opposing influences is a steady and minimized strain on the machine structure, reducing friction and tending toward a long period of service without breakdowns or interruptions and with the lowest repair cost.

The seventh advantage appears in the probability that an accident to a duplex machine will affect only one side; and in emergency the injured half may be disconnected and the other half run as a simple straight-line machine with reduced output, often tiding over a critical period. Occasionally, also, it is desirable to install only one side of a duplex compressor; and later the other half can be added—compounded if necessary—when a complete double cross-compound is had.

Only one disadvantage has ever been rightly urged against the duplex construction, and that is its dependence upon its foundation; but even this point is usually exaggerated, and certain modern duplex compressors have a continuous sub-base under the entire machine, making them as self-contained and as independent of their foundation as the simplest straight-line. While it is a fact that a duplex compressor occupies somewhat more floor space than a straight-line machine of equivalent capacity, yet the question of floor space has not as yet become of sufficient importance in compressor practice to make this point serious when compared with the many well-known advantages of the duplex type.

WATER JACKETS

The necessity for some means of cooling the air during compression, has already been referred to (see page 8). Compressor practice of to-day uses one or both of two methods for securing this result—namely, (1) *water-jacketing;* (2) *compound* or *stage compression.*

In modern air-compressors, a space is provided around the barrel, and, if possible, around the heads of the air cylinders, through which water as cool as can be had is circulated in order to carry away the heat produced in compression, with the object of bringing the compression curve more nearly to the isothermal line. If the piston were made to move very slowly, so that this water could absorb and carry off all the heat, it is evident that this jacketing process would save all the work represented on the indicator diagram by the area between the actual compression line and the isothermal curve. In practice, however, a compressor cannot be run so slowly, and water-jacketing is only a partial solution of the problem of cooling.

While temperatures increase with the pressure in the cylinder, it is evident that the advancing piston steadily reduces the area of cool cylinder walls with which the air is in contact. Near the end of the stroke, when pressure and temperature are highest, only a very little of the walls is exposed, and the greater part of the cooling area remaining is in the cylinder-head. This emphasizes the importance of head jacketing, and the superiority of those constructions in which the air-valves are elsewhere than in the heads, thus affording the maximum percentage of cooled head area.

Moreover, air is a poor conductor of heat, so that even under the best conditions the heat from the interior of the air volume is not given up. On the contrary, only the external film in contact with the cylinder walls and head is cooled in any degree.

All things considered, therefore, water-jacketing an air cylinder realizes only a part of the saving of power which in theory it promises. Some beneficial effect in this respect, it certainly has—particularly in slow-speed, long-stroke machines; but the real problem of cooling during compression must be solved by other means, which will be discussed in the next section.

Cylinder jacketing is nevertheless essential in air or gas compression. Aside from the economical consideration just explained, it has other important and beneficial effects upon the satisfactory operation—and, indeed, upon the economy—of the compressor.

Not the least important of these is its effect upon the lubrication and therefore upon the wearing qualities of a machine. The compression of air from sea-level pressure and 60° F., to six atmospheres or 73.5 lbs. gauge, produces (unless some means of cooling is provided) a temperature of about 419° F. Such a temperature as this is destructive of all ordinary lubricants and packings. The former would be charred to a hard, coke-like substance, and the latter would be burned out. The inevitable result would be a rapid wear of piston rings, cylinder, rods, and valves, with extravagant leakage, loss of capacity, increased friction, and excessive power con-

COMPRESSED AIR ON THE PANAMA CANAL
Sullivan Air Channeler at Bas Obispo

sumption. But the water jacket keeps all these parts reasonably cool, resulting in tight joints, free lubrication, little wear, reduced friction, easy running, and sustained capacity and efficiency.

An attempt to run continuously without jackets would heat the cylinder, piston, valves, rods, etc., to such a degree that dangerous expansion and contraction strains might be produced, resulting in distortion of parts, added friction, probable binding, possible breakage, and almost certain loss of capacity. The alternate expansion and contraction between the extreme temperatures of operation and shut-down, would make it practically impossible to maintain tight joints, and heavy leakage loss would certainly follow.

It has already been stated that it is almost as important that air should be admitted cool into the cylinder as that it should be kept cool during compression. If a water jacket is not used, the heating of the cylinder, piston, rods, valves, ports, head, etc., means that the air will be heated during admission, first, while entering through hot valves and ports; and second, by contact with a hot cylinder and piston inside the cylinder. This means a rarification of the entering air, reduced cylinder capacity, lower efficiency of compression, and excessive terminal temperature. Since jacketing is at best only partially effective, this latter argument suggests the advantage of those constructions which admit air through one, or at most only a few, large openings where only the outer film of air is heated, rather than through a number of small openings through which the air is finely subdivided and greatly increased in temperature.

The water-jacketing of compressor cylinders is thus seen to be essential on the grounds of power saving, large capacity, better wearing qualities, and a generally improved performance.

COMPOUND AIR-COMPRESSION

If at several points in the stroke the compressing piston could be stopped, and the air already partially compressed and heated could be withdrawn long enough to be cooled by some external means to its initial temperature (or even lower), and then returned to the cylinder to be further compressed, heated, and again cooled, it is evident that a fairly uniform temperature could be maintained in this air volume over the full range of pressures, and the result would be very close to isothermal compression.

Evidently practical considerations forbid such a repeated starting and stopping of the piston; but the same results may be obtained by carrying on the process of compression in several cylinders, in the first of which moderate pressure and temperature are reached, and the air in this condition discharged through a cooling device which restores its initial temperature, then admitted to another cylinder

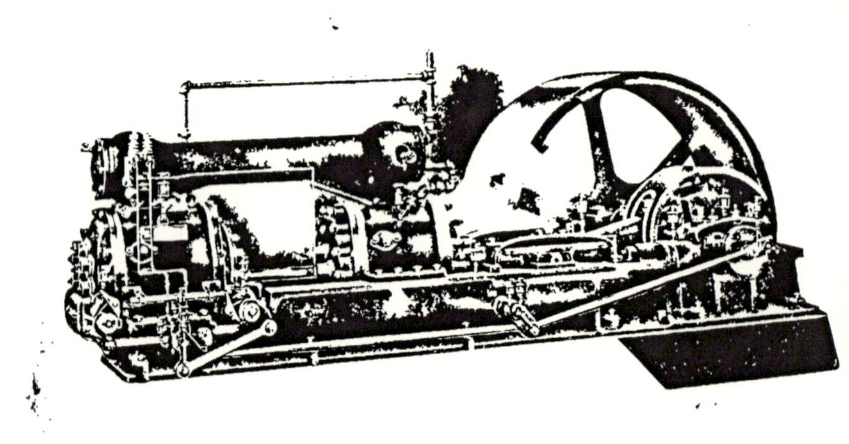

Fig. 60. Straight-Line Belt-Driven Two-Stage Air-Compressor. Norwalk Iron Works Company, South Norwalk, Conn.

where it is further compressed and heated, again withdrawn and cooled, again compressed—and so on until the desired terminal pressure is reached. Such a process, developed to a practical working cycle of operations, constitutes *compound* or *stage compression*, which is almost universally employed to-day when pressures are fairly high.

While the limiting pressures for which various degrees of stage compression are advisable, are not definitely fixed, an average of the practice of the leading builders seems to give the following:

For pressures up to 70 lbs., and under some conditions to 100 lbs., single-stage compression.

For pressures of 75 to 500 lbs., compound or two-stage compression.

Fig. 61. Class WH-2 Straight-Line Two-Stage Belt-Driven Air-Compressor.
Equipped with mechanically operated Corliss air-inlet valves and removable poppet discharge valves. Air cylinder water-jacketed.
Sullivan Machinery Co., Chicago, Ill.

For pressures of 500 to 1,500 lbs., three- or four-stage compression, preferably with duplex construction.

For pressures of from 1,500 to 3,000 lbs., or even higher, four-stage compression.

A three-stage straight-line machine should be used only for small or moderate capacities. Beyond that, the duplex type is to be preferred because of its balanced construction and of the readiness with which it lends itself to compounding. Whether three or four stages shall be used for 500 to 1,500 lbs. pressure, is largely a matter of preference with different builders; but since four air cylinders are required for the duplex in either case (two intermediate cylinders being used in duplex three-stage work), probably the four-stage cycle is to be preferred for from 1,000 to 3,000 lbs. pressure.

The primary object of compound compression is a saving of power by eliminating what may be called the *heat element* in the total power required for compression. Other gains by compounding will be discussed later; but at this point it is to be noted that successful stage compression depends primarily upon the efficiency of the cooling devices used. It has already been shown that a water jacket is essential, but hardly to be considered as a great factor in the cooling process. The main burden of cooling, therefore, falls upon the device employed for cooling the air between compressions. This device is called the *intercooler*. Before taking up this feature, however, the gains by various degrees of compound compression will be investigated in a specific case.

Let the problem be to determine the best means of compressing 1,000 cubic feet of free air at sea-level to 100 lbs. gauge pressure. By reference to Table VII (page 27), it is found that, if this could be done isothermally, the power required would be 1,000 × .1317, or 131.7 H.P. The power required for this duty in single-stage adiabatic compression is 180 H.P.; for two-stage adiabatic compression, 154 H.P.; for three-stage adiabatic, 146 H.P.; and for four-stage adiabatic, 142 H.P. It is to be remembered that the values in this table assume adiabatic compression in the cylinders, but perfect cooling to initial temperature in the intercooler. Neither assumption is correct in practice, but the relative power requirements in the several cases are fairly shown by the values given. As compared with ideal isothermal compression, therefore, the efficiency of average single-stage compression in this case is the ratio of 131.7 to 180, or 73.2 per cent; of two-stage compression, 85.5 per cent; of three-stage compression, 90.2 per cent; and of compression in four stages, 92.7 per cent. At first glance, four-stage compression would seem to get first choice, three-stage second choice, and two-stage third choice. As a matter of fact, however, two-stage compression would probably be used; for while in theory there is undoubtedly better economy in three- and four-stage compression, yet in actual practice it is found that the added complication of design in the two latter cases, owing to the greater number of cylinders, valves, and working parts, with additional intercoolers, produces losses probably offsetting any gain by higher compounding. The limiting pressures for various stages of compression represent averages determined by the long experience of builders of compressors, and may therefore be safely accepted.

The Intercooler. The ideal intercooler is one which, at the full rated speed of the compressor of which it is a part, reduces the temperature of the air passing through it to the temperature of the cooling water which it uses. This probably can never be fully realized; but a good, practical standard for an intercooler is that it shall deliver air at a temperature not more than ten degrees above that of the cooling water. If the water is cold enough, this will bring the air down to, or even below, atmospheric temperature; and it is just as important that cool air shall enter the high-pressure cylinder as the low-pressure cylinder, for the reasons already given in considering the intake air temperature.

The successful intercooler, considered now as a heat-removing device, involves four fundamental requirements:

1. It must provide for a complete and minute subdivision of the air volume passing through it, in order that the heat may be dissipated without reliance upon the heat-conducting properties of the air itself. In other words, the air should be split up into thin sheets or streams so that the internal heat can be fully given up.

2. It must present an ample cooling surface to this subdivided air stream.

3. The circulation of the cooling water must be free and at such a velocity that the maximum amount of heat can be carried away. Furthermore, the flow of the water should be properly directed in relation to the flow of the air.

4. An ample cross-section must be provided in all air-passages, in order that the rate of flow of the air may be slow enough to give a length of contact with the cooling surfaces sufficient for the complete removal of the heat.

The first two requirements are met by the same means—namely, by the use of a nest of tubes inside the intercooler casing or shell, through which tubes the water circulates, and over, about, and between which the heated air flows. Baffle-plates should be so disposed as to cause the air stream to pass back and forth and between the tubes, splitting the air up into thin sheets. The prevailing practice puts the water inside the tubes, not the air; for in the latter case each tube would contain a column of air of which

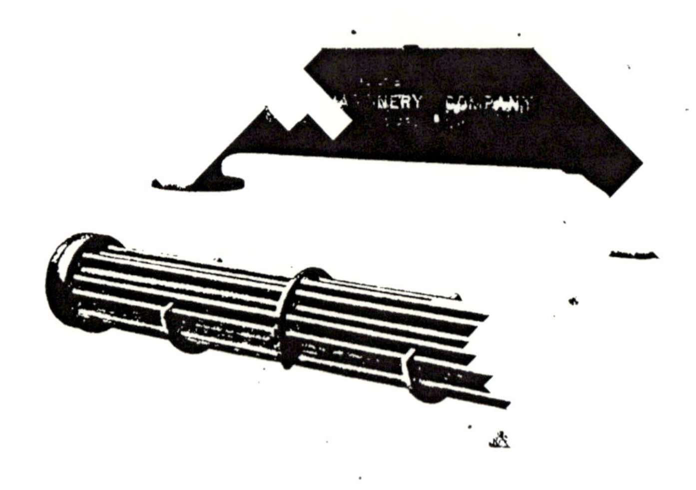

Fig. 62. Intercooler for Two-Stage Straight-Line Belt-Driven Air-Compressors. Views showing shell, and tubes removed from same. Sullivan Machinery Co., Chicago, Ill.

only the outer layer or film would be cool, the interior retaining its heat. This general arrangement gives the maximum cooling from a given flow of water.

The third requirement is met by providing plenty of tubes giving an ample flow with a low velocity of the water; and by making the water flow through the intercooler in a direction contrary to that of the flow of air. The hottest air thus encounters the warmest water, and the air as cooled steadily comes in contact with cooler water.

The fourth requirement can evidently be met by making the intercooler shell large enough, and the tubes sufficiently numerous to provide an ample passage for the air without loss of pressure due to friction and without an undue velocity due to restricted area of passage.

One consideration of a purely mechanical nature, in the intercooler, is nevertheless as important as those just mentioned in its bearing on intercooler efficiency. Expansion and contraction must be provided for; otherwise there results either a leakage of air, with a loss of capacity and a waste of power, or a leakage of water into the air chamber, which may be carried over to the next cylinder, with disastrous results.

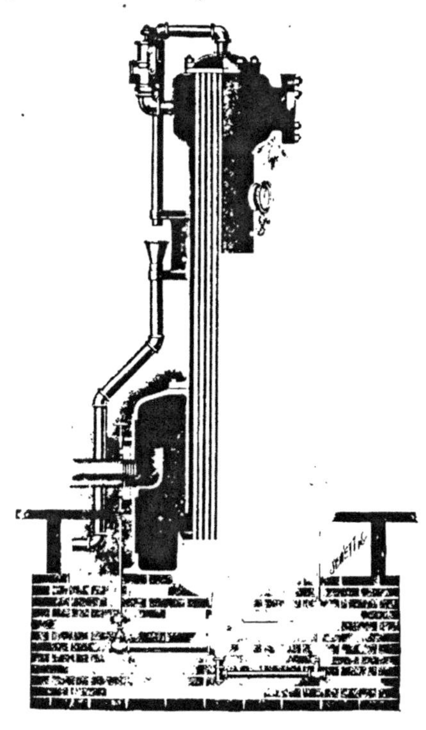

Fig. 63. Vertical Intercooler for Duplex Compressors.
Also used as an aftercooler.
Ingersoll-Rand Co., New York City.

This suggests also another important point—namely, the proper drainage of the intercooler. The lowering of the air temperature condenses the water vapor in the air, and this water accumulates in the intercooler until it may cover some of the tubes, reducing the air passage as well as the cooling area. But the greatest danger is that this water, if not removed from the intercooler through the drain, will be caught in the sweep of air and carried in a mass into the next cylinder, when a broken cylinder-head or piston, or a bent piston-rod, may result.

Other Advantages of Compound Compression. The mechanical structure of an air-compressor must be equal to the maximum stresses which may come upon it; and this determines the amount and disposition of materials in it. The endurance of the machine depends upon its factor of safety; and good design exacts that this factor shall be large without extravagant use of metal. In other words, while the maximum stresses must be provided for, the average stresses must be kept well below this maximum if the machine is to operate satisfactorily in long, steady service.

Compound air-compression reduces the average structural stresses in the compressor, resulting in a higher factor of safety and longer life of the machine, with a given weight of metal. It has been seen that in a single-stage machine the piston starts against no resistance (except friction), but rapidly builds up pressure to the maximum, which is held until end of stroke. A 24-inch piston compressing to 100 lbs. gauge, single-stage, meets a total resistance of 45,239 lbs.,

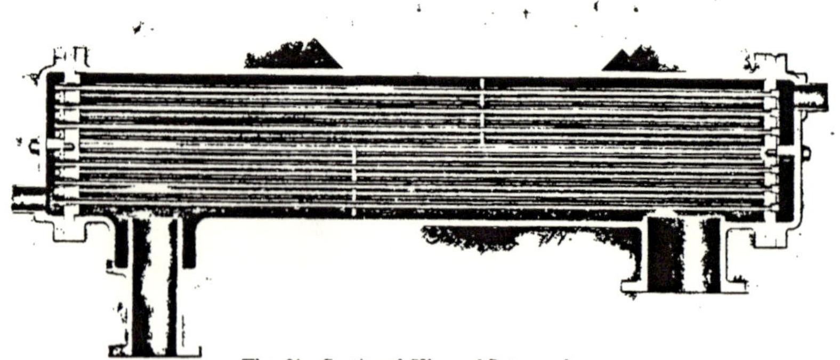

Fig. 64. Sectional View of Intercooler.
Laidlaw-Dunn-Gordon Company, Cincinnati, Ohio.

or nearly 23 tons; and at 90 R.P.M. this is encountered and relieved 180 times each minute. It takes the best of metal, and plenty of it, to stand up under such service as this. But if this same work is done in two cylinders, with two stages, the condition is very different. The maximum total pressure on the low-pressure piston will depend upon the cylinder ratios; and the pressures in high- and low-pressure cylinders will partially balance and equalize. As a matter of fact, without going into the mathematics of it, it is found that the maximum stresses encountered in compression in two stages to a given pressure are only 55 or 60 per cent of what they would be in single-stage compression to the same pressure. In the case just cited, this means a reduction of terminal strains from 180 × 23 or 4,140 *ton-minutes* to 2,277 or 2,484 *ton-minutes*.

This difference in maximum pressure is found not only in the machine structure, but also in the air-valves, which are the most vital parts of a machine and which show a notably easier operation under these improved conditions.

But a very vital point must not be overlooked. Though the maximum stresses of average operation may be reduced and brought well within the limit of safety, yet the maximum possible stresses must be provided for in the design. For the failure of the supply of cooling water, preventing the cooling between stages, or the breaking of a high-pressure valve, may at any moment throw the full high-pressure load on the low-pressure piston, in which case disaster follows if the machine has not been designed with such a contingency in view. It is not safe to assume, therefore, that a multi-stage machine can safely be built lighter than a single-stage machine of equivalent capacity.

Another advantage of compound compression lies in the resultant improved steam economy if the compressor is steam-driven. This may be best illustrated by reference to the example just cited, where it was seen that the terminal stresses on the machine, and therefore the terminal power required, are reduced 40 to 45 per cent by compounding the air end. It was seen, also, that the load was more equably distributed throughout the stroke. Evidently this means a lower M.E.P. in the steam cylinder to do the work, secured by a shorter cut-off and consequently less steam. More of the work of passing dead center can be intrusted to the fly-wheel, instead of being provided for by a later cut-off. The mechanical gain by reduced terminal stresses also permits a higher piston speed with safety, this again improving the steam economy by reducing condensation losses and the possibility of leakage in the steam cylinder.

Higher volumetric efficiency comes with the use of compound compression, a gain in this respect being the result of three different causes. The first of these is the lower clearance volume; and it is to be remembered that, as affecting the free-air capacity of the compressor, the clearance in only the low-pressure or intake cylinder is involved. The air in the low-pressure clearance, at the end of the stroke, is at maximum pressure in that cylinder; and before free air can enter, this clearance air must re-expand to atmosphere. In a single-stage compressor, clearance pressure is the full terminal pressure of the machine. If this is (say) 7 atmospheres, and the clearance percentage is 1.5 per cent, the space occupied by this expanded air will be 7 × 1.5, or 10.5 per cent of the cylinder volume. If, however, the compression is to 7 atmospheres in two stages, the terminal pressure in the low-pressure cylinder will be somewhere around 3 atmospheres. In this case, therefore, the reduction in cylinder capacity due to the expansion of the clearance air (the percentage of clearance in the low-pressure cylinder being assumed the same as before) will be 3 × 1.5, or 4.5 per cent. The volumetric gain in this respect alone by compounding becomes evident.

The second cause of gain in volumetric efficiency by compounding, is the lower maximum temperatures in the cylinders, corresponding to the lower pressures. These being much lower in a compound machine than in a single-stage compressor, the pistons, cylinder walls and heads, valves, and ports are kept much cooler, and the entering air is not so much heated. This means a greater density of air at the beginning of compression, with a corresponding increase in volumetric capacity.

The third element in volumetric gain by compounding is the reduced leakage past valves, pistons, and rods in the compound machine, consequent upon the reduced extremes of pressure in each cylinder.

Compound compression results in more efficient lubrication of the machine, meaning easier running, improved mechanical efficiency, less wear, lower repair costs, reduced leakage losses, and longer life. The lower maximum of temperature which holds in each compound cylinder is not sufficient to affect seriously a good oil, and all parts are thoroughly and freely lubricated.

The last advantage of compounding is in the delivery of drier air. It was noted that a properly designed intercooler would condense and remove much of the moisture in the air. This reduces the liability to freezing-up at the exhaust ports of machines using the compressed air. It also largely obviates the accumulation of water in pipe-lines, where the latter are long, reducing their cross-section or even choking them entirely at low points.

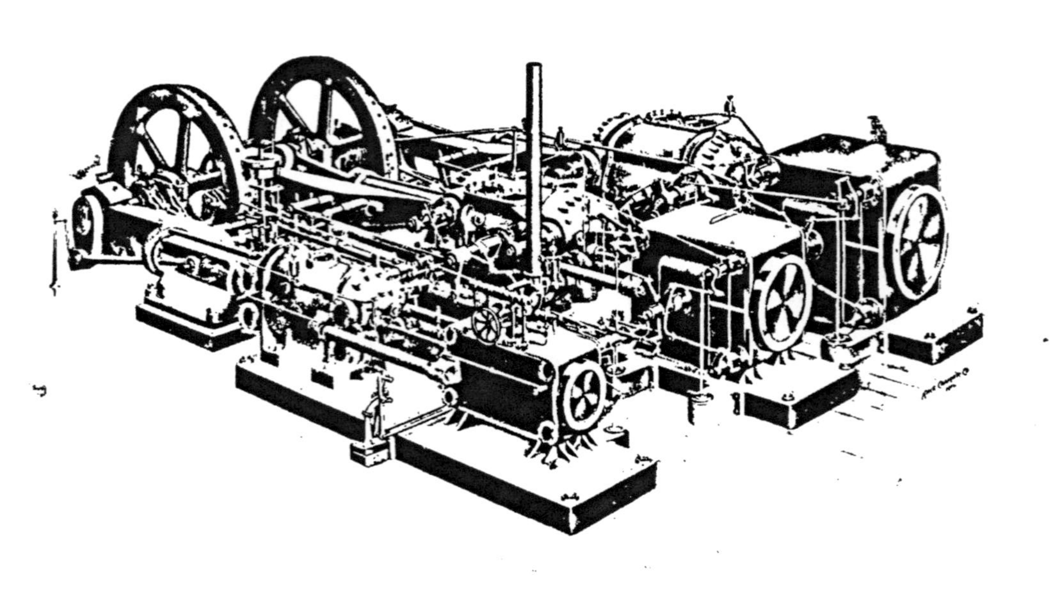

GENERAL VIEW OF THREE-STAGE TRIPLE EXPANSION AIR COMPRESSOR.
Nordberg Manufacturing Co.

COMPRESSED AIR

PART II

AIR-COMPRESSOR VALVES

Two sets of air-valves are essential on each air cylinder of a compressor. The first are the *inlet valves*, their function being the admission of air to the cylinders; the second are the *discharge* or *outlet valves*, which govern the discharge of the air compressed.

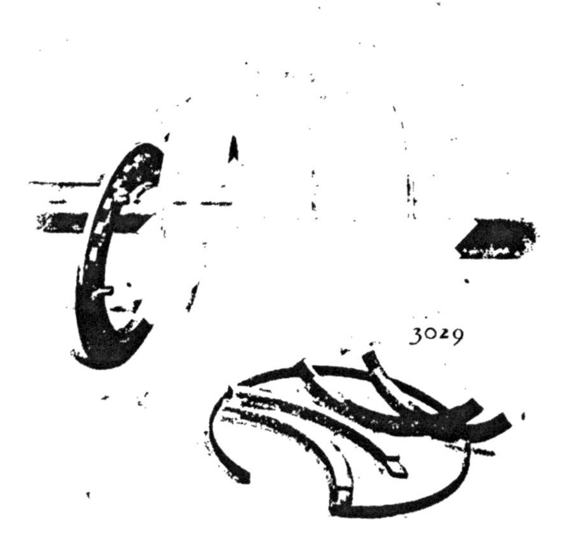

Fig. 65. Details of "Hurricane" Piston-Inlet Air-Valve. Ingersoll-Rand Co., New York City.

While the number of different types of air-valves is as great as the number of compressor builders, yet an analysis of all types will show a broad classification into two groups: *Mechanically actuated* valves, which have a positive motion derived through some mechanical connection from the rotating or reciprocating parts of the compressor; and *poppet* or *automatic valves*, operated wholly by differences of pressure. There are some air-valves in use to-day which cannot be exactly placed in either class, notably the *piston inlet* valve of a prominent builder, which is operated by its own inertia. In still other cases the valve seems to combine the features of both classes, an automatic adjustability supplementing a primary mechanical control.

Mechanically actuated valves are frequently of the Corliss type,

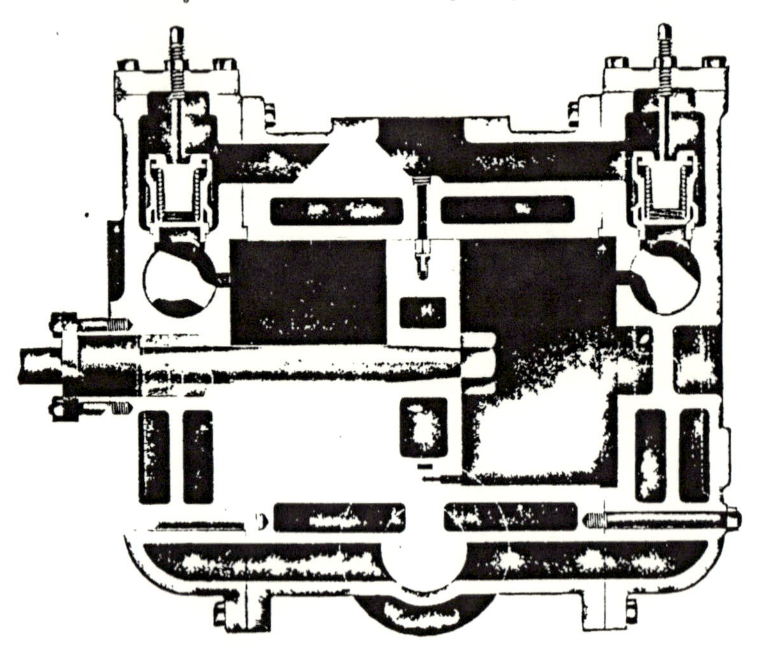

Fig. 66. Section of Air-Compressor Cylinder with "Cincinnati" Valve Gear for Inlet and Discharge, with Poppet Valves Supplementing the Corliss Valve on Discharge. Laidlaw-Dunn-Gordon Co., Cincinnati, Ohio.

operated by levers from a wrist-plate driven by an eccentric or return-crank on the main shaft. Usually there is nothing in this valve movement corresponding to the release mechanism of the Corliss steam valve gear, since there is nothing in the air card corresponding exactly to the phenomenon of cut-off on the steam card. Another form of mechanical valve is similar in general appearance to the regular poppet valve, but its movement is mechanically produced. A third form of valve which may be considered in this class is the *air-thrown* poppet valve of one leading manufacturer, in which the movement of a poppet valve is produced by air-pressure from the receiver or intercooler on a small piston on the stem of the valve.

Mechanical air-valves are practically limited in their application to the inlet valves of single-stage compressors, or the low-pressure inlet valves of compound machines. There are instances, however, in large sizes, where mechanical valves are used on the intake of the high-pressure cylinder as well. Almost the only instance in which

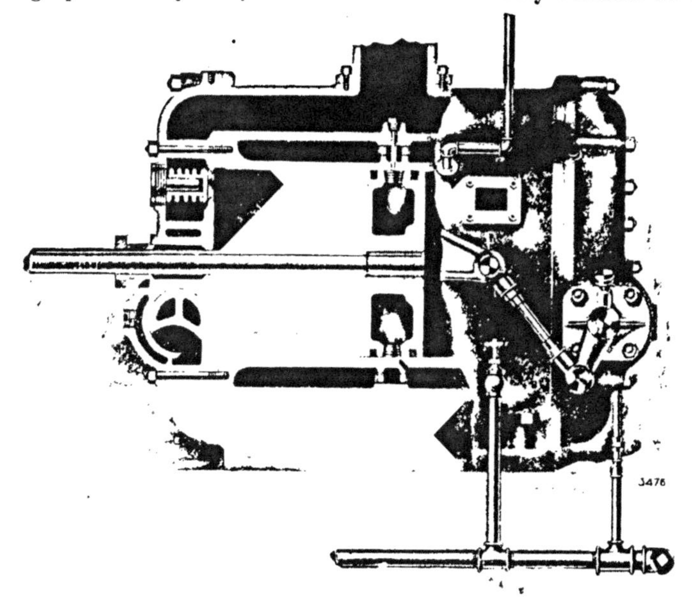

Fig. 67. Section of Compressor Cylinder with Rand-Corliss Inlet Valves and Direct-Lift Poppet Discharge Valves. Ingersoll-Rand Co., New York City.

mechanical valves are used for discharge purposes is found in the case of some blowing engines, which are simply compressors designed for very low pressures. The function of an atmospheric inlet valve is simply to open full when the clearance air has expanded to atmosphere, and to remain open to the end of the stroke. This duty is light, for no pressure higher than atmospheric pressure is encountered while the valve is in motion. The same valve used for the high-pressure inlet of a compound machine, while subject to higher pressures—namely, the discharge pressure of the low-pressure cylinder—is nevertheless not compelled to move while under this pressure.

Fig. 68. Section of Compressor Cylinder with Corliss Inlet Valves and Mechanically Operated Poppet Discharge Valves
Allis-Chalmers Company, Milwaukee, Wis.

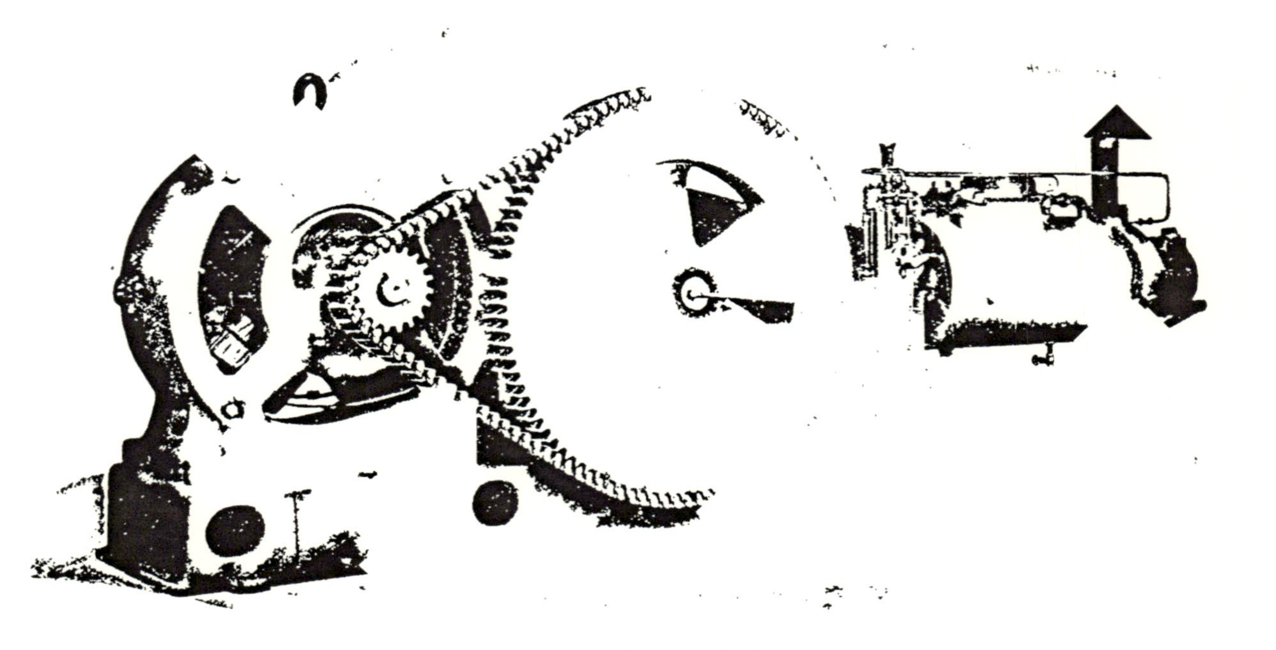

MOTOR-DRIVEN AIR COMPRESSOR.

Morse Chain Company.

However, when used for controlling the air discharge, mechanical valves suffer under the necessity of opening at a fixed point of the

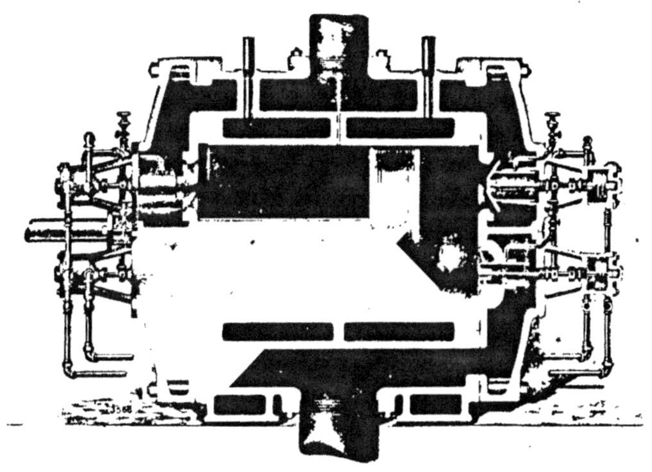

Fig. 69. Section of Compressor Cylinder with Air-Thrown Inlet and Discharge Valves. Ingersoll-Rand Co., New York City.

stroke, while the correct point of discharge will vary in the stroke with varying receiver pressures, even where automatic pressure-regulators are employed. In this event, one of two things will happen.

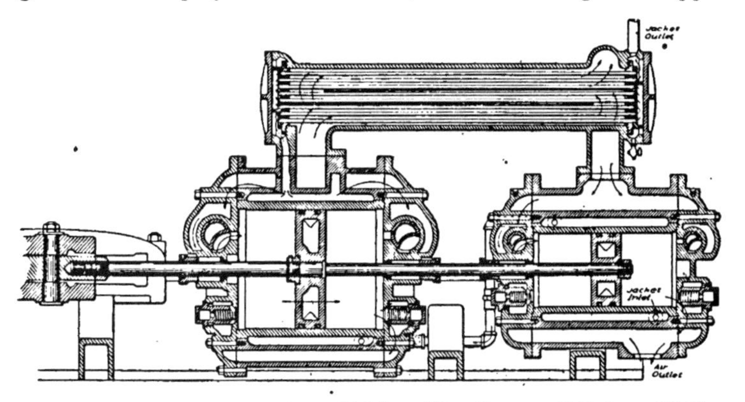

Fig. 70. Section through Inter-cooler and High- and Low-Pressure Cylinders of Sullivan Straight-Line Compressor. Corliss Inlet and Poppet Discharge Valves on Both Air Cylinders.

If the point of discharge comes *before* the point of discharge valve opening, due to a drop in receiver pressure, the valve and valve gear must move under a heavy, unbalanced load from the excess of cylinder pressure over receiver pressure. If the point of discharge falls *after* the point of valve opening, the mechanism is again under heavy strain when it moves, due to the excess receiver pressure. In either case the duty is hard on the valve and gear, wear is rapid, and leakage almost inevitable.

Poppet valves are simply held to their seats normally by a spring, and move when the air-pressure upon their face overcomes the spring tension. As applied to atmospheric inlet purposes they are open to the objection that sufficient vacuum must be produced within the cylinder on the admission stroke to give an excess of atmospheric pressure sufficient to overcome the spring tension. This brings the admission line below atmosphere, and reduces the volumetric efficiency. This objection, of course, does not hold on the high-pressure inlet of compound compressors, where the pressure is all that is required for the prompt opening and closing of the valve. The poppet type is very generally used for high-pressure inlet purposes, on this account.

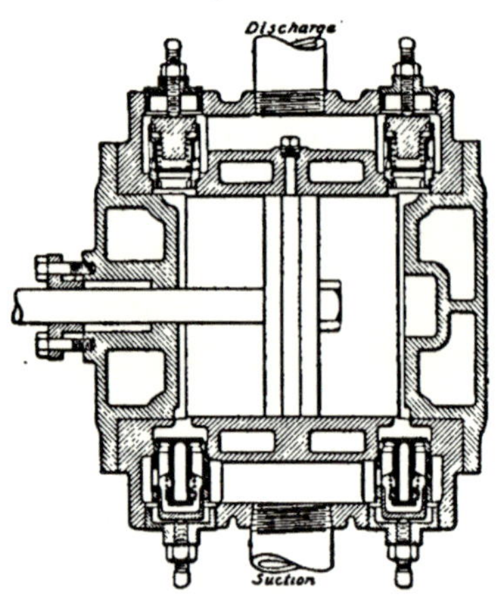

Fig. 71. Section of Air Cylinder of Laidlaw-Dunn-Gordon "Climax" Compressor, with Poppet Inlet and Discharge Valves.

The poppet valve is also almost universally applied for discharge purposes, on both high- and low-pressure cylinders. In this service, there is plenty of pressure to operate such a valve against its spring; and since it will not open until the relation of pressures on its front and back is exactly what it should be, it automatically adjusts itself to varying discharge pressures.

Air-Inlet Valves. Without going into a detailed discussion of the many different types of air-inlet valves, some general observations may be made as to the functions and requirements of these important parts. The functions of the air-inlet valve (or valves) are: To admit the maximum volume of cool, clean air to the cylinder, with the least loss or expenditure of power; to keep this air in; and to continue to perform this duty indefinitely, at the least possible cost. An analysis of these fundamental facts reveals the following requirements of a successful air-inlet valve:

1. It must admit a volume of air which will entirely fill the compressor, this air being as nearly as possible at atmospheric pressure. This calls for an instantaneous and complete opening of the valve, so timed that there shall be no escape of the compressed air in the clearance space. This full opening should be held to the end of the stroke. This first consideration calls further for short, direct, and unobstructed passages through the ports and valves.

2. It must admit clean air, suggesting a design permitting ready connection to an intake passage or conduit supplying air from a place free from dust and grit.

3. The air admitted must be as cool as possible. This is partially met by the second consideration, whereby cool air as well as clean air should be

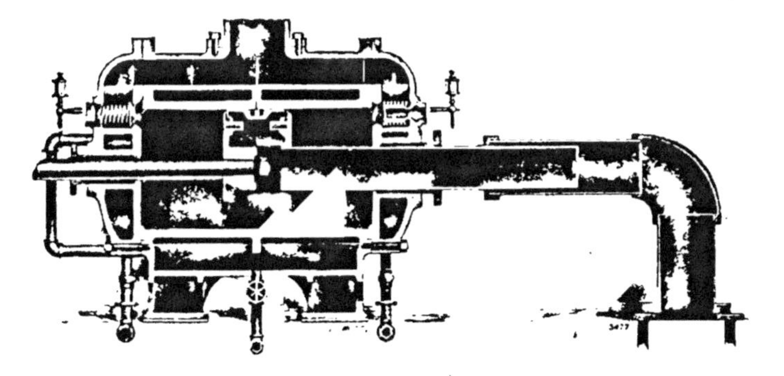

Fig. 72. Section of Compressor Cylinder with Hurricane-Inlet and Direct-Lift Poppet Discharge Valves.
Ingersoll-Rand Co., New York City.

supplied; but, more than this, it demands that this cool air supplied shall nowhere be heated in its course into the cylinder, suggesting that valves, ports, and air-passages should be so placed as to be cooled by the cylinder jacket. Moreover, the air should be admitted through one or only a few large openings, rather than sieved through many small ones.

4. All the air admitted must be kept in the cylinder, calling for instantaneous and complete closing of the valve at the end of the stroke, with a perfect seating and with no leakage.

5. All the processes hitherto outlined must be carried on with the least possible expenditure of power for the manipulation of the valve gear and the overcoming of friction—a condition necessitating nice adjustments, simplicity of mechanism, perfect lubrication, and as few bearing and wearing surfaces as possible. This exacting performance must be continued indefinitely, twice each revolution and hundreds of times per minute, without undue wear or loss of adjustment. This condition is fundamentally one of correct design, high-class materials and workmanship, and generous lubrication.

The ideal inlet valve has here been outlined. Probably no compressor on the market to-day exactly fulfils all these conditions, but many types represent the judgment of their several builders applied in a wise compromise. In the foregoing statement of primary essentials, it is to be remembered that these apply to high-pressure as well as to low-pressure valves, and they can be studied in relation to high-pressure work by substituting *intercooler pressure* for atmospheric pressure wherever the latter has been used in defining these essentials.

Fig. 73. Details of Poppet Discharge Valve. Blaisdell Machinery Co., Bradford, Pa.

Air-Discharge Valves. The functions of the air-discharge valve (or valves) of an air-compressor are: To release the full volume of compressed air from the cylinder, with the least power expenditure; to keep it all out after discharge; and to continue indefinitely, without undue expense, to maintain this duty. The primary essentials of such a valve are as follows:

1. The opening of the valve must be full and instantaneous, occurring at just the right point in the stroke; and this opening must be held until the piston comes to rest. The valve opening must be ample, so that the velocity of discharge may not be excessive; and the ports and discharge passages should be short and free from obstruction.

2. As few parts as possible—and these correctly adjusted—should provide for the necessary valve movement, with as little power consumed in friction as possible.

3. The valve should close instantly and completely as soon as the piston comes to rest at the end of the stroke, in order that there may be no return of the air discharged when the piston starts on its back-stroke. Such a return of compressed air would act as an increased clearance to reduce the capacity of the cylinder.

4. The continued maintenance of this performance—more arduous upon the discharge valves than upon the inlet valves, because of the higher pressures encountered—suggests the advisability of a simple construction with as few bearings as possible, with careful workmanship, the use of good materials, complete lubrication, and a careful distribution of materials to give the best results.

It is always to be remembered that air-discharge valves and ports are continually exposed to a flow of heated air, even where the most careful jacketing is used. On the other hand, a requirement of inlet valves and ports is that they should be kept cool. The latter should therefore be as far removed as possible from the former; nor should the same ports be used for both inlet and discharge.

Areas of Valves, Ports, and Passages. It might at first glance seem that the area of the discharge valves, ports, and passages could be much less than the corresponding inlet areas, since the former must pass air at a higher pressure and therefore in less volume than the latter; but it is to be remembered that while the inlet areas are open during the full stroke, the discharge areas are open for only a part of the stroke—usually about one-third. In order, therefore, to make the velocity of discharge about the same as that of intake or admission, it is the general practice to make the inlet and discharge areas approximately equal. It is important that the velocity of flow through valves, ports, etc., shall not be excessive; and this depends upon the piston speed, which is a widely variable quantity. The higher the piston speed, the larger should be the valve and port areas. In actual practice, the inlet area varies from 3 to 15 per cent of the cylinder area, with an average probably around 9 or 10 per cent. The corresponding discharge areas run from 7 to 15 per cent, with an average in the neighborhood of 10 or 12 per cent. Owing to the increased friction through a number of small passages, as compared with that through a single large one, the actual *measured* area in the former case should be 50 to 100 per cent larger than in the latter. This explains why the measured inlet area of machines with Corliss or other large-area inlet valves is so often much less than the measured discharge area of many small poppet discharge valves, while still giving good results.

Inlet and Discharge Port Areas. While reference has been repeatedly made to the clearance of an air-compressor, it may be well to recall at this point that clearance is really made up of three elements: (1) the *inlet port clearance,* which is all the space left between the closed inlet valve and the inner surface of the cylinder-head; (2) the *discharge port clearance,* consisting of all the space between the closed discharge valves and the inner surface of the cylinder-head or walls; (3) the *piston clearance,* which is the space allowed for safety between piston and cylinder-head at the end of the stroke. The latter element is an essential in all compressors; but it is not unusual for the other two elements together to constitute by far the larger proportion of the total clearance. This suggests the superiority of those valve designs in which inlet and discharge port clearances are reduced to the lowest practical limit.

MECHANICAL EFFICIENCY OF AIR-COMPRESSORS

It will be remembered (see page 32) that an air-compressor can be considered as made up of three elements—a driving element, a compressing element, and a power-transmitting element, which latter constitutes the mechanical structure of the machine. The first two have been discussed; the last remains for consideration.

It has been seen that the mechanical efficiency of a compressor—the ratio of power output to power input, which is the ratio of the I.H.P. of the air cylinder to the I.H.P. of the steam cylinder—can be exactly determined by a comparison of the steam and air indicator cards. The difference between the I.H.P.'s of these two cards represents power consumed in the friction of the machine. Good mechanical efficiency, therefore, is seen to depend upon the extent to which friction is reduced. No attempt will here be made to enter into the details of compressor design. It will be enough simply to enumerate the points which have a vital bearing on this subject:

1. Power should be applied to resistance—steam pressure to air pressure—in as direct a manner as possible, so that bending or deflecting strains may be avoided as far as possible, with the resultant binding of bearings and loss of correct alignment.

2. Ample metal should be used, properly disposed to afford the rigidity necessary as a further precaution against the attendant binding stresses just mentioned. A light construction is always to be avoided.

3. Lest this correct proportioning of materials to meet stresses should result in an excessive weight and an extravagant cost, these materials should be of the best quality and intelligently selected for specific purposes, affording the maximum strength per unit of weight.

4. All bearings should be of ample area, so proportioned as to give a safe, moderate pressure per inch of bearing surface.

5. Wherever possible, bearings should be supplied with removable linings (preferably of anti-friction metal). In all cases, whether thus lined or otherwise, they should be scraped smooth and true, properly adjusted and arranged for the ready distribution of lubricant—all of this tending toward a lower coefficient of friction for these bearing surfaces.

6 Complete provision should be made for a plentiful supply of lubri-

cant to all bearing surfaces, furnished preferably by an automatic system which will be effective and efficient at all speeds.

7. The utmost simplicity should be sought throughout, reducing the number of parts, the number of joints (each an element of weakness), and the number of bearings, with their attendant friction and their possible loss of economical adjustment.

8. The design should afford a ready accessibility at every point, inviting frequent and careful attention to the maintaining of correct adjustment of all parts for most efficient operation.

9. Steam compounding and compound air-compression have important effects upon the mechanical efficiency, which have already been discussed in the proper place.

Friction has been aptly defined as "that portion of the power consumed by a machine in wearing itself out." This definition throws on the question of compressor endurance a light in which the importance of a high mechanical efficiency appears in bold relief. True economy is a question, not of momentary results, but of results covering months and years of continuous service under all conditions; and a compressor in which the percentage of "wearing-out power" is least promises not only the highest, but also the longest, record of sustained economy and satisfactory service.

IGNITIONS AND EXPLOSIONS

Ignitions and explosions in compressor cylinders, air-receivers, and pipe-lines, while comparatively rare, are yet of sufficient frequency to warrant attention. Such instances evidence a condition which should not exist in a well-designed and properly managed air-power plant; and several causes may account for them. In all cases it is to be remembered that air itself is not inflammable; and combustion or explosion can occur only when a foreign inflammable substance is present in the air. This is usually an explosive vapor produced by the volatilization of the lubricant used in the compressor; and ignition or explosion is always the result of some abnormal increase in temperature above the flash-point of the oil.

A common mistake in the operation of air-compressors is to use more oil than is necessary in the air cylinders. In steam cylinders more oil is required, because the tendency of the steam is to cut or wash away the oil; but in air cylinders there is no such action, and a drop of oil now and then is all that is necessary. Where too much oil is used, there is a gradual accumulation which may clog the valves and passages, interfering with the valve action and air discharge, and contributing toward a dangerously high temperature. Only the best oil—*never cylinder oil*—should be used in the air cylinders.

Benzine, naphtha, kerosene, and other light oils should never under any circumstances be introduced into a compressor cylinder. While they are unquestionably good for cleaning the cylinder walls, valves, ports, etc., they are an element of great danger, because of the ease with which they are volatilized and ignited. Soap and water is practically as good for cleaning purposes, and there is no danger attendant upon its use. A good practice is to fill the cylinder lubricator with strong soap and water once or twice a week—or even oftener, if there is a pronounced tendency to clog up—feeding it into the cylinder just as oil is fed.

Several causes may contribute toward the higher temperatures necessary for ignition and explosion. There may be an excessive temperature produced by an abnormal pressure not shown by the gauge, or there may be a very high temperature produced without any increase in pressure.

In the first case a poor oil may cause deposits of carbon in ports, passages, pipes, etc., which will so reduce the discharge area that a momentary pressure sufficient to produce an ignition temperature may be reached, even though this pressure may not extend to the receiver and be shown there on the gauge. Insufficient valve and port area may accentuate this trouble, and numerous bends in the pipe between the compressor and cylinder may still further contribute. In the second case, if a discharge valve should stick and remain open because of the carbon deposit upon it, the heated discharge air may return and be mingled with the free air entering. This air, already at a high temperature, would then be compressed with a further heating, and the final temperature might well be above the flash-point of the oil used, though the pressure might not exceed normal.

Heated intake air may also produce a dangerously high discharge temperature in the same manner as that just described. For instance, if the air for the compressor is drawn from a heated engine-room at a temperature of 80 or 90 degrees, its final temperature on compression may easily be above the danger point.

Explosions are much less likely to occur with compound compressors having intercoolers; and an aftercooler is an additional safeguard. Freedom from explosion and ignition can be practically assured by the following means: By providing for the admission of cool intake air; by the use of only the best oils, in the minimum quantity possible; by never using kerosene, benzine, naphtha, gasoline, etc., in the air cylinders; by frequent inspection and cleaning of air-valves, ports, cylinder bore, etc.; by the use of an aftercooler; by making the pipes between compressor and receiver of large area

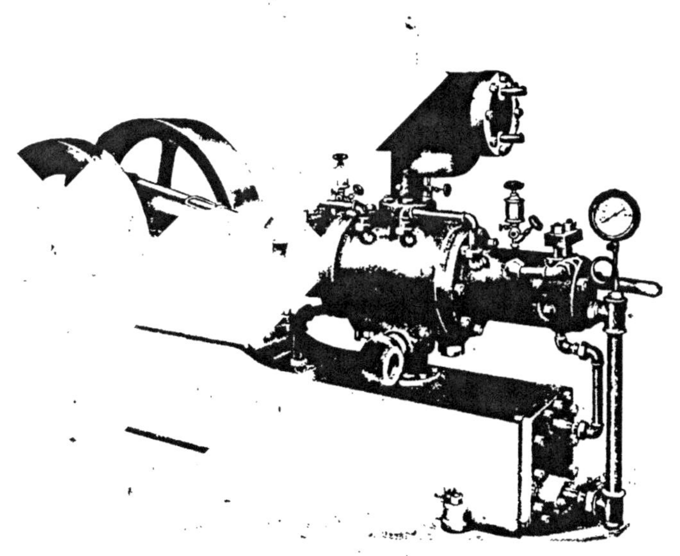

Fig. 74. Class EE-3 Straight-Line Three-Stage Belt-Driven Compressor for High Pressures.
Ingersoll-Rand Co., New York City.

and as short and free from bends as possible; and by selecting a compressor with ample valve and port areas and with direct air-passages.

HIGH-PRESSURE COMPRESSORS

The development of pneumatic haulage requiring the storage of air at high pressure, the use of storage air-brake systems, torpedo work, etc., have developed the modern high-pressure compressor for pressures up to 3,000 lbs. These are essentially multi-stage machines, and may be straight-line or duplex in type. The requirements of this line of work are particularly exacting, because of the high pres-

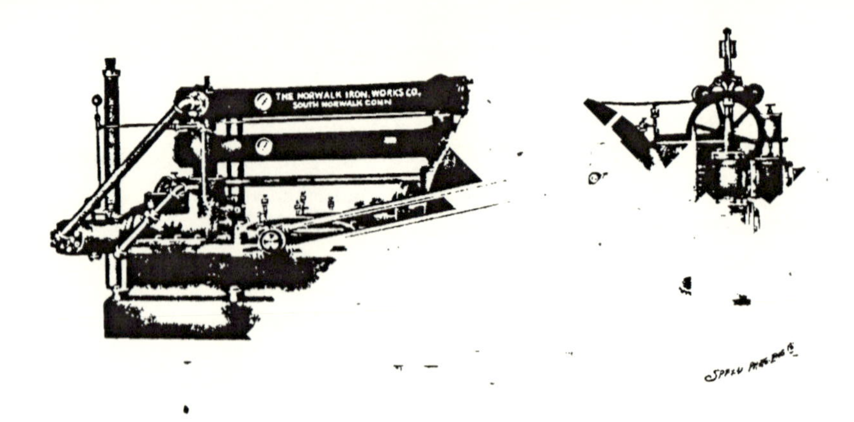

Fig. 75. Straight-Line Three-Stage Compressor for High-Pressure Locomotive Charging. Norwalk Iron Works Company, South Norwalk, Conn.

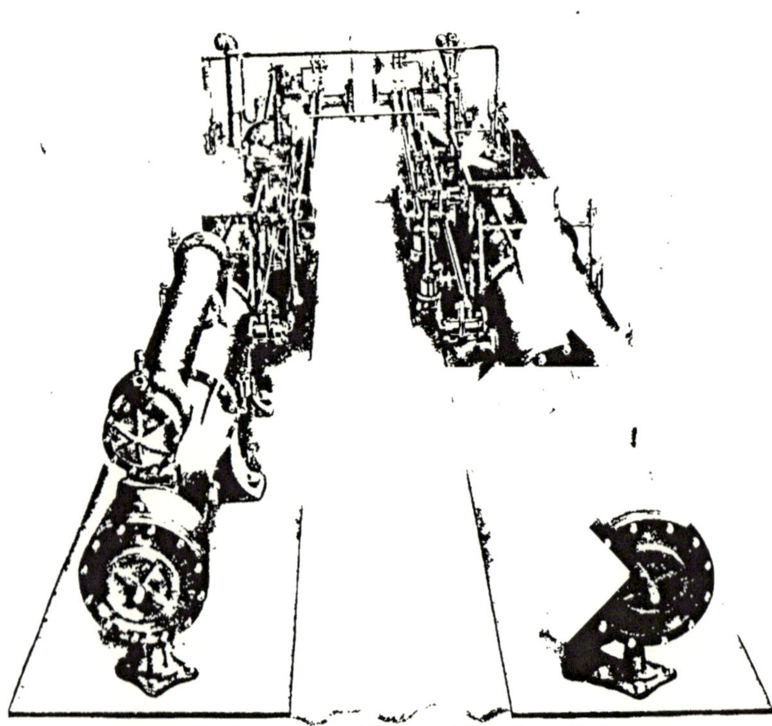

Fig. 76. Class WX-4 Cross-Compound Steam and Four-Stage Air-Compressor. Sullivan Machinery Co., Chicago, Ill.

sures involved; and the greatest care in design, workmanship, and materials of construction is necessitated.

The greatest difficulty is in the handling of the high temperatures which may be encountered, and the necessity for very perfect and complete cooling devices. It will be evident that the ordinary tubular intercooler cannot always be used for this work in the higher

Fig. 77. Four-Stage Air Cylinders for Double Cross-Compound High-Pressure Air-Compressor.
Ingersoll-Rand Co., New York City.

stages, because the volume to be handled becomes so small. The various high-pressure machines on the market, therefore, differ principally in the details of construction of intercoolers, high-pressure valves, high-pressure packings, cooling devices, etc. Practice varies so that no details can here be taken up without going into a discussion of several machines of specific builders; and this has hitherto been avoided in this treatise, which deals only with fundamental principles. The catalogues of the various builders may be referred to for more complete information on this point.

GAS-COMPRESSORS

The modern gas-compressor differs but little from the standard air-compressor; and the general principles already laid down are equally applicable to the compression of air and gas. It is, of course, to be understood, that the physical properties of various gases differ from those of air, making necessary the statement that the tabulated data presented in these pages cannot be used for gas-compression. For instance, the exponent of the equation for adiabatic compression of gas differs from that of air; but the fundamental principles of design, stage compression, cooling, etc., apply to gas as well as to air. Indeed, with some gases, the question of cooling is even more important than with air as affecting the thermal or illuminating power of those gases.

Another peculiarity of gas compression is the necessity for an absolutely closed intake and for the elimination of all leakage, for the escape of gas may seriously affect the health of the attendants, or may even result in explosion where the conditions are suitable. Occasionally a gas must be handled which chemically affects the metal of the ordinary compressor, in which case the cylinder bore, piston, rods, valves, and other parts in contact with the gas must be made of a special metal to resist this action. Gas-compression is a subject for special study, and requires a broad experience and a more or less specialized training.

DESIGN OF THE AIR-POWER PLANT

Compressed-air power plants may be broadly divided into two general classes. The first of these is the plant for temporary or semi-permanent use, which supplies power for a specific job, after the completion of which the plant may be sold or moved to another location. In this class will come the average plant for mining and contracting, for running rock-drills in tunnel work, operating pneumatic tools for bridge or structural work, for the development of a mining property, etc.

The second class includes the plant for permanent and continuous service, such as for running drills, coal-cutters, and pumps in a developed mine, operating a quarry, supplying air for railway switch and signal work, factory or municipal water supplies, railway and manufacturing shop service, and kindred work.

For work covered by the first class, the straight-line compressor is usually preferred, because of its self-contained character, its ease of transportation and installation, its simplicity, and the readiness with which it lends itself to the care of attendants of only ordinary skill. In this class of service, high operating economy is usually of less importance than the ability to "keep going" day after day, often for weeks at a time, without a shut-down. Frequently the work must be completed within a given time, or other considerations demand the utmost speed in the prosecution of the work. Under such conditions, it is of primary importance that the air-power plant shall be absolutely dependable, this point having even greater weight than the lowest steam consumption. Steam valves of plain slide or of Meyer cut-off type are usually preferred to more refined valve movements. Steam pressures are usually those which can be furnished by the ordinary return-tubular boiler, which can easily be transported and erected. There is, however, a notable tendency in recent years among the most successful contractors, to install high-grade duplex Corliss machines and water-tube boilers with high steam pressures and condensing apparatus, even for jobs which are more or less temporary in character. The saving in fuel and other operating charges by this advanced practice is evidently an important consideration in large enterprises where competition is keen and where ultimate profit depends upon close attention to every economy, however seemingly small.

In power plants of the second class, practice recommends the use of strictly high-grade compressors; and preference leans toward the double cross-compound type with Meyer cut-off or Corliss steam valves, the use of high steam pressures, a condensing equipment, and other accessories, making up a plant of the highest fuel and steam economy. Such plants are usually in charge of skilled engineers, and repair facilities are immediately at hand in emergency. It is not to be understood from this that the better class of compressors are more liable to breakdown. On the contrary, all things considered, they are fully up to the standard of the simpler machines; but any mechanical refinement adopted with a view to higher economy is likely to introduce elements requiring closer attention and demanding the maintaining of exactly correct adjustments. This is the price which must be paid for better efficiency, but it need not mean weakness or liability to breakdown. The very fact that a machine does better work, justifies its claim for better care and closer attention.

As to the boilers, steam piping, feed-water heaters, condensers, etc., involved in the high-grade air-power plant, no extended discussion is necessary here. The fundamental principles of steam-power plant design apply in this case.

As to the layout of the compressor plant proper, probably the first consideration is that of adequate protection against total loss of air supply by accident to the compressor installation. Obviously this lies in the subdivision of the plant into two or more units, in proportions to be determined by a careful study of the conditions under which the machines must work, the character of their load, etc.

It is always to be remembered that an air-compressor has no overload capacity. Its full rated speed as determined by the manufacturers is the limit of safe operation, and it can be exceeded only at a risk. The economy of the duplex steam-driven machine is practically constant at all loads up to full rating; and therefore nothing is lost in economy by having one compressor capable of carrying the full load, while still automatically regulating itself under fractional loads. But in such a case a breakdown of that machine would cripple the entire system. Moreover, provision must be made for occasional inspection and overhauling of the machine without stopping the operations depending upon it. It is safer, therefore, to have two compressors, each of one-half the maximum required capacity. One machine would then run at full capacity so long as the load on the plant did not fall below half of full load, the other compressor meantime caring for the intermediate fluctuations. For loads between half-load and no load, one machine would be shut down, saving oil, wear and tear, and attention. In case of accident to one machine, the other could continue to supply half the demand. Conditions may readily be conceived—and, indeed, are frequently met in practice—where three, four, or even eight duplex compressors afford the best solution of the problem of economical operation and regulation.

But where the compressors are power-driven—usually by electric motors—mere duplication of units is not sufficient, for the electric motor falls off in efficiency at fractional loads, and it is wise to keep all the compressors running as far as possible at full load. A theoretical case will best illustrate the method of handling such conditions. Let the problem be to install an electrically driven compressor plant to work under load variations of from 75 H.P. to 500 H.P., with an average normal load of 400 H.P. Probably the best solution would be the use of three motor-driven compressors, two of 200 H.P. and one of 100 H.P. The smaller unit would have a high-grade unloading device. Such a plant would afford the following combinations; 500, 400, 300, 200, and 100 H.P.—in all of which the machines in use would operate under full load and at maximum efficiency. When load fell below 100 H.P., the unloader of the smaller unit would safely care for a reduction of 25 or 30 per cent.

TRANSMISSION OF COMPRESSED AIR

The compressed-air transmission and distribution system is here understood as including all apparatus taking the air from the compressor discharge and delivering it to the machine where it is applied. With this understanding, the transmission system will be seen to include the following: (1) the *aftercooler,* if one is used; (2) the *primary air-receiver* or *receivers*; (3) the *secondary receivers* which may be used along the line; (4) the *pipe-line* proper; (5) the *reheater,* if one is used. These several divisions will be taken up in order; but before their discussion, the general theory of compressed-air transmission will be considered.

It will be remembered that the discharge from any air-compressor is more or less heated; but the compressed air quickly gives up its heat whether an aftercooler is used or not; and thereafter the transmission of this air may be considered as isothermal, or at a constant temperature approximately the same as that of the surrounding atmosphere. The initial cooling of the air after discharge and on just entering the transmission system, represents, of course, a loss of power; but inasmuch as the transmission system proper is not responsible for this loss, the latter cannot be charged against the transmitting devices.

With the air cooled to normal temperature in the system, and assuming for the present that there is no leakage, there can be no loss of power during transmission. A drop in pressure there will be, in an amount determined by the physical characteristics of the transmission system. But since the conditions are isothermal—that is, PV = a constant—this drop in pressure is accompanied with a corresponding increase in volume, and the intrinsic energy of the air in transit remains practically the same.

The drop in pressure accompanying transmission (assuming that there is no leakage) is the result of friction in the pipe-line and accessories; and there is no known formula which adequately covers even average conditions. Authorities differ widely, and even the results of actual tests have revealed no basis for accurate computation. However, the drop in pressure, due to friction, depends upon the following factors: the volume of air; the pressure of the air; the velocity of the flow; the diameter of the pipe; the length of the line and the condition of its interior surface; the number and the angle of elbows or bends; and the number and style of valves used. Moreover, the velocity of flow is not uniform, for the steadily reducing pressure and the steadily increasing volume produce a gradual acceleration in velocity, in order that this larger volume may be passed by the pipe of fixed diameter.

It is considered best in this treatise, therefore, to omit any formulæ for compressed-air transmission, which, at the best, would be only approximate for any particular case.

Tables of Air-Power Transmission. The problem usually confronting a practical man dealing with compressed air is the selection of a pipe of suitable size which will deliver the output of a given compressor plant over a given distance, with a given initial pressure, and with a drop in pressure not to exceed a certain number of pounds which may be arbitrarily decided upon. It will often be useful, also, to know the equivalent volume of compressed air delivered under these conditions.

Tables XI, XII, XIII, and XIV, which are copyrighted by the Ingersoll-Rand Company and here used with their permission, give all the data above referred to for four common initial pressures—

COMPRESSED AIR ROCK DRILLS IN THE GUNNISON TUNNEL OF THE UNITED STATES RECLAMATION SERVICE, COLORADO
Sullivan Machinery Co., Chicago, Ill.

TABLE XI

Loss of Pressure in Pounds, by Friction, in Transmission of Air through Pipes 1,000 Feet Long

Copyright, 1906, by Ingersoll-Rand Company

Initial Air-Pressure, 60 Pounds Gauge

(Wightman)

Size Pipe	Equivalent Delivery in Cubic Feet of Compressed Air per Minute																									
	9.84	14.73	19.64	24.60	29.45	34.44	39.35	49.20	58.90	68.6	78.6	88.4	98.4	118.1	137.5	156.6	176.5	196.4	294.5	393.7	492	589	686	786	884	984
	Delivery in Cubic Feet of Free Air per Minute																									
	50	75	100	125	150	175	200	250	300	350	400	450	500	600	700	800	900	1,000	1,500	2,000	2,500	3,000	3,500	4,000	4,500	5,000
1 in.	18.24	....	....	....	....	....	....	....	....	....	....	....	....	....	....	....	....	....	....	....	....	....	....	....	....	....
1¼ "	5.06	11.34	20.16	....	....	....	....	....	....	....	....	....	....	....	....	....	....	....	....	....	....	....	....	....	....	....
1½ "	1.95	4.33	7.79	12.23	17.53	....	....	....	....	....	....	....	....	....	....	....	....	....	....	....	....	....	....	....	....	....
2 "	.42	.95	1.69	2.65	3.80	5.17	6.77	10.61	15.20	....	....	....	....	....	....	....	....	....	....	....	....	....	....	....	....	....
2½ "	.13	.29	.52	.81	1.16	1.58	2.09	3.24	4.65	6.31	8.28	10.47	....	....	....	....	....	....	....	....	....	....	....	....	....	....
3 "	.05	.11	.19	.30	.44	.59	.78	1.22	1.78	2.37	3.11	3.94	4.88	7.03	9.52	....	....	....	....	....	....	....	....	....	....	....
3½ "	....	.05	.08	.13	.19	.26	.36	.55	.78	1.07	1.40	1.77	2.20	3.17	4.29	5.57	7.08	8.77	....	....	....	....	....	....	....	....
4 "	....	....	.04	.07	.09	.13	.17	.27	.38	.53	.69	.88	1.08	1.56	2.12	2.75	3.49	4.33	9.73	....	....	....	....	....	....	....
4½ "	....	....	....	.03	.05	.07	.09	.15	.21	.29	.39	.48	.60	.87	1.17	1.52	1.94	2.40	5.39	9.65	....	....	....	....	....	....
5 "	....	....	....	....	.03	.04	.06	.08	.12	.17	.22	.28	.34	.49	.67	.87	1.17	1.37	3.08	5.51	8.61	....	....	....	....	....
6 "	....	....	....	....	....	.01	.02	.03	.05	.06	.08	.11	.14	.19	.27	.34	.43	.54	1.20	2.16	3.36	4.82	6.54	....	....	....
7 "	....	....	....	....	....	....	.01	.01	.02	.03	.04	.05	.06	.09	.12	.15	.19	.24	.55	.98	1.53	2.19	2.97	3.91	4.94	6.19
8 "	....	....	....	....	....	....	....	....	.01	.01	.01	.02	.03	.04	.06	.08	.09	.12	.27	.41	.77	1.11	1.50	1.98	2.51	3.10
9 "	....	....	....	....	....	....	....	....	....	....	....	.01	.01	.02	.03	.04	.05	.06	.15	.27	.42	.61	.83	1.08	1.36	1.69
10 "	....	....	....	....	....	....	....	....	....	....	....	....	....	.01	.02	.03	.03	.04	.09	.16	.25	.36	.48	.63	.79	.99
12 "	....	....	....	....	....	....	....	....	....	....	....	....	....	....	....	.01	.01	.02	.03	.06	.09	.14	.19	.25	.32	.39
14 "	....	....	....	....	....	....	....	....	....	....	....	....	....	....	....	....	....	.01	.01	.03	.04	.06	.09	.11	.15	.18
16 "	....	....	....	....	....	....	....	....	....	....	....	....	....	....	....	....	....	....	....	.01	.02	.03	.04	.05	.07	.09

For longer or shorter pipes, the friction loss is proportional to the length—that is, for 500 feet, ½ of the above; for 4,000 feet, four times the above; etc.

TABLE XII

Loss of Pressure in Pounds, by Friction, in Transmission of Air through Pipes 1,000 Feet Long

Copyright, 1906, by Ingersoll-Rand Company

Initial Air-Pressure, 80 Pounds Gauge

(Wightman)

Size Pipe	Equivalent Delivery in Cubic Feet of Compressed Air per Minute																									
	7.74	11.3	15.2	19.4	23.2	27.2	31.0	38.7	46.5	54.2	62.0	69.7	77.4	92.9	108.2	124.0	139.5	152	232	310	387	465	542	620	697	774
	Delivery in Cubic Feet of Free Air per Minute																									
	50	75	100	125	150	175	200	250	300	350	400	450	500	600	700	800	900	1,000	1,500	2,000	2,500	3,000	3,500	4,000	4,500	5,000
1 in.	14.31	....	....	....	....	....	....	....	....	....	....	....	....	....	....	....	....	....	....	....	....	....	....	....	....	....
1¼ "	3.96	8.46	15.31	....	....	....	....	....	....	....	....	....	....	....	....	....	....	....	....	....	....	....	....	....	....	....
1½ "	1.53	3.26	5.92	9.64	13.79	....	....	....	....	....	....	....	....	....	....	....	....	....	....	....	....	....	....	....	....	....
2 "	.33	.71	1.28	2.09	2.99	4.09	5.34	8.32	12.01	....	....	....	....	....	....	....	....	....	....	....	....	....	....	....	....	....
2½ "	.10	.21	.39	.64	.91	1.25	1.63	2.54	3.67	4.99	6.53	8.25	10.81	....	....	....	....	....	....	....	....	....	....	....	....	....
3 "	.03	.08	.14	.24	.34	.47	.61	.96	1.38	1.88	2.45	3.13	3.83	5.61	7.46	9.86	....	....	....	....	....	....	....	....	....	....
3½ "	.01	.03	.06	.11	.15	.21	.27	.43	.62	.84	1.11	1.40	1.73	2.46	3.37	4.42	5.61	6.64	15.41	....	....	....	....	....	....	....
4 "	....	.01	.03	.05	.07	.10	.13	.21	.30	.41	.54	.69	.85	1.22	1.66	2.18	2.77	3.29	7.62	13.62	....	....	....	....	....	....
4½ "	....	....	.02	.03	.04	.06	.07	.12	.17	.23	.30	.38	.47	.68	.92	1.19	1.54	1.82	4.24	7.58	11.79	....	....	....	....	....
5 "	....	....	.01	.01	.02	.03	.04	.07	.09	.13	.17	.22	.27	.39	.53	.69	.88	1.04	2.43	4.32	6.88	9.72	13.25	....	....	....
6 "	....	....	....	....	.01	.01	.01	.02	.03	.05	.06	.08	.10	.15	.20	.27	.34	.40	.95	1.69	2.64	3.79	5.27	6.78	8.54	10.55
7 "	....	....	....	....	....	....	....	.01	.01	.02	.03	.04	.05	.06	.09	.12	.15	.18	.43	.77	1.19	1.73	2.35	3.07	3.89	4.79
8 "	....	....	....	....	....	....	....	....	....	.01	.01	.01	.02	.03	.04	.06	.08	.09	.22	.39	.60	.87	1.19	1.55	1.97	2.46
9 "	....	....	....	....	....	....	....	....	....	....	....	....	.01	.02	.02	.03	.04	.05	.12	.21	.33	.48	.65	.85	1.08	1.33
10 "	....	....	....	....	....	....	....	....	....	....	....	....	....	.01	.01	.02	.02	.03	.06	.12	.19	.28	.37	.49	.66	.77
12 "	....	....	....	....	....	....	....	....	....	....	....	....	....	....	....	.01	.01	.01	.02	.04	.07	.11	.15	.19	.25	.30
14 "	....	....	....	....	....	....	....	....	....	....	....	....	....	....	....	....	....	....	.01	.02	.03	.05	.06	.09	.11	.14
16 "	....	....	....	....	....	....	....	....	....	....	....	....	....	....	....	....	....	....	....	.01	.01	.02	.03	.04	.05	.07

For longer or shorter pipes, the friction loss is proportional to the length—that is, for 500 feet, ½ of the above; for 4,000 feet, four times the above; etc.

TABLE XIII

Loss of Pressure in Pounds, by Friction, in Transmission of Air through Pipes 1,000 Feet Long

Initial Air-Pressure, 100 Pounds Gauge

(Wightman)

Size Pipe	Equivalent Delivery in Cubic Feet of Compressed Air per Minute																									
	6.41	9.61	12.81	15.81	19.22	22.39	25.62	31.62	38.44	44.78	51.24	57.65	63.24	76.88	89.56	102.5	115.3	126.5	192.2	256.2	316.2	384.4	447.8	512.4	576.5	632.4
	Delivery in Cubic Feet of Free Air per Minute																									
	50	75	100	125	150	175	200	250	300	350	400	450	500	600	700	800	900	1,000	1,500	2,000	2,500	3,000	3,500	4,000	4,500	5,000
1 in.	11.89	....	....	....	....	....	....	....	....	....	....	....	....	....	....	....	....	....	....	....	....	....	....	....	....	....
1¼ "	3.29	7.42	13.20	....	....	....	....	....	....	....	....	....	....	....	....	....	....	....	....	....	....	....	....	....	....	....
1½ "	1.28	2.87	5.11	7.75	11.42	....	....	....	....	....	....	....	....	....	....	....	....	....	....	....	....	....	....	....	....	....
2 "	.27	.62	1.15	1.68	2.48	3.36	4.43	6.72	9.95	13.41	....	....	....	....	....	....	....	....	....	....	....	....	....	....	....	....
2½ "	.08	.19	.34	.52	.76	1.03	1.36	2.06	3.04	4.11	5.40	6.85	8.21	12.21	....	....	....	....	....	....	....	....	....	....	....	....
3 "	.03	.07	.12	.19	.29	.39	.51	.77	1.14	1.54	2.06	2.57	3.08	4.58	6.19	8.13	10.23	12.39	....	....	....	....	....	....	....	....
3½ "	.01	.03	.05	.08	.13	.17	.23	.35	.51	.69	.92	1.16	1.39	2.14	2.79	3.67	4.64	5.60	12.81	....	....	....	....	....	....	....
4 "	....	.01	.02	.04	.06	.09	.12	.17	.25	.34	.45	.57	.68	1.03	1.38	1.81	2.29	2.76	6.68	11.35	....	....	....	....	....	....
4½ "	....	....	.01	.02	.03	.04	.06	.09	.14	.19	.25	.32	.38	.57	.77	1.00	1.27	1.23	3.51	6.61	9.56	14.04	....	....	....	....
5 "	....	....	....	.01	.02	.03	.04	.05	.08	.11	.15	.18	.22	.33	.44	.57	.76	.88	2.03	3.62	5.51	8.11	10.95	14.48	....	....
6 "	....	....	....	....	.01	.01	.02	.02	.03	.04	.05	.07	.08	.12	.17	.22	.28	.34	.78	1.41	2.14	3.16	4.26	5.59	7.04	8.51
7 "	....	....	....	....	....	....	.01	.01	.01	.02	.03	.03	.04	.05	.07	.10	.13	.16	.36	.67	.97	1.44	1.93	2.55	3.22	3.88
8 "	....	....	....	....	....	....	....	....	....	.01	.01	.02	.02	.03	.04	.05	.06	.08	.18	.33	.49	.76	.98	1.30	1.84	1.98
9 "	....	....	....	....	....	....	....	....	....	....	....	.01	.01	.02	.02	.03	.04	.04	.09	.18	.27	.39	.53	.72	.89	1.07
10 "	....	....	....	....	....	....	....	....	....	....	....	....	....	.01	.01	.02	.02	.03	.05	.10	.16	.23	.31	.41	.52	.63
12 "	....	....	....	....	....	....	....	....	....	....	....	....	....	....	....	.01	.01	.01	.02	.04	.06	.09	.12	.16	.21	.25
14 "	....	....	....	....	....	....	....	....	....	....	....	....	....	....	....	....	....	....	.01	.02	.03	.04	.05	.07	.09	.11
16 "	....	....	....	....	....	....	....	....	....	....	....	....	....	....	....	....	....	....	....	.01	.01	.02	.03	.04	.05	.06

For longer or shorter pipes, the friction loss is proportional to the length—that is, for 500 feet, ½ of the above; for 4,000 feet, four times the above; etc.

TABLE XIV

Loss of Pressure in Pounds, by Friction, in Transmission of Air through Pipes 1,000 Feet Long

Initial Air-Pressure, 125 Pounds Gauge

(E. F. Schaefer)

Size Pipe	Equivalent Delivery in Cubic Feet of Compressed Air per Minute																									
	5.26	7.89	10.51	13.15	15.79	18.41	21.05	26.30	31.58	36.81	42.10	47.30	52.60	63.20	73.70	84.20	94.70	105.1	157.9	210.5	263.0	315.8	368.1	422.0	473.0	526.0
	Delivery in Cubic Feet of Free Air per Minute																									
	50	75	100	125	150	175	200	250	300	350	400	450	500	600	700	800	900	1,000	1,500	2,000	2,500	3,000	3,500	4,000	4,500	5,000
1 in.	9.88	22.20	39.50	....	....	....	....	....	....	....	....	....	....	....	....	....	....	....	....	....	....	....	....	....	....	....
1¼ "	2.70	6.07	10.82	16.88	24.33	33.05	....	....	....	....	....	....	....	....	....	....	....	....	....	....	....	....	....	....	....	....
1½ "	1.05	2.37	4.22	6.58	9.47	12.90	16.84	26.30	37.90	....	....	....	....	....	....	....	....	....	....	....	....	....	....	....	....	....
2 "	.23	.51	.91	1.42	2.04	2.78	3.63	5.68	8.18	11.08	14.51	18.38	22.68	....	....	....	....	....	....	....	....	....	....	....	....	....
2½ "	.07	.16	.28	.43	.63	.85	1.11	1.73	2.51	3.39	4.44	5.61	6.95	10.00	13.60	17.80	....	....	....	....	....	....	....	....	....	....
3 "	.03	.06	.10	.16	.23	.32	.42	.65	.94	1.27	1.67	2.11	2.61	3.76	5.11	6.68	8.45	10.42	23.48	....	....	....	....	....	....	....
3½ "	.01	.03	.05	.07	.11	.14	.19	.29	.42	.58	.75	.95	1.18	1.69	2.31	3.01	3.81	4.71	10.59	18.81	29.40	....	....	....	....	....
4 "	....	.01	.02	.04	.05	.07	.09	.15	.21	.28	.37	.47	.58	.84	1.14	1.49	1.88	2.32	5.23	9.30	14.52	20.90	28.51	....	....	....
4½ "	....	....	.01	.02	.03	.04	.05	.08	.12	.16	.21	.26	.32	.46	.63	.83	1.04	1.29	2.90	5.15	8.05	11.59	15.78	20.61	26.10	32.20
5 "	....	....	....	.01	.02	.02	.03	.05	.07	.09	.12	.15	.18	.27	.36	.47	.60	.74	1.65	2.94	4.60	6.63	9.01	11.80	14.90	18.45
6 "	....	....	....	....	.01	01	.01	.02	.03	.04	.05	.06	.07	.10	.14	.18	.23	.29	.64	1.15	1.80	2.59	3.53	4.61	5.83	7.20
7 "	....	....	....	....	....	....	....	.01	.01	.02	.02	.03	.03	.05	.06	.08	.11	.13	.29	.52	.82	1.18	1.61	2.19	2.65	3.27
8 "	....	....	....	....	....	....	....	....	....	.01	.01	.01	.02	.02	.03	.04	.05	.07	.15	.26	.41	.60	.81	1.06	1.34	1.65
9 "	....	....	....	....	....	....	....	....	....	....	....	.01	.01	.01	.02	.02	.03	.04	.08	.15	.23	.33	.45	.58	.73	.90
10 "	....	....	....	....	....	....	....	....	....	....	....	....	....	.01	.01	.01	.02	.02	.05	.08	.13	.19	.26	.34	.43	.53
12 "	....	....	....	....	....	....	....	....	....	....	....	....	....	....	....	.01	.01	.01	.02	.03	.05	.07	.10	.13	.17	.21
14 "	....	....	....	....	....	....	....	....	....	....	....	....	....	....	....	....	....	....	.01	.02	.02	.03	.04	.06	.08	.10
16 "	....	....	....	....	....	....	....	....	....	....	....	....	....	....	....	....	....	....	....	.01	.01	.02	.02	.03	.04	.05

For longer or shorter pipes, the friction loss is proportional to the length—that is, for 500 feet, ½ of the above; for 4,000 feet, four times the above; etc.

TABLE XV

Globe Valves, Tees, and Elbows

The reduction of pressure produced by globe valves is the same as that caused by the following additional lengths of straight pipe, as calculated by the formula,

$$\text{Additional length of pipe} = \frac{114 \times \text{Diameter of pipe}}{1 + (3.6 \div \text{Diameter})}:$$

Diameter of pipe	1	1½	2	2½	3	3½	4	5	6	inches
Additional length	2	4	7	10	13	16	20	28	36	feet
Diameter of pipe	7	8	10	12	15	18	20	22	24	inches
Additional length	44	53	70	88	115	143	162	181	200	feet

The reduction of pressure produced by elbows and tees is equal to two-thirds of that caused by globe valves. The following are the additional lengths of straight pipe to be taken into account for elbows and tees. For globe valves, multiply by $\frac{3}{2}$:

Diameter of pipe	1	1½	2	2½	3	3½	4	5	6	inches
Additional length	2	3	5	7	9	11	13	19	24	feet
Diameter of pipe	7	8	10	12	15	18	20	22	24	inches
Additional length	30	35	47	59	77	96	108	120	134	feet

These additional lengths of pipe for globe valves, elbows, and tees must be added in each case to the actual length of straight pipe. Thus a 6-inch pipe 500 feet long, with 1 globe valve, 2 elbows, and 3 tees, would be equivalent to a straight pipe $500 + 36 + (2 \times 24) + (3 \times 24) = 656$ feet long.

60, 80, 100, and 125 lbs.—for volumes of from 50 to 5,000 cubic feet of free air per minute, and for a distance of 1,000 feet. For longer or shorter distances, the drop in pressure will be practically proportional—that is, one-half as much for 500 feet, and four times as much for 4,000 feet. The delivered volume of compressed air given in these tables is the equivalent volume at initial pressure, and is therefore not strictly correct at the point of delivery, since the pressure will be decreased by the amount given, and the volume increased accordingly. However, since these are minimum volumes, they may be safely used where the reduced pressure is taken into account. These tables show that it is possible to get almost any pressure drop wanted, and that this loss of pressure can be practically eliminated if a pipe large enough is used. But here the question of cost of pipe and fittings enters in, and decision in any case will be a compromise for a moderate drop of pressure in connection with a moderate cost of line.

Effect of Valves, Tees, and Elbows. Supplementing the four tables just referred to, is Table XV, also furnished by the Ingersoll-Rand Company, giving the effect of valves, tees, and elbows in a pipe-line, expressed in terms of additional lengths of pipe. The data given in these five tables will furnish a solution for almost any problem of transmission which will be encountered in average practice.

Necessity for Dry Air. There are many reasons why air should be as dry—that is, as free from water vapor—as possible, during transmission. At the outset, let it be clearly understood that air cannot "freeze up." It is the water in the air which freezes when the conditions are favorable.

The capacity of air for carrying water vapor increases with its temperature. This means that the cooler the air, the less moisture it will contain; and as temperature falls, the water vapor is condensed to water, and accumulates wherever opportunity offers. If this water is in the pipe-lines, it will freeze up in cold weather, clogging and ultimately choking the line, and frequently bursting the pipe. Or, the small particles of ice may be carried along with the movement of the air until they encounter a valve or other obstruction, where they will gather and eventually fill the pipe, closing the air-passage. Even in warm weather, the accumulation of water in the pipe results in *water-hammer* and leaky joints. It flows to the low points in the line, and fills the pipe, when it must be pushed ahead by the air-pressure, resulting in an added loss of pressure.

Water in the air is the cause of freezing at exhaust ports of drills, pumps, etc., since the sudden expansion of the air on exhaust produces such a low temperature that ice is formed and the exhaust is clogged, oftentimes even in warm weather. There can be no difficulty of this kind where the air is dry

Water carried with the air into the cylinders of drills, pumps, pneumatic tools, and other compressed-air appliances, condenses on the walls and excludes the lubricant, resulting in an added friction load and more rapid wear and leakage.

All of these points emphasize the importance of removing the water from the air, so far as this is practically possible. The logical sequence of operations in which this should be accomplished, is: *first,* before compression; *second,* during compression; *third,* after compression and before transmission; *fourth,* during transmission. In all of these cases the problem is one of adequate cooling.

The Antecooler. Although not properly a part of the transmission system, the antecooler may here be considered in its relation to the question of dry air. It is not generally used, and has not received the attention its importance deserves. It is very similar in construction to the intercooler of a compound compressor, and is placed on the air intake to the compressor, so that all air entering the compressor must pass through it. Of course it must be larger in proportion, to provide for the larger volume and lower velocity of the atmospheric air passing through it, which must not be perceptibly rarified in passing. If the intake air is warm and moist, and water sufficiently cool can be had for the antecooler, this device will condense a large part of the moisture before it enters the compressor. Aside from the drying effect, the antecooler may have an important effect upon the economy of the air-power system, when it is remembered that for every five-degree reduction in the temperature of the intake air there is an increase of about one per cent in volumetric efficiency. As to the design of the antecooler, it need only be said that it should conform to the requirements already outlined in connection with the intercooler.

Removal of Moisture during Compression. The removal of the moisture from the air during the compression process has already been considered in the discussion of Compound Compression, in which it was seen that the intercooler played an important part in this process by condensing the water vapor between the stages of compression. No further reference need here be made to this point.

The Aftercooler. The aftercooler is practically identical with the intercooler, it being important to remember, however, that it must deal with higher pressures and smaller volumes. It should be placed as close as possible to the air-compressor, and provided with a suitable means of draining it. The function of the aftercooler is to reduce the temperature of the compressed air as low as possible, condensing as much as possible of the moisture and delivering the air to the line at maximum density and minimum temperature.

Aside from the withdrawal of moisture, the aftercooler has a most important effect on pipe-line efficiency. Nothing about the air-power plant will waste more power than a badly leaking pipe-line; and nothing is so conducive to a leaky line as wide extremes of temperature, producing large strains of expansion and contraction. Without an aftercooler, air enters the pipe-line hot, while the pipe itself is cold; but the latter gradually takes up heat from the air, and the heated portion extends along the line the longer the load is on the line; and the whole system expands accordingly. When the plant is shut down—probably at night, when the atmosphere is cooler—there is a corresponding contraction all along the line. This goes on day after day, with alternate expansions and contractions; and the resultant difficulty in keeping tight joints and valves and in preventing leakage becomes evident, even where the best mechanical precautions are taken along the line. The use of an aftercooler obviates these extremes of temperature, and leaves the pipe-line subject only to the expansion and contraction consequent upon changes in atmospheric temperature.

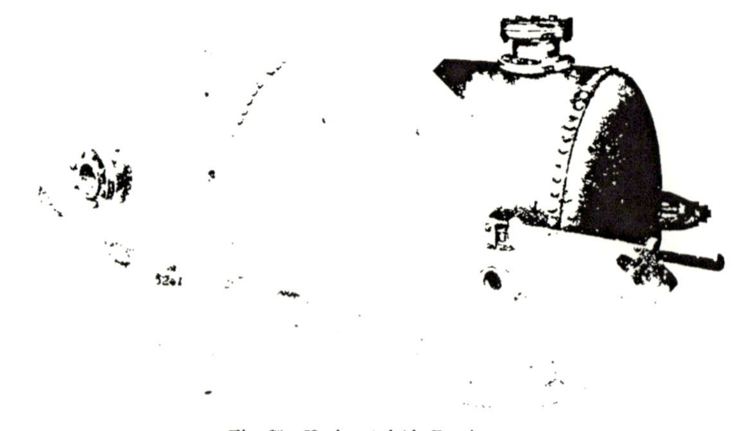

Fig. 78. Horizontal Air-Receiver.
Ingersoll-Rand Co., New York City.

Primary Air-Receivers. The discharge from the compressor is more or less pulsating in character, corresponding to the strokes of compression; and an air-receiver is a *rectifier*, so to speak, which receives and absorbs the pulsations and delivers a steady flow of air to the line. It is also in a very small degree an *accumulator*, in which excess energy is momentarily stored and momentarily withdrawn; but it cannot be relied upon as a power storage in this respect.

The air-receiver cannot be too large, nor can there be too many of them, provided that leakage is properly guarded against. Ample receiver capacity is especially useful on work of intermittent character, such as running rock-drills and pneumatic tools. It makes the problem of regulation simpler, the work of the governor or unloader easier, and helps out on overloads. In a broad sense, the air-receiver is the *balance wheel* of the air-power system, equalizing supply and demand and smoothing out the minor fluctuations in the load curve.

Preference should be given to the *vertical* receiver as against the *horizontal.* It should be placed as close as possible to the compressor or aftercooler, and a pipe amply large in section should be used for connecting it up. It is well to have the pipe entering the receiver from the compressor a size larger than the discharge pipe from the receiver to the line. Elbows should be avoided as far as possible in the former section of piping, and preference given to wide-sweep bends. No valve should be placed between the compressor and receiver unless it is protected by a relief-valve on the side nearest the compressor. It is perhaps unnecessary to state that each receiver should have a relief- or safety-valve; and where the receiver is out of doors, its safety-valve should be piped back into a warm place, to prevent its freezing up. The piping for the main or primary receiver should enter near the top, and leave near the bottom.

Fig. 79. Vertical Air-Receiver. Ingersoll-Rand Co., New York City.

There is, of course, some cooling of the air in its passage through the receiver; and on this account several small receivers are better than one large one, as offering a larger cooling surface. Each receiver should be provided with a drain-cock at its lowest point, and this cock should be opened regularly and frequently for the withdrawal of water and oil. The best location for a receiver is out of doors, where it is exposed to the coolest atmosphere. A pressure-gauge on each receiver is convenient.

Secondary Receivers. Secondary receivers are essentially moisture traps, and, in large transmission systems, should be placed at every low point in the line, and always *below* the line, so that they may collect the water which condenses in the piping and flows to this point. They need not be very large, and are preferably of horizontal pattern; and the pipe-lines should enter and leave from the top or upper side, so that the flow of the air will not be impeded by the accumulated water. Each receiver should have a drain-cock at its lowest point, and should be drained regularly and often. Where only one secondary receiver is used, it should be placed far enough away from the primary receiver, so that the air may have a chance to cool to atmospheric temperature before entering it. All branches from the main transmission line should be taken off *beyond* the first secondary receiver. It is not necessary that these small tanks should have a safety-valve.

The Pipe-Line. The size of pipe to be used having been determined from the tables already given, or otherwise, the problem of transmission efficiency becomes one of care in the smaller details of installation. The most important consideration is that the pipe-line *shall not leak.* A transmission system which will not hold gauge pressure over night with all outlets and valves closed, cannot be considered a first-class system. More power can be wasted through a few seemingly insignificant leaks in the line than can be saved by the most refined methods of air-compression.

It is better to lay pipe on the surface of the ground than underground, for, in the former case, not only is there more cooling, and therefore more complete removal of moisture, but a leak is more readily discovered and stopped. If it is not desirable to lay the line on the surface, it should be laid in a trough or box so that it can be readily examined. A good way of testing a line for leaks is to shut down all the air outlets provided with valves, and observe how many strokes of the compressor are necessary to maintain gauge pressure. This will give a basis for figuring piston displacement which is necessary to carry the *leakage load*; and the results are frequently surprising. A pipe-line should be regularly inspected for leaks, by going over it with a brush and a pail of soapy water, "painting" each joint, branch, valve, or connection on pipe-line, receivers, etc., with this solution. A leak will be revealed by the soap bubbles formed, even when it is not sufficient to make a sound. Flanged joints are to be preferred to screwed joints, as more easily packed and kept tight. In making up a pipe-line, care should be exercised to see that none of the packing in the joints projects into the pipe, offering an obstruction; and each length of pipe should be cleaned of all obstructions before connecting up. Suitable supports should be provided at safe intervals, so that the line will not sag, with danger of sprung joints. Expansion and contraction must be provided for, and consequent leakage guarded against, either by the use of packed slip joints, or, preferably and more cheaply, by offsetting the line at intervals of a few hundred feet with wide-angle bends. Wide-sweep bends should be preferred to short elbows. Branches should be taken off with angle fittings pointing in the direction of flow, rather than with tees. Valves should be of gate pattern, rather than of globe type, as offering more direct passage and less friction. Rising-stem valves have the advantage over others, of showing at a glance whether the valve is closed or open, and the degree of opening. Every low place in the line where water might accumulate should have a secondary receiver, or at least a drain-cock, which should be opened frequently and the water blown off. Mr. Frank Richards is authority for the statement that "the velocity of air in main transmission lines should never exceed twenty feet per second; and in small branch pipes it should be still lower."

The Loop System. The usual air-transmission system consists of a main or trunk line, with branches and sub-branches taken off at intervals. Evidently a break at any point in the main line means the shutting down of all the machines connected up beyond the break, while repairs are in progress. To avoid this shut-down of a part of the air-transmission scheme, the *loop system* of distribution was devised. It is illustrated in the diagram, Fig 80. The main supply-pipe from the primary receiver is divided into two branches which later unite in a loop or closed circuit. From this main loop, other secondary loops are formed by intersecting and uniting branch pipes. Valves are placed as indicated by the cross marks in the diagram.

A little study of this sketch will show that nothing short of a break in the main between the receiver and the first branch forming the loop can shut down this system or interfere with its workings. Any section in any loop or sub-loop can be cut out for alterations or repairs, without in any way affecting the other sections. While any one branch of the main or secondary loops might be overloaded while supplying the whole load on that loop, yet the shut-down of one side of the loop would be only temporary, and the work could go on at slightly less efficiency in the transmission system while changes or repairs were in progress.

The loop system is probably not justified in any but comparatively large installations, such as air-power supply for shops, quarries, mines, and contracts where a large number of compressed-air appliances are in use continuously, and where it is important that the work shall not suffer interruption. This system will probably cost more to install than a straight trunk-and-branch system; but its extra cost may be looked upon as an insurance against total shut-down during repairs or alterations in the mains.

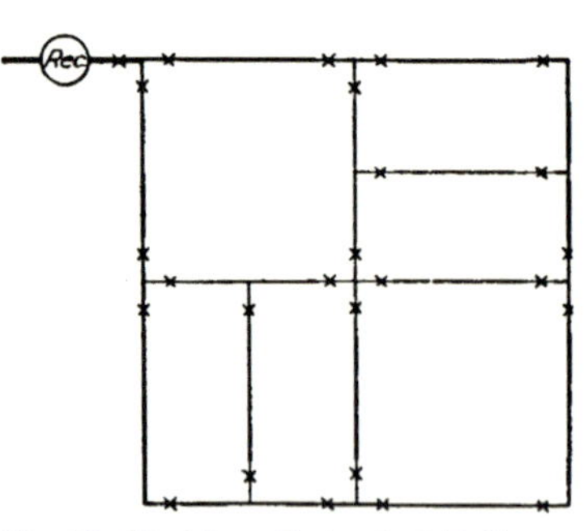

Fig. 80. The "Loop System" of Air Power Distribution.

The Dense-Air System. The usual practice in ordinary air-compression and application is to compress from atmosphere to a given pressure, transmit at this pressure, apply at this pressure, and exhaust to atmosphere. In the *dense-air* or *return-pipe system*, on the contrary, the exhaust from the air-motor is above atmosphere, and is piped back to the compressor intake, the cycle of compression, transmission, and expansion thus occurring in a closed circuit, and pressures always being well above atmosphere.

The dense-air system is based upon the fact that the power required in the first stages of compression, and for comparatively low pressures, is greater than that required in the higher stages of compression. For instance, it takes the same power to compress a given volume of free air from atmosphere to 30 lbs. gauge as to continue to compress it from 30 lbs. gauge to 90 lbs. gauge. In the first case, if the air at 30 lbs. is used in a motor discharging to atmosphere, the maximum effective pressure is 30 lbs. In the second case, if the air at 90 lbs. is used in a motor exhausting at 30 lbs., and this exhaust is piped back to the compressor intake for re-compression to 90 lbs., there will be a maximum effective pressure of 90–30, or 60 lbs., at the motor. The cost of compression in the latter case will be the same as in the first, but the maximum effective pressure is twice as great.

The higher the compression is carried, the greater the comparative advantage of the dense-air system. For example, the power required to compress from atmosphere to 60 lbs. gauge is practically the same as that needed to compress from 60 lbs. to 350 lbs. gauge. The maximum effective pressures are in these two cases 60 lbs. and 290 lbs., respectively, assuming that the dense-air system is used with the higher pressure. But the ordinary mechanical devices on the market are not adapted to more than 125 or 150 lbs. pressure, and this fixes the practical limit for the use of the dense-air system. It is to be remembered that even though the effective difference of pressures be only 125 lbs., the air supply line from compressor to motor must carry the full highest pressure of the system. This is another fact setting a limit to the maximum pressures to be employed in this system, as standard pipe and fittings are always to be preferred.

Aside from the saving in compressor power, the dense-air system also eliminates all losses due to clearance, since the air is in a closed circuit.

The complete dense-air system consists of a compressor (usually single-stage, since the ratio of intake and discharge pressures is low, even though these pressures themselves may be high), two pipe-lines, and a motor using the compressed air. There will be some leakage in operation, so that a small *booster* compressor is ordinarily used also, supplying air to the low-pressure side of the system to make good these leakage losses and maintain the proper ratio of pressures.

It must be confessed that the dense-air system does not fully realize all the economies which it promises; but the actual working-out of the system defining these shortcomings demands a mathematical investigation which is beyond the scope of this book. It is evident that the system is not adapted to intermittent service or widely fluctuating loads. Its best field of application is in the operation of direct-acting pumps by compressed air.

The general scheme of compression and expansion in a closed circuit, however, is used in three very practical devices on the market —namely, the *return-air pumping system*, the *electric-air rock-drill*, and the *electric-air channeler*, reference to which will be made later.

Fig. 81. Sergeant Air-Reheater.
Ingersoll-Rand Co., New York City.

Reheating Compressed Air. The reheating of compressed air might not seem properly to belong in a section on air-transmission; but it is here introduced because the definition of the transmission system here adopted covers all devices for delivering the compressed air from the compressor to the machine using it.

Reheating is, in brief, a process of restoring to the compressed air the heat which was withdrawn from it by the cooling devices during compression and transmission. If it were economically desirable and practically possible to retain all the heat generated in the air during compression and up to the point of application, reheating would be unnecessary; but it has been seen that economical compression and transmission demand the removal of all the heat possible.

It will be remembered that, according to Charles's law, the volume of air at a constant pressure varies directly as its absolute temperature. The reheating of compressed air at a constant pressure, therefore, produces an increase in volume at that pressure proportional to the degree of heating. In that respect it is equivalent to an increase in compressor capacity; but a vital difference is that this increased volume by reheating is secured much more cheaply than by increased compressor capacity. Mr. Frank Richards, in his book on "Compressed Air," has explained this very clearly; and the comparison which follows is used here with his permission, the figures all being theoretical:

"Weight of 1 cu. ft. of steam at 75 lbs. gauge, 0.2089 lb.

"Total heat units in 1 lb. of steam at 75 lbs. gauge, from water at 60° F., 1,151.

"Total heat units in 1 cu. ft. of steam at 75 lbs. gauge, 1,151 × 0.2089 = 240.44.

"To produce by compression with a steam-driven compressor 1 cu. ft. of compressed air at 75 lbs. gauge and 60° F., about 2 cu. ft. of steam at the same pressure are required; or the heat units employed in producing 1 cu. ft. of compressed air (at this pressure) will be about 240.44 × 2, or 480.88 heat units, which is the thermal cost of 1 cu. ft. of compressed air at the above temperature and pressure. . . . . The difference in the thermal cost of any volume of compressed air thus produced by mechanical compression and the cost of any additional volume that may result from the subsequent reheating of the air, is very striking.

"Weight of 1 cu. ft. of free air at 60° F., 0.076 lb.

"Weight of 1 cu. ft. of compressed air at 75 lbs. gauge and 60° F., 0.456 lb.

"Units of heat required to double the volume of 1 lb. of air at 60° F., 123.84.

"Units of heat required to double the volume of 1 cu. ft. of compressed air at 75 lbs. and 60° F., 123.84 × 0.456 = 56.47.

"Cost of 1 cu. ft. of reheated compressed air at 75 lbs., compared with the cost of 1 cu. ft. as produced by ordinary compression in a steam-driven compressor, 480.88 to 56.4771, or 1 to 0.1174.

"Here we see that the cost in heat units of the volume of air produced by reheating is less than one-eighth of the cost of the same volume produced by compression."

To double the volume by reheating, it will be necessary to double the absolute temperature. In the case just cited, the initial absolute temperature is 60 + 461, or 521° absolute; and doubling this gives 1,042° absolute, or 1,042 − 461 = 581° F. But air loses its heat rapidly, and it is probable that to secure a double volume *in the motor*, the air would have to be heated to at least 800° F. This temperature cannot be practically handled, and probably the best that should be attempted in practice is to increase the volume by reheating 50 per cent. In this latter case, assuming initial temperature of 521° absolute, the final temperature theoretically necessary to add 50 per cent to the volume will be $1\frac{1}{2} \times 521$, or 728° absolute, corresponding to 321° F. A safe allowance for loss by cooling between a reheater and motor, would bring this up to 450° F.—an increase of 450 − 60, or 390° F. It was seen that 56.47 heat units were used in raising one cubic foot of compressed air at 75 lbs. from 60° to 581° F., the temperature increase being 521°. The heat units needed to reach 390 F. are found by the proportion 521 : 390 = 56.4 : 42.27.

In other words, the first cubic foot of compressed air at 75 lbs. required 480.88 heat units, and the additional half-foot secured by reheating required 42.27 heat units, making the total thermal cost of $1\frac{1}{2}$ cubic feet of compressed air at 75 lbs. gauge 523.15 heat units, or at the rate of 348.78 heat units per cubic foot. The relative cost per cubic foot of compressed air at 75 lbs. gauge by compression alone, and by compression and reheating, is thus 480.88 : 348.78 = 1 : 0.72.

From the above it appears that the gain by reheating compressed air at 75 lbs. gauge to increase its volume one-half, is 28 per cent. Stating the relation otherwise, 0.72 : 1 = 1 : 1.38. In other words, the total fuel used in a system using compression and heating will give 38 per cent more work than can be had from the same amount of fuel by compression alone, without reheating.

There are reheaters of many different types on the market, and some which are not generally available. The more common ones may be divided into two classes: (1) Internal-combustion reheaters, in which the compressed air passes directly over the flame from burning oil or coke, the products of combustion going through into the motor; (2) reheaters in which the air is heated by passing through closed tubes or shells above or surrounding a fire of coke or coal. The former method is the more efficient, but it contaminates the air. The latter delivers the air clean and free from ash, soot, and foreign sub-

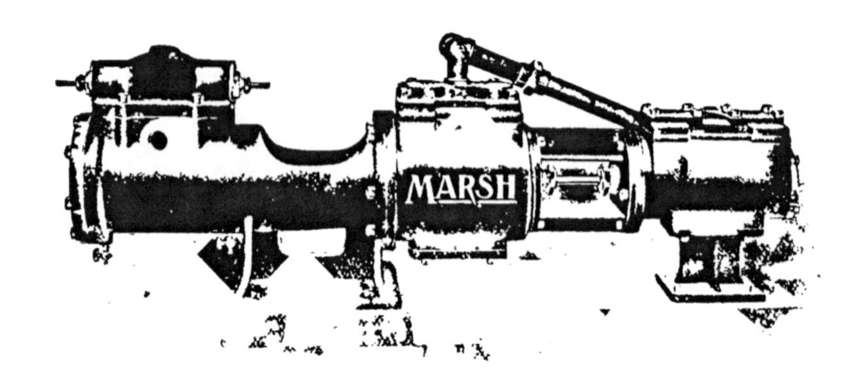

MARSH TWO-STAGE AIR COMPRESSOR
American Steam Pump Company, Battle Creek, Mich.

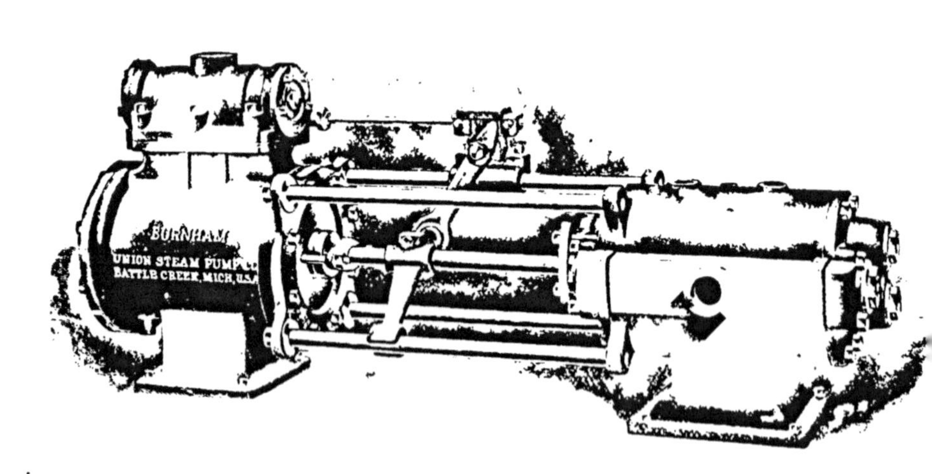

BURNHAM AIR COMPRESSOR
Union Steam Pump Co., Battle Creek, Mich.

stances which may collect in the motor. Other types not in general use employ an electric heating coil in the air passage; or the air is heated by hot water; or steam coils furnish the necessary heat.

As to the construction of reheaters, the first essential is that they shall be made of a material which will not burn out, for it is to

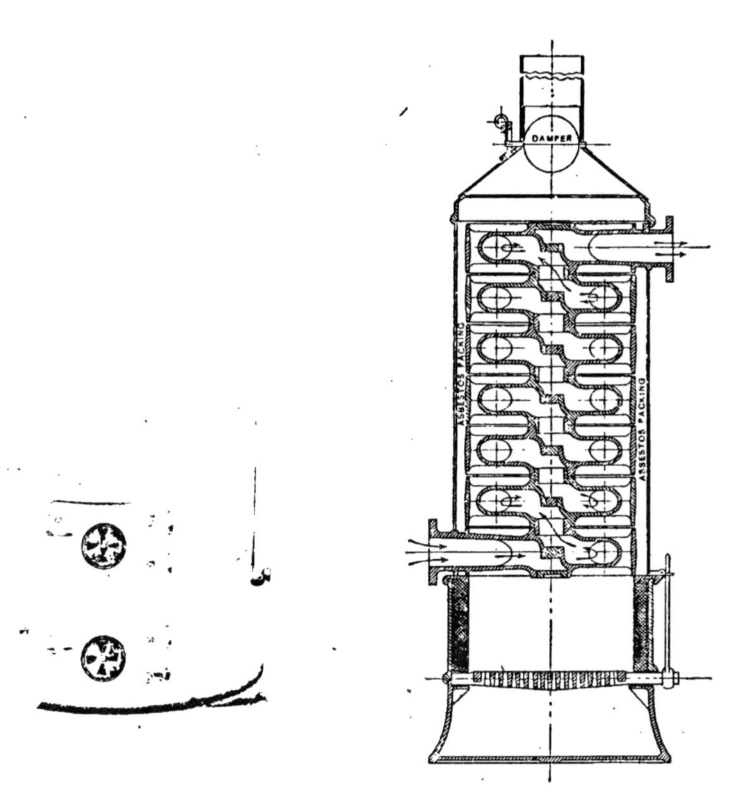

Fig. 82. General and Sectional Views of Stationary Air-Reheater. Sullivan Machinery Co., Chicago, Ill.

be remembered that they are not protected by water as are the tubes and plates of steam boilers. Cast iron seems to give the best results in this respect. Another requirement is that they shall give an ample heating surface without materially interfering with the free flow of air. A third essential is that leakage shall not develop under the expansion and contraction strains.

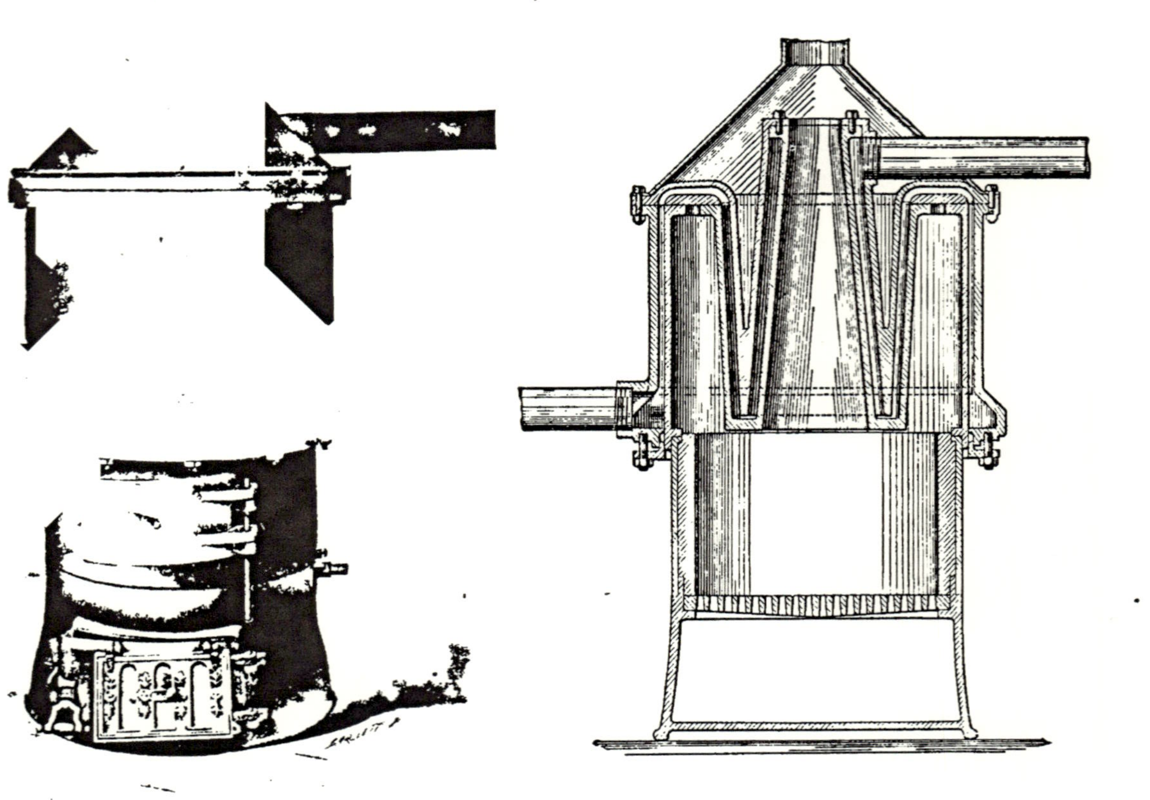

Fig. 83. Rand Air-Reheater. Sectional view shown at right. Ingersoll-Rand Co., New York City.

The reheater should be placed as close as possible to the work it supplies, and the connecting pipes should be well protected against loss of heat by radiation. Where the reheated air is used in a motor without expansion, particular care must be exercised in the matter of lubrication at this high temperature. Where the air is used expansively, the mean cylinder temperature in the motor will be lower and safer.

It is probable that reheating seldom pays, except where the air is used in large volumes and for comparatively steady service, such as in pumps, hoists, stone channelers, and air-motors of large power.

APPLICATION OF COMPRESSED AIR

Elementary Theory. The phenomena of the expansion of compressed air are governed by the same laws as those holding for the compression of air. It was seen in the beginning of this treatise, that the compression of air produces an increase in temperature due to intermolecular action. Conversely, by a reversal of this molecular activity, the expansion of air produces a drop in temperature.

Isothermal expansion is expansion at constant temperature, according to Boyle's law, PV = a constant. It is impossible of attainment, as is isothermal compression; but it is the ideal to be sought in the practical working of expansion air engines or motors.

Adiabatic expansion is expansion with a total proportionate drop in temperature, according to Charles's law governing the relation of pressures, volumes, and temperatures. Practical work approaches this, though it is the process to be avoided as far as possible. It is never fully realized, because there is always some absorption of heat from the cylinder in which expansion occurs.

Air-Motor Indicator Cards. In Fig. 84, the theoretical phenomena of compression and expansion are shown in an ideal indicator card. A volume of air compressed adiabatically will follow the curve AB, and, after cooling to initial temperature, will occupy the volume represented by EC. If it had been possible to compress it isothermally, the line of compression would have been the curve AE, and a considerable saving of power would have resulted, as represented by the area ABE, as already explained. If this volume of air could now be expanded isothermally to initial pressure, the expansion line would retrace the curve AE; and the original volume, at original temperature and pressure, would be recovered, with a restoration of all the work expended in isothermal compression and represented by the area *AECD*. If it should expand adiabatically, however, it would follow the curve *EF* with a loss of work, as compared to isothermal, represented by the area *EAF*, the volume at initial pressure and at the lower temperature represented by the line *AD* showing a decrease in volume represented by *AF*. The total loss, therefore, in adiabatic compression and expansion (assuming a perfect intermediate cooling to initial temperature), is represented by the area *ABEF*.

As a matter of fact, however, compression will more nearly follow the line *AJ*, on account of the cooling devices used on the compressor; and expansion will possibly follow the line *EK*, owing to the absorption of heat from the motor. The loss of work, therefore, between practical compression and practical expansion, is probably more correctly represented by the area *AJEK*, so that the conditions are not quite so bad as they might at first appear.

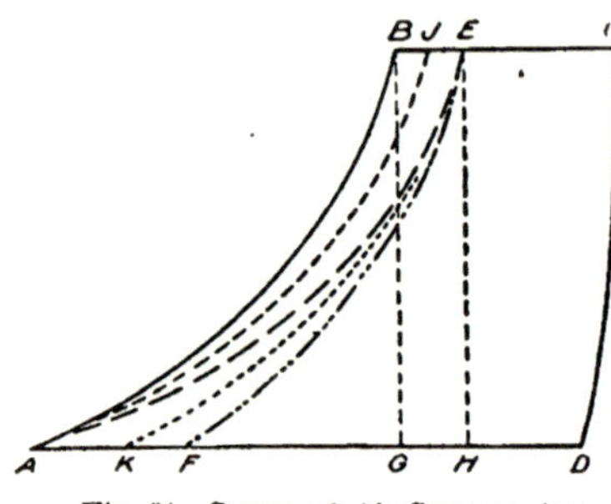

Fig. 84. Curves of Air Compression and Expansion.

The operation thus far discussed has assumed that the air after compression was allowed to cool to its initial temperature before it was expanded. If it were possible, however, to compress adiabatically, and then, without any loss of the heat of compression, to expand adiabatically, it is evident that both compression and expansion would follow the curve *AB*; and while more work was required in compression, still the full equivalent of this would be returned in expansion. This is plainly impossible and impracticable.

But assume that compression follows the practical curve *AJ*; that the air is cooled to initial temperature and transmitted, its volume then being represented by *EC*; that enough heat is applied to this transmitted volume to re-expand it to the volume *JC*; and that at this temperature and volume the air is allowed to expand along the practical curve *JA*. This is a cycle of operations quite possible of attainment, and represents good, everyday compression, cooling, and transmission, reheating to increase the volume, and expansion along practical lines. Reheating of the air to increase its volume from *EC* to *JC* has saved the power (as compared with cool air) shown by the area *AJEK*. In the discussion of reheating in a previous section, it was shown that this gain is secured at a relatively small cost.

It was seen, in discussing air-compression, that the ideal method of cooling would be that in which the heat was removed as fast as produced, compression following the isothermal curve; but since this is impossible in practice, the next best arrangement is the division of compression into stages, with cooling to original temperature between each compression.

Similarly, the ideal expansion air-motor would be one in which the isothermal curve was followed by supplying, during expansion, the heat necessary to maintain a constant temperature. This being impossible in practice, the nearest approach to the process would be one of stage expansion and interheating, which has been proposed, and which, indeed, has been used to a limited extent. But the problem of supplying heat is more difficult than that of withdrawing it. In this suggested process, the air would be heated before each stage of expansion; and actual expansion in each cylinder would follow the intermediate practical curve between isothermal and adiabatic.

It is to be regretted that there are no indicator cards from actual expansion air-motors at hand for reproduction here, showing what results are really obtained; but these cards are singularly hard to secure. It is quite probable, however, that they would show a striking resemblance to the ordinary steam indicator card, except that their M.E.P. would be lower than that of the steam cards with the same initial back-pressure and with the same cut-off. This is explained as follows:

Mean Effective Pressures for Air-Expansion. In calculations for steam engines, it is generally assumed that steam expands in a cylinder in accordance with Boyle's law, or isothermally. In calculations for air-expansion, on the contrary, it is safest to assume that the expansion follows the adiabatic curve.

Thus, in Fig. 85, used by permission from Richards' "Compressed Air," a diagram is shown representing a volume of steam and the same volume of air, both at 100 lbs. gauge, and both expanded to below atmosphere. It will be noted that the steam curve is everywhere above the air curve, so that the M.E.P. of the steam during this expansion will be higher than that of the air. It is also to be observed that the air curve reaches the atmospheric line after expansion to about 4¼ times its original volume, while the steam curve reaches atmosphere after expanding 6¾ times. It thus appears that, volume for volume, at equal pressures, steam has more intrinsic power than air.

Fig. 86 shows the relations in another form. Here are a steam and an air curve, both expanding to atmosphere at the same point. They are really portions of the curves from Fig. 85, placed so that their points of intersection with the atmospheric line coincide. The air curve is now seen to be everywhere above the steam curve, showing a higher M.E.P. for the air than for the steam. But it is to be observed that while this means more power from the air, it calls for a greater initial volume of air than of steam, in the proportion of the distance *CB* to *CE*.

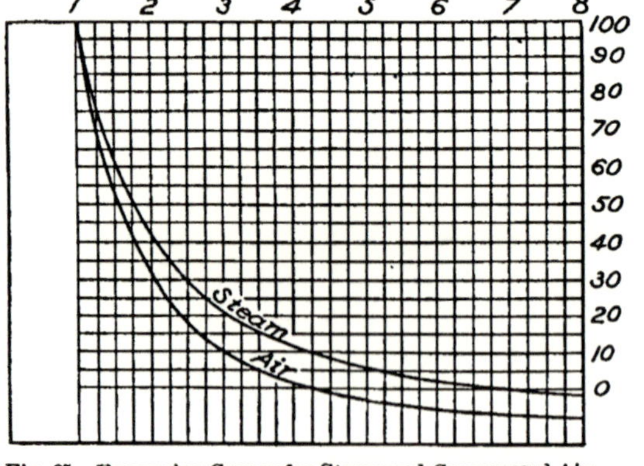

Fig. 85. Expansion Curves for Steam and Compressed Air.

From the foregoing, it is evident that the great distinction between an engine or motor driven by air and the same motor driven by steam, is that, the initial pressure and the required power being the same in both cases, the air-motor will require a later cut-off, and therefore a greater weight of working fluid per I.H.P., than the steam motor.

This fact will effectively condemn the use of air as against steam in all cases except where other conditions—particularly long transmissions where steam would condense with a great loss and an excessive steam consumption—justify the application of compressed air, even in the face of this disadvantage. Reheating may largely reduce the difference in favor of steam; but it can never fully wipe it out, because of the limits imposed by practical temperature conditions.

Temperatures of Expansion. The adiabatic compression of air from 60° F. at atmospheric pressure to 100 lbs. gauge, produces a terminal temperature of 484° F. If this air could be immediately expanded adiabatically to atmosphere, its temperature would fall to 60° F. This difference in temperature of 484 - 60, or 424 degrees corresponds to any expansion or compression with this ratio of pressures. If the air after compression is allowed to cool to initial temperature of 60° F., and then expanded adiabatically, the temperature will fall in practically the same degree—namely, 424 degrees, giving a terminal temperature at atmospheric pressure of - 364° F. If there is moisture in the air, the result in this case is very evident, even remembering that these temperatures are theoretical only and would never be actually realized. Difficulties in lubrication are also encountered in dealing with these low temperatures; and in consequence of these facts, it is very seldom attempted to use compressed air expansively in an engine or motor without reheating. The use of a reheater not only gives the added volume and improved economy already referred to, but it also gives an initial temperature before expansion begins, sufficiently high to avoid impracticably low temperatures at the motor exhaust.

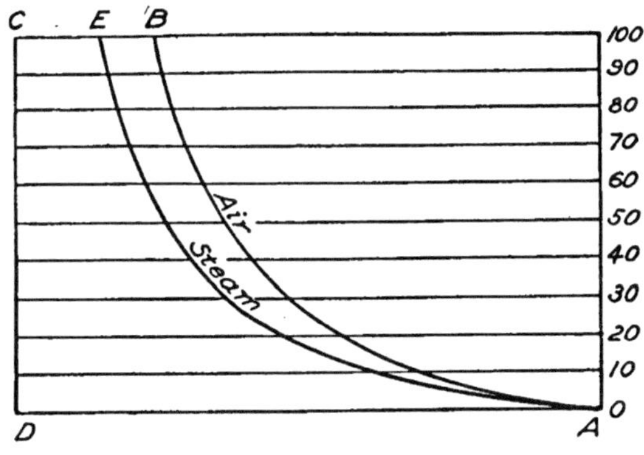

Fig. 86. Volumes of Steam and Air for Equivalent Expansions.

Use of Compressed Air without Expansion. While there have been successful air-power installations where the full advantages of expansion have been realized by means of reheating and careful extraction of the moisture in the air, it nevertheless remains a fact that probably 95 per cent of the applications of compressed air use the air without expansion. The greatest field of compressed-air application is that where reciprocating mechanisms are to be used, such as rock-drills, stone channelers, coal-cutting machines, pneumatic chipping and riveting hammers, etc. In all of these machines the operation is simply that of driving a piston forward and back, there being no attempt to convert this reciprocating motion into rotary motion as in the air-motor. It is practically impossible to use compressed air expansively in such machines, because there is no fly-wheel effect to come into play after cut-off. In all devices of this kind, the thing sought above all else is a heavy, powerful blow at the end of the piston stroke; and expansion would not give this desired effect. The indicator card from such machines, therefore, would be practically a rectangle with the corners rounded where the movement of the valves began a little before the end of the stroke, probably throttling slightly the intake and exhaust passages at these points.

In the small air-motors used on pneumatic drilling machines for wood and metal, there are usually several sets of cylinders arranged to give effects on the shaft equivalent to that of several cranks placed at quarters or thirds of a circle. This is not, as it might at first appear, to permit the expansive use of the air, but to give the maximum turning moment on the shaft throughout each revolution —a result which could not be secured from a single cylinder equivalent in power to all the smaller cylinders combined.

It is to be remembered that in all such machines as are here under discussion, where the air is not used expansively, the standard of economy is not in any way similar to that used for measuring the efficiency of a steam engine or of an air-driven rotary motor. The latter machines are rated in horse-powers, and their economy depends upon the steam or air consumption per horse-power-hour. But in the former machines of straight reciprocating type, the object sought is not so much a great economy of air as a great convenience and a great capacity for doing the class of work called for.

Such machines must be as light as possible, and must be capable of standing a large amount of abuse or neglect, while still doing good work. This is only one of many considerations which, in the aggregate, overbalance the importance of a high economy of air to be secured only by a greater refinement of design. With a rock-drill, for instance, the proper standard is the air consumption per foot of hole drilled; and this evidently includes many other factors than the amount of air used per stroke. With a pneumatic metal-boring drill, a proper standard should be the air consumption per cubic inch of metal drilled. Similar practical standards might be enumerated for all the special applications of compressed air.

It may be well to emphasize here the fact that the air consumption of the average compressed-air appliance, though not a thing to be ignored, can yet be much exaggerated. In the operation of a large air engine or air-driven pump working under more or less steady conditions and usually in a place where it will be looked after, carefully lubricated, packings adjusted, etc., it is well worth while to adopt all reasonable means for getting maximum power with minimum consumption of air. But with drills, coal-cutters, pneumatic hammers, and other small air-devices, which are seldom given the care and attention they deserve, the ability to "keep going" and to do good work without frequent breakdown and repairs, becomes of more importance than the air consumption. With machines of the former class, the cost of power far outweighs the cost of up-keep or maintenance; with machines of the latter class, it is not at all unusual for the repair or up-keep charges to far exceed the power charge. Questions of sturdy design and wearing power, therefore, become more important than the question of air economy.

APPLICATIONS OF COMPRESSED AIR

It is entirely beyond the scope of this paper to deal with the details of construction of the many mechanical devices for the application of compressed air to various specific duties. Every builder of such devices has his own specialized designs, each of which is usually found to have some point of merit over others. For details of these machines, the student should refer to the catalogues of various manufacturers, the majority of which are clear and explicit, and which indeed constitute an invaluable addition to the library of the practical engineer. This paper, dealing as it does only with compressed air as a power-transmitting medium, is not concerned with the mechanical details of any air-power devices, except as they bear upon the method of control of the power which actuates them. The remainder of this paper, therefore, will contain only brief discussions of the fundamental principles of some of the more common applications of com-

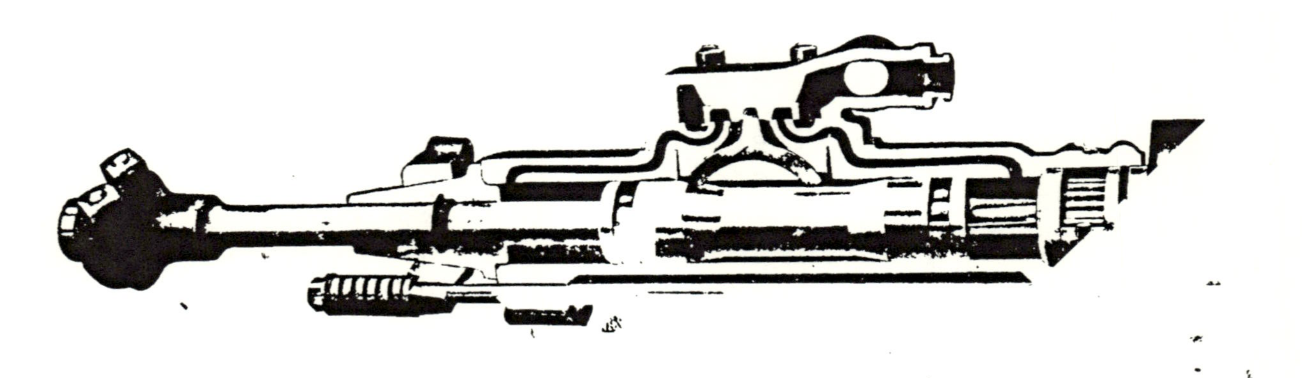

Fig. 87. Section of Tappet Valve Rock-Drill, Showing Relative Position of Working Parts. Sullivan Machinery Co., Chicago, Ill.

pressed air. No attempt will be made to cover the entire field of compressed-air application.

The Rock-Drill. The rock-drill was the precursor and the cause of the air-compressor. The rock-drill was built first, and then air-compressors were designed to supply it with air. In this respect the rock-drill is the origin of all the pneumatic development of to-day.

The rock-drill is in essentials a reciprocating engine designed to give a heavy, powerful, forward stroke (the *blow*) and a very quick back stroke (the *return*). On the forward stroke, all the energy ordinarily should be applied to penetration at the drill bit, though there are certain qualities of rock which need a cushioned blow rather than a dead, stunning blow. On the return stroke, the function of which is simply to draw the steel back ready for another blow, and to *mud* the hole, a cushion must be provided to prevent the piston striking the back head of the drill. This cushion is provided by throttling the exhaust and causing compression between piston and back head, this compression helping to start the piston on its next forward stroke. But as a safeguard, in case leakage destroys this air-cushion, springs or buffers are also usually employed to absorb, by their elasticity, an accidental blow of the piston on the head. As the steel in feeding forward may sometimes encounter a hole or *pocket*, or the drill may not be fed forward rapidly enough, the front head must also be protected by some means. Frequently there is a partial air-cushion on the forward end also; but the principal reliance in this case is to be placed upon buffers or springs.

Rocks of various characters require blows of different qualities. Some are drilled most rapidly by long, swinging, powerful blows with the full stroke of the drill. Others require many short, rapid blows with only part of a full stroke. This variation in stroke is secured by manipulating the feed-screw, crowding the drill forward for short strokes, and keeping it well back for longer strokes.

Evidently this variability of stroke demands a method of valve movement equally effective on long or short piston strokes; and it entirely prevents any expansive use of the air. The rock-drill, therefore, receives and discharges practically a cylinder full of air at working pressure each stroke.

Rock-drill valve movements are of two general classes. The

first form is the *tappet valve,* in which a slide-valve, balanced usually for higher pressures, is thrown by means of a tappet or rocker, which projects into the cylinder, and which is struck or moved by shoulders on the piston in its travel. This is a positive valve movement especially useful where a wide stroke variation is not necessary.

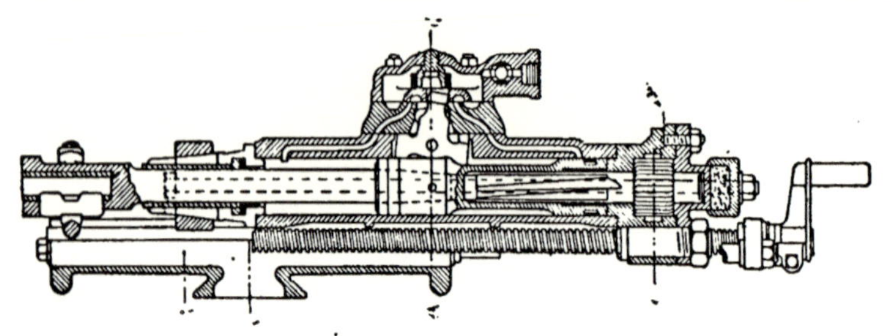

Fig. 88. Section of "Little Giant" Tappet Valve Rock-Drill, with Balanced Valve. Ingersoll Rand Co., New York City.

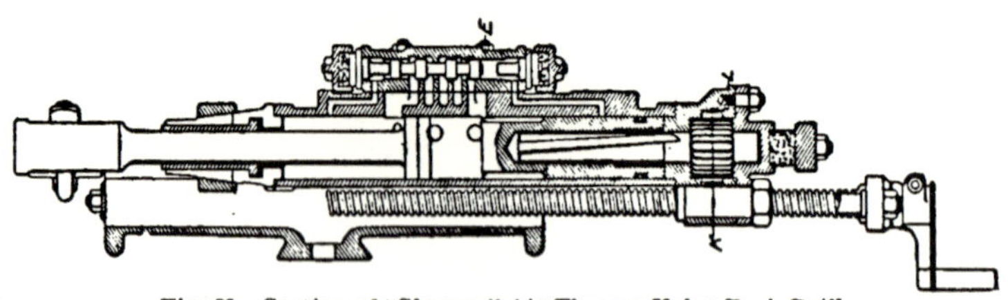

Fig. 89. Section of "Slugger" Air-Thrown Valve Rock-Drill. Ingersoll Rand Co., New York City.

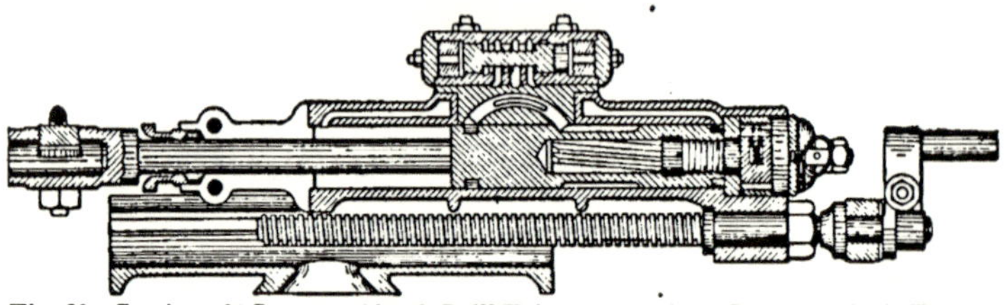

Fig. 90. Section of "Sergeant" Rock-Drill Using a Combined Tappet and Air-Thrown Valve Movement. Ingersoll Rand Co., New York City.

Being in no way dependent for its action upon the fit of the piston in the cylinder, it will continue to operate even after wear has passed the point of good economy. The tappet action has been generally employed on drills designed to be run by steam where condensation may result in low pressures and much water in the line.

The second type of valve movement is the *air-thrown valve,* which is a spool or piston valve moved by unbalanced pressures on its two ends. A lower pressure on one end is produced by the piston uncovering in its travel a small port which exhausts part of the air in the valve-chest; full pressure on the other end of the valve then throws it, and reverses the piston movement. This valve action usually gives a more *lively* drill than the former movement, and it permits a wider variation of stroke than the tappet movement; but it is more or less dependent for its effectiveness upon the air-tight fit of all parts, and therefore suffers more under wear. This, however, is in a sense an advantage, for the action of the drill thus becomes somewhat of an index of the economy with which it is operating.

In one well-known drill on the market, these two types of valve action are combined. A tappet movement controls the admission and exhaust of air to the main valve, which in turn governs the movement of the piston. These drills combine some of the best features of both original types.

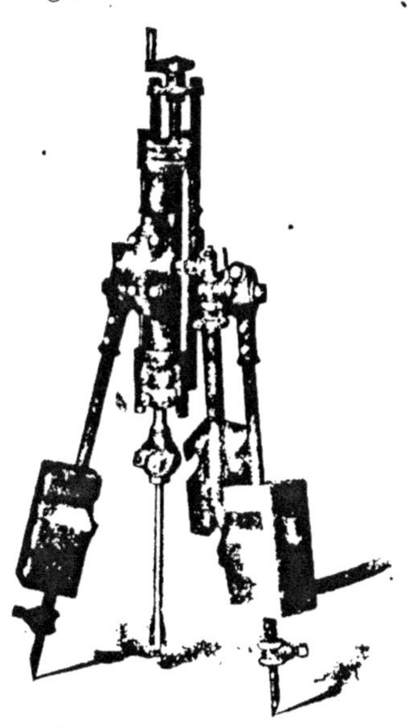

Fig. 91. Rock-Drill on Tripod. Sullivan Machinery Co., Chicago, Ill.

Hammer Drills. In the ordinary rock-drill or piston drill, the cutting steel is attached to, and reciprocates with, the piston. A modern development from this scheme is the *hammer drill*, in which the steel remains stationary (except for its rotation) and is struck by the rapidly moving piston or hammer. The piston, being freed from the weight of the steel and the friction of the steel in the hole, moves much more rapidly than in the ordinary rock-drill, and the action of the hammer drill more nearly approaches that of hand-drilling. While some large hammer drills are on the market designed for the ordinary work of the piston drill, this type has found its greatest application for drilling small and comparatively shallow holes.

Hammer drills may be divided into two classes. The first has a separate valve controlling the piston, this valve usually being of the air-thrown type, but with many mechanical variations. These valve hammers have practically no cushioning, strike a very powerful blow, and make a very large number of blows per minute. The second class includes valveless hammers, in which the moving piston or hammer itself controls the admission and exhaust of air by cover-

Fig. 92. Class D-21 Hammer Drill with Air Feed.
Sullivan Machinery Co., Chicago, Ill.

ing and uncovering in its travel certain air-ports. The valveless hammer will probably not strike so hard a blow, nor so many of them, as a valve hammer of the same diameter; but it is often preferred because of its greater simplicity, there being but one moving part—the piston or hammer—in the machine. Choice between the

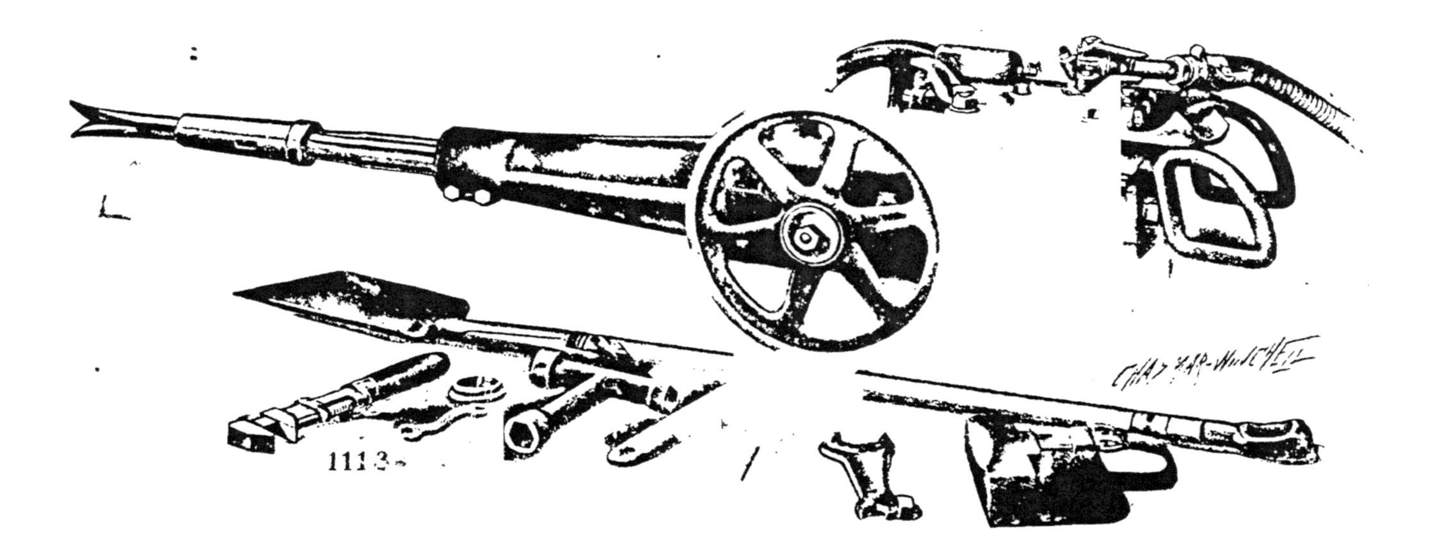

Fig. 93. "New Ingersoll" Coal Undercutter, with Complete Equipment. Ingersoll-Rand Co., New York City.

valve and valveless hammer drills seems to be largely a matter of personal preference and of local precedence in different sections of the country.

Except for the large hammer drills already mentioned as competing with the piston drills, the hammer drill is generally used without any fixed mechanical mounting. For certain classes of work, where downward holes only are required, it is fitted with a hand grip or handle, the operator pressing the tool downward against the rock; but for upward pointing holes, an air feed is used, consisting of a long piston and cylinder in which air-pressure forces the drill up against the rock. The back end of the air feed has a pointer or spud which is simply butted against the floor or wall. In certain rare classes of work, the air-feed drill is held in a mounting similar to those used with the heavier piston drills. It is seldom attempted, except in the large sizes, to provide mechanical rotation for the hammer-drill steel. Usually the steel is rotated by means of a lever or handle swung by the operator.

Coal-Mining Machines. The *coal-undercutting machine* or *puncher* is another reciprocating machine carrying a coal pick on its extended piston-rod. It is mounted on wheels, and is fed downward as the cut deepens, along an inclined board upon which it rolls. In this machine, cushioning is important in taking up the jar and recoil which would otherwise come upon the operator, as well as in protecting the machine itself. The valve movement may be independent, in which case the valves will continue to keep on operating, even when the pick sticks in the coal and the piston stops; or it may be mechanically dependent upon the movement of the piston, so that when the piston stops the valve stops. The successful puncher must have a ready means of regulating the force, number, and quality of the blows it strikes, for coals of varying hardness and character. A governing device is also essential to prevent the "racing" of the machine when drawn away from the coal, to shift its position. For ordinary undercutting, the puncher is mounted on wheels of comparatively small diameter. It is sometimes used for shearing, however, or making a vertical cut in the coal face, in which case the puncher is mounted on the trunnions of wheels of large diameter, the cutting engine swinging in a vertical plane around this axis.

SULLIVAN PNEUMATIC PICK MACHINE AT WORK IN OPEN COAL PIT

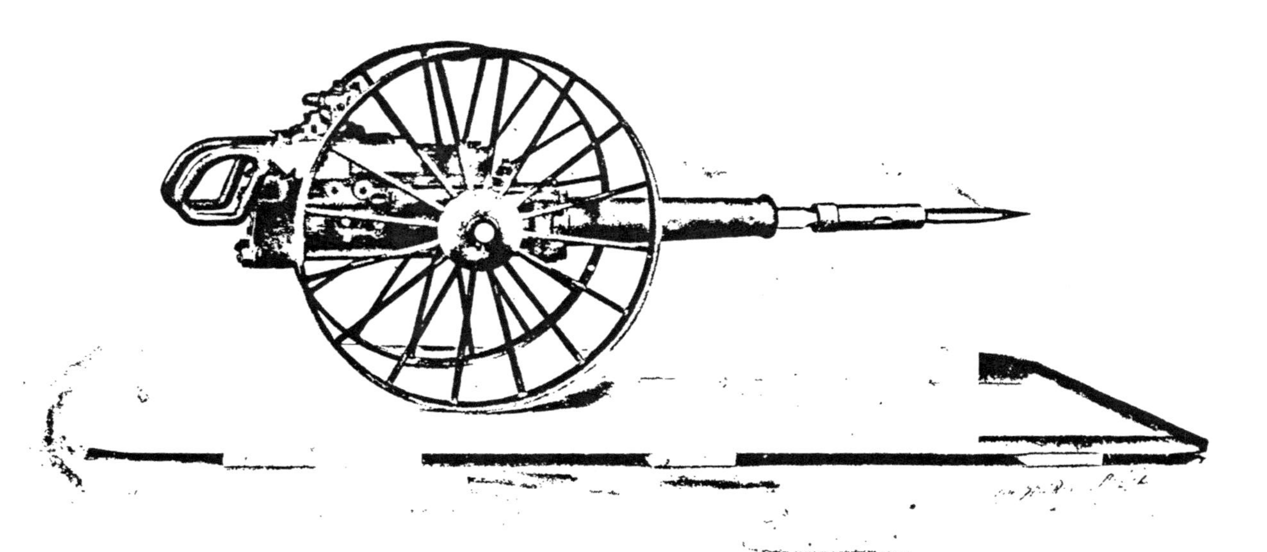

Fig. 94. Pick Machine or Puncher Mounted for Shearing. Sullivan Machinery Co., Chicago, Ill.

Fig. 95. Pick Machine at Ordinary Work of Undercutting in a Pennsylvania Coal Mine. Sullivan Machinery Co., Chicago, Ill.

Fig. 96. Compressed-Air Chain-Type Coal-Mining Machine. Sullivan Machinery Co., Chicago, Ill.

Coal-Shearing Machines. A comparatively recent development in compressed-air coal-mining machines, is one for *shearing*, or making a vertical cut, in the coal vein. This device is in essentials a rock-drill with a long feed arranged on a column mounting, with a gear for swinging it in a vertical plane. Thus, as the machine is fed forward and swung around its axis, a deep, narrow, arc-shaped cut is made in the coal. This shearing cut, in combination with the undercut of the puncher, makes the shooting down of the coal easier and safer, produces a larger percentage of lump coal, and reduces the amount of powder required to bring down the face. This type of shearing machine also has a limited application as an under-cutter for use in coal veins having so steep a pitch that the ordinary puncher cannot be used. In this case the machine is arranged to swing horizontally instead of vertically.

Track Channelers. The track channeler is an essential in the quarrying of stone cut to exact dimensions. It consists of a heavy, reciprocating, cutting engine mounted on a truck which moves along a track, cutting a channel in the stone parallel to the track, to any depth or at any angle required. Frequently track channelers are operated by steam, each machine carrying its own boiler. But the tendency in modern quarry practice is toward the use of air-driven channelers supplied with power from a single compressing plant, this production of power in a large centralized station resulting in a greater economy of fuel and power. Not only is the cutting engine operated by air; but the engine providing the movement along the track, and frequently the motor providing for the vertical feed of the cutting steels, are also air-driven.

Channeler valve movements are always positive in character, governed by some mechanical connection with the main piston of the cutting engine. A wide variation of stroke must be provided for; and the cushioning is important because of the large diameter and heavy weight of the reciprocating parts involved. In certain qualities of rock—notably in different grades of marble—a *dead* blow shatters the material and injures it. The air-cushion in the channeler cylinder provides for the necessary variation in the character of the blow which this condition demands; and usually adjustable cushioning is provided by means of valves at both ends of the cylinder of the cutting engine.

Some modern track channelers have a *swivel-head* which permits swinging the cutting engine in the plan of the cutting steels, the maximum swing permitted being about 45 degrees either side of vertical. In addition, a *swing-back* is usually furnished, by which the cutting engine can be swung at any angle from vertical to horizontal in a plane at right angles to that permitted by the swivel-head. Such machines are useful in cutting in materials lying in diagonal strata or layers, in transferring cuts at corners, and in tunneling.

Fig. 97. H-8 "Monitor" Track Channeler, with Air-Reheater. Ingersoll-Rand Co., New York City.

Undercutting Track Channelers. Where the rock which is to be quarried has no well-defined horizontal cleavage lines so that the blocks cut by the ordinary track channeler cannot be split from their bed by wedges, it is necessary to *undercut* the block. For this purpose, the *undercutting track channeler* has been designed, which is a special modification of the regular track channeler. Its cutting engine works in a horizontal plane, with an angular adjustment for corner cuts in this plane up to about 45 degrees on either side. The undercutting channeler is not so heavy a machine as the vertical track channeler, since it is never called upon for such heavy work or such deep cuts as the latter machine.

Bar Channelers. In general construction the cutting engine of the bar channeler is similar to that of the track channeler, but much smaller on account of its lighter duty. Instead of the whole machine moving on a track, the cutting engine of the bar channeler traverses a frame mounted on adjustable legs, movement being provided by an air-motor operating a pinion engaging a rack on the frame or bar.

Fig. 98. Class Z Swivel-Head Swing-Back Track Cnanneler. Sullivan Machinery Co., Chicago, Ill.

This type of machine is more limited in its scope than the track channeler, but has a wide range of usefulness as an accessory to the larger machine in quarries and contract work.

Pneumatic Hammers. Pneumatic hammers for chipping metal or stone and for driving rivets, are simply reciprocating machines in which the piston is the hammer or striking part, which travels perfectly free in the cylinder. Two general classes prevail—the *valve tool,* in which air-thrown valves in almost endless variety control the movement of the piston; and the *valveless tool,* in which the piston itself regulates the admission and exhaust of air. Very light pneumatic hammers are used for dressing and carving stone, these latter usually being known as *stone tools.* Heavier, short-stroke hammers are used for chipping and trimming metal. Long-stroke hammers prevail for riveting work. The piston or hammer of these machines is usually cushioned on air on the back stroke, to relieve the shock on the operator; but on the forward stroke a dead blow is desirable.

Rotary Pneumatic Drills. Rotary pneumatic drills for boring in wood or metal, for grinding and similar light work, are furnished in a great variety of sizes and patterns. But in essentials

Fig. 99. "Crown" Pneumatic Chipping Hammer.
Ingersoll-Rand Co., New York City.

they all consist of a small air-motor—either of reciprocating or of rotary type—geared to a spindle carrying the working tool. No further description can be offered here without going into details of special constructions which would be out of place at this point.

Compression Riveters and Pneumatic Punches. In these machines, air-pressure on a relatively large piston is applied, through toggles or other power-multiplying devices, to the compression of red-hot rivets, or to the punching of holes in plates or other metal forms. These are comparatively heavy machines, and their use is limited to shop and manufacturing purposes where the ready portability of the pneumatic hammer or drill is not essential.

Pneumatic Hoists. Pneumatic hoists are made in two general classes. The first is the *direct-lift* or *plunger hoist,* in which air

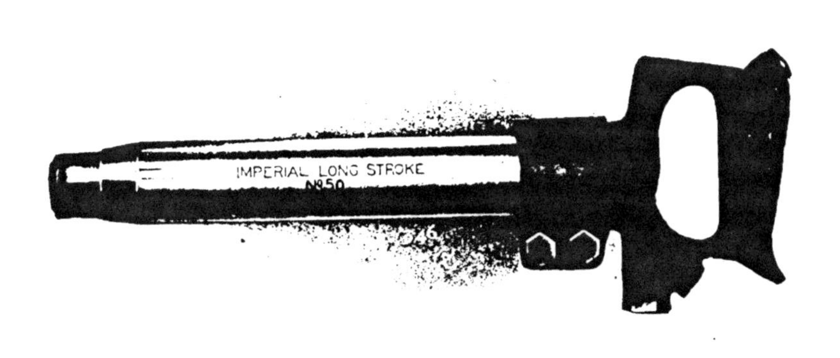

Fig. 100. "Imperial" Pneumatic Riveting Hammer.
Ingersoll-Rand Co., New York City.

Fig. 101. "Imperial" Rotary Pneumatic Drill.
Ingersoll-Rand Co., New York City.

is admitted behind the piston of a very long cylinder, power thus applied directly lifting the weight attached to the piston-rod. These machines are necessarily limited to work where a comparatively short lift only is needed, as the longer lifts necessarily demand an inconveniently long cylinder. However, this arrangement is sometimes used in connection with cables and pulleys for providing long lifts at slow speeds, as in ice elevators and similar work. The second class is the *motor-hoist,* in which an air-motor is geared to a chain or rope drum lifting the weight. The latter machines are fully as economical and reliable as the former, and their lift is limited only by the winding capacity of their drum or sheave.

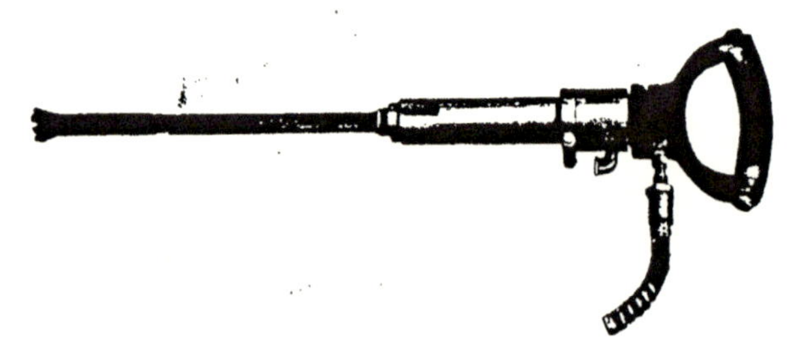

Fig. 102. Class D-19 Hand Hammer Drill.
Sullivan Machinery Co., Chicago, Ill.

The Sand Blast. A pneumatic sand blast for cleaning castings and building faces, for removing paint, for etching glass, etc., consists simply of a jet of air under pressure, carrying with it a stream of sharp sand directed against the surface to be operated upon. The air imparts to the sand the requisite velocity under which it simply cuts away the surface that it encounters. The air-pressure used varies among various builders of the apparatus.

Pumping by Compressed Air. There are three general methods of applying compressed air to the pumping of water or other fluids. In this class of work, compressed air has a distinct advantage over steam, in that there is no loss by condensation in long pipe-lines, and, though pressure may fall in transmission, there is a corresponding increase in volume, so that the net amount of energy delivered at the pump is practically the same as that put into the transmission line. The three methods of pumping by compressed air are: (1) The *direct-acting plunger pump;* (2) the *air-lift system;* and (3) the *direct-displacement pump.*

Direct-Acting Pumps. The ordinary direct-acting steam pump is probably the most wasteful machine in practical use to-day. It has been said of it that "it is the only method of pumping which requires as much steam (or air) to pump water down hill as to pump it up hill." The extravagance of this device is due primarily to the fact that it uses a full cylinder of steam at full pressure each stroke, with no expansion whatever. Duplex non-compounded pumps with mechanical valve movements, may use actually more than a full cylinder of steam per stroke; for the piston on one side may throw the steam valve on the other side before the piston on the latter side has finished its stroke. In such a case the live steam is discharged directly into the exhaust. All of these defects of the steam pump are found to hold when this machine is run by compressed air. It is to be noted here, however, that if an air-driven direct-acting pump is excessively extravagant of power, the fault lies in the pump itself, and not in compressed air, which stands ready to do its work economically if given a proper machine to work in.

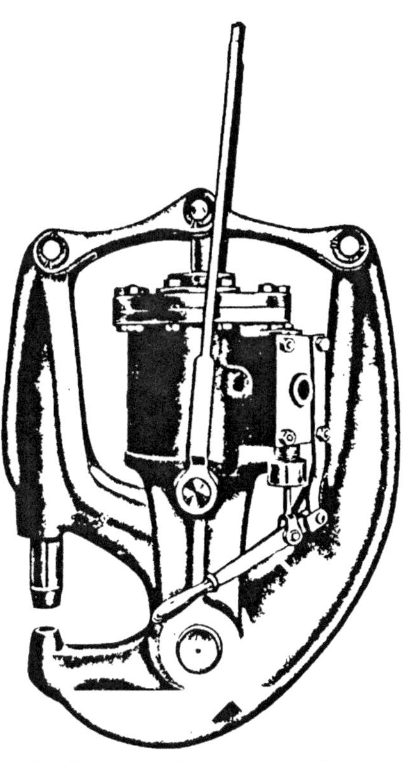

Fig. 103. Pneumatic Compression Riveter. Chicago Pneumatic Tool Company, Chicago, Ill.

Another cause of extravagant air-consumption in the direct-acting air-driven pump, is the fact that such pumps are usually stock machines, bought without due consideration of the work they are to be used for. For instance, a pump with air and water cylinders so proportioned that air at 100 lbs. pressure will just handle a cylinder-

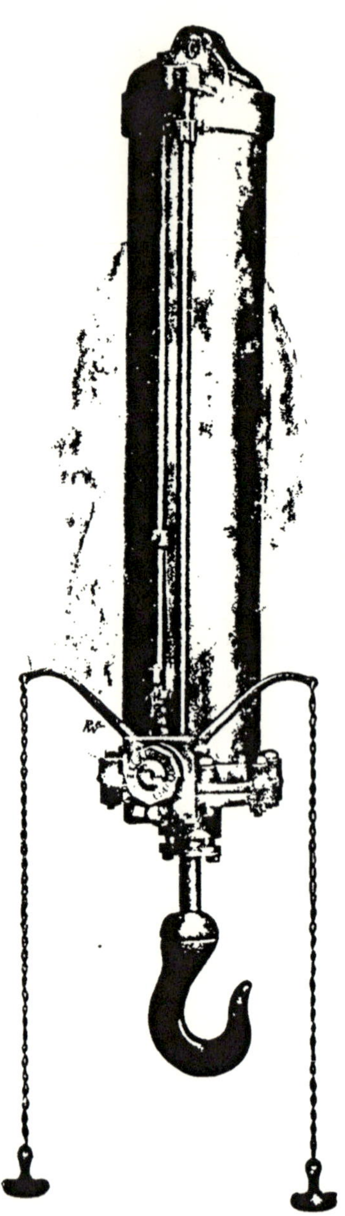

Fig. 104. Direct-Lift Pneumatic Hoist.
Curtis & Co. Mfg. Co., St. Louis, Mo.

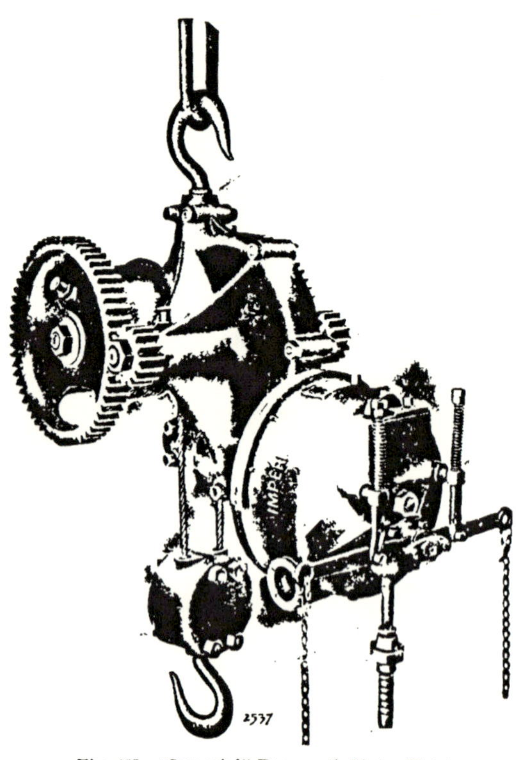

Fig. 105. "Imperial" Pneumatic Motor-Hoist,
Ingersoll-Rand Co., New York City.

ful of water under a head of 200 feet, may be put at work where it pumps against a head of only 50 or 100 feet. Evidently much smaller air-cylinders could be used in the latter case than in the former; but usually the old air-cylinder is retained, using at each stroke two to four times the amount of air actually needed, simply because there is no cut-off and no means of adjusting air-consumption to the load.

Table XVI, used here by permission of the Ingersoll-Rand Company, affords the means of figuring the air-pressures and air-volumes required for a pump with the given ratios of air and water cylinders and for various lifts up to 500 feet. The method of using this table is explained by the example in the note below it.

In compound direct-acting pumps driven by air, there are four methods of utilizing the air with more or less expansion. In this connection it is to be remembered that any attempt to use compressed air expansively without reheating will result in terminal temperatures so low as to make this scheme difficult or even impossible. Some method of heating, therefore, is essential to the compounding of pump cylinders.

The first plan contemplates the heating of the air before admission to the high-pressure cylinder, to such a degree that the terminal temperature of discharge from the low-pressure cylinder is within a practical limit. This is open to the objection that so high an initial temperature would be required that proper lubrication of the high-pressure cylinder would be almost impossible. Moreover, since there is no fly-wheel on the pump, cut-off in either cylinder, with the subsequent expansion, would mean a reduction in power at the end of the stroke, and a loss of speed, so that the movement would be irregular.

The second plan utilizes both preheating before high-pressure admission and interheating between the cylinders. While the heating between cylinders permits a reduction of the initial temperature over that required by the first method, the second objection cited in the first method would still hold.

The third method uses air without heating and without cut-off or expansion in the high-pressure cylinder; but this air is expanded after high-pressure exhaust into the low-pressure cylinder, where it is used either expansively or with no cut-off. Unless interheating is used, a troublesome drop in temperature will probably result from

TABLE XVI

Compressed-Air Table for Pumping Plants

Reasonable allowances have been made for loss due to clearances in pump and to friction in pipe.

Ratio of Diameters	Perpendicular Height, in Feet, to which the Water is to be Pumped															
	25	50	75	100	125	150	175	200	225	250	300	350	400	450	500	
1 to 1	13.75	27.5	41.25	55.0	68.25	82.5	96.25	110.0	.....	.....	.....	.....	.....	.....	.....	Air-pressure at pump
	0.21	0.45	0.60	0.75	0.89	1.04	1.20	1.34	.....	.....	.....	.....	.....	.....	.....	Cubic feet of free air per gallon of water
1½ to 1	.....	12.22	18.33	24.44	30.33	36.66	42.76	48.88	55.0	61.11	73.32	85.4	97.66	.....	.....	Air-pressure at pump
	.....	0.65	0.80	0.95	1.09	1.24	1.39	1.53	1.68	1.83	2.12	2.41	2.70	.....	.....	Cubic feet of free air per gallon of water
1¾ to 1	.....	.....	13.75	19.8	22.8	27.5	32.1	36.66	41.25	45.83	55.0	64.16	73.33	82.5	.....	Air-pressure at pump
	.....	.....	0.94	1.14	1.24	1.30	1.54	1.69	1.84	1.99	2.39	2.59	2.88	3.19	.....	Cubic feet of free air per gallon of water
2 to 1	.....	.....	.....	13.75	17.19	20.63	24.06	27.5	30.94	34.38	41.25	48.13	55.0	61.88	68.75	Air-pressure at pump
	.....	.....	.....	1.23	1.37	1.52	1.66	1.81	1.96	2.11	2.40	2.69	2.98	3.28	3.57	Cubic feet of free air per gallon of water
2¼ to 1	.....	.....	.....	.....	13.75	16.5	19.25	22.0	24.75	27.5	33.0	38.5	44.0	49.5	55.0	Air-pressure at pump
	.....	.....	.....	.....	1.533	1.68	1.83	1.97	2.12	2.26	2.56	2.85	3.15	3.44	3.73	Cubic feet of free air per gallon of water
2½ to 1	.....	.....	.....	.....	.....	13.2	15.4	17.6	19.8	22.0	26.4	30.8	35.2	39.6	44.0	Air-pressure at pump
	....	.....	.....	.....	.....	1.79	1.98	2.06	2.104	2.34	2.62	2.88	3.18	3.36	3.23	Cubic feet of free air per gallon of water

To find the amount of air and pressure required to pump a given quantity of water a given height, find the ratio of diameters between water and air cylinders, and multiply the number of gallons of water by the figure found in the column for the required lift. The result is the number of cubic feet of free air. The pressure required on the pump will be found directly above, in the same column. For example: The ratio between cylinders being 2 to 1, required to pump 100 gallons, height of lift 250 feet. We find under 250 at ratio 2 to 1, the figures 2.11: $2.11 \times 100 = 211$ cubic feet of free air. The pressure required is 34.38 pounds delivered at the pump piston.

the expansion between cylinders — a difficulty further increased where expansion in the low-pressure cylinder is attempted.

The fourth method uses the air at full stroke in both cylinders; but between the cylinders the exhaust from the high-pressure cylinder is reheated to a point where its original pressure is restored. By this method there is no expansion or drop in temperature in the cylinders, and speed is uniform. This method has been successfully applied in triple-expansion pumps using three air cylinders with two interheaters, giving a very high economy.

To get the best economy (never very good, however) from the simple direct-acting air-driven pump, the clearances should be made as small as possible and the cylinder ratio made right for the work in hand. From air-driven compound direct-acting pumps, no great degree of success can be expected unless one of the four methods above described is used; and the fourth method gives the best results.

The mechanical features of direct-acting air-driven pumps are similar to those of direct-acting steam pumps.

Air-Lift Pumping System. The air-lift was originally designed for deep-well pumping, and still finds its greatest application in that field; but it has also been successfully applied for mine pumping and for industrial work, particularly in handling solutions of varying chemical possibilities. A small air-pipe is carried down the well, and compressed air is discharged in the water discharge or *eduction* pipe at the bottom. This air, mingling with the water in the eduction pipe, gives this column of mixed air and water a lower specific gravity than that of the solid water column outside the pipe. There is, therefore, an unbalanced condition of affairs; and the greater weight of the outer mass of water forces the column of mingled air and water in the pipe upward and out. This is the principle of the air-lift in its simplest terms.

There are no generally accepted rules for designing air-lift systems. The question is one involving the depth of water below the surface, the *submergence* or length of eduction pipe below the water level during pumping, the total lift or distance water must be elevated, the diameter of the eduction pipe, the volume and pressure of the air-supply, etc. Different companies furnishing this system appear to have different rules for manipulating these several factors. Various *foot-pieces* are also supplied by different builders, for securing the proper mingling of the air and water at the bottom of the eduction pipe.

A properly designed air-lift system will probably pump from a driven well at higher efficiency than any other method of pumping, particularly if the cost of up-keep, as well as of power, is taken into account. The ordinary deep-well pump has most of its working mechanism at the bottom of the well, submerged, inaccessible, and

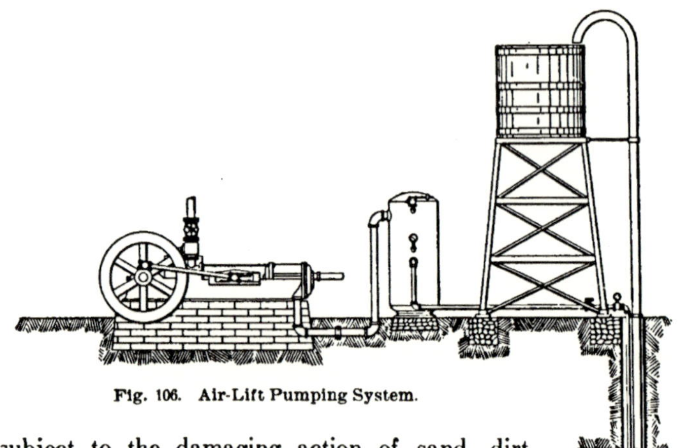

Fig. 106. Air-Lift Pumping System.

subject to the damaging action of sand, dirt, and grit. The air-lift, on the contrary, has nothing in the well to wear out, nothing but some pipes exposed to the action of the fluid; and all its working mechanism is the compressor, always under the eye of the operating engineer and presumably kept at highest efficiency. One compressor plant, moreover, can pump any number of wells scattered over the water-bearing area; and any or all of these wells can be instantly controlled from the compressor plant. The air-lift system is distinguished by a very low cost for wells and equipments. All of these advantages are equally applicable where the system is used for other purposes than deep-well pumping. The fact that this device has no working parts exposed to the fluid pumped, is of particular value where acid or corrosive solutions are to be handled.

Direct-Displacement Pump. The pneumatic direct-displacement pump consists of one or more tanks submerged in the fluid to be pumped, so that they are filled by gravity. When a tank is filled with the fluid, air under suitable pressure is admitted at the top of the tank, forcing out or displacing the volume of liquid. As soon as the liquid has been displaced, the air-volume is discharged or exhausted to atmosphere, and the tank again fills by gravity and is again discharged under air-pressure. This process continues automatically as long as the pressure is supplied. Usually the tanks are arranged in pairs, so that while one is being discharged the other is being filled. Alternate admission and discharge of compressed air are controlled by valves operated by floats or buckets in the tanks, so disposed that as soon as a tank is completely filled air is admitted, and the air-admission closed as soon as the tank is emptied. The volume discharged is limited only by the size of the tanks, and the height of the lift is limited only by the air-pressure available.

The displacement pump, even at its best, is open to two objections, which, however, for limited lifts, are often more than offset by the perfect simplicity and sturdy reliability of the device. The first objection is that the valve mechanism must be on the tanks, and therefore submerged with them. This means that this mechanism is inaccessible for examination and adjustment, and is fully exposed to the action of the fluid pumped. In some of the better constructions, however, the valve movement is enclosed so that the latter difficulty is minimized. The second objection is that the discharge of a tank full of air at full pressure is a great waste of power, and evidently this objection increases as the lift, and therefore the pressure required, increases. However, as stated, the direct-displacement pump has distinct advantages often recommending it for use, even in spite of these shortcomings.

Return-Air Pumping System. One of the large builders of compressed-air machinery has produced a modified direct-displacement pumping system designed to obviate the two objections just cited in the case of the plain displacement pump. This system is known as the *return-air system,* and is an application of the closed-pipe or dense-air system of power transmission mentioned earlier in this paper. With this device, twin tanks are used; but the valve mechanism governing the admission and discharge of air, instead of being mounted on the tanks, is located in the compressor room. The air, after displacing the fluid, is not discharged to atmosphere, but is carried back through a return pipe to the intake side of the compressor. Thus its energy of expansion is applied on one side of the compressor piston, helping to do the work of compression on the other side. In this arrangement a single-stage compressor is ordinarily used, since the ratio of compression in the cylinder is small, though the terminal pressure in the machine itself may be relatively high. This system, therefore, removes the valve mechanism from the action of the fluid pumped, and places it in the compressor room where it can be kept in proper adjustment. The air is also used with higher economy. There is, of course, some leakage of compressed air in the system; and this is compensated by a check-valve on the piping, which opens to admit free air when pressure falls below a certain point.

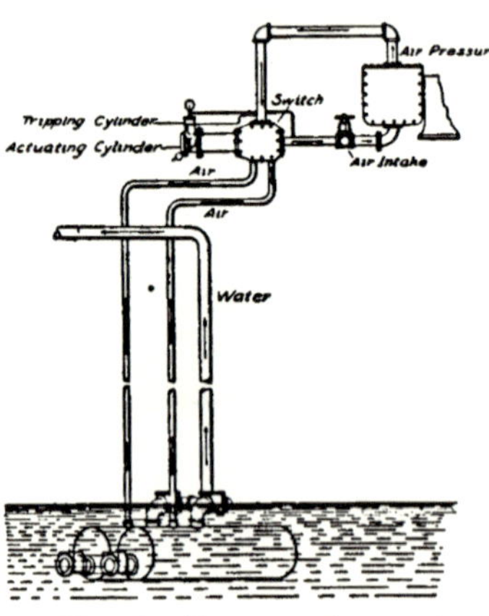

Fig. 107. Diagram of Return-Air Pumping System.
Ingersoll-Rand Co., New York City.

The manufacturers of the return-air system also offer the direct-displacement pump, and in their literature fix the distinction between the proper applications of these two systems as follows: The plain direct-displacement pump is recommended for lifts up to 50 feet; for lifts of 50 feet to 250 feet, the plain return-air system is recommended; and for lifts of above 250 feet, a compounded return-air system is offered.

The Electric-Air Principle. Another example of the application of the dense-air system is found in the *electric-air drill* and *electric-air channeler* produced by a prominent builder of pneumatic machinery. In these machines each device consists of a small, complete air compressing, transmitting, and applying system. Air is compressed by a small motor-driven duplex air-compressor designated as the *pulsator*. From each of the pulsator cylinders, a length of hose or pipe leads to one end of the cylinder of the reciprocating engine of the drill or channeler. The air, however, is not compressed and discharged to atmosphere, but is compressed and expanded back into the pulsator, where it is recompressed and re-expanded; and so on, indefinitely. Air compressed in one pulsator cylinder passes through a length of

SULLIVAN COMPRESSED AIR DIAMOND DRILL

hose to one end of the reciprocating engine, driving the piston in the latter forward. At the same time, the air on the other side of the reciprocating piston is withdrawn from the cylinder through the other length of hose and into the other pulsator cylinder. On the return stroke, this process is reversed. Thus there is had in effect a compressor and a reciprocating engine without any inlet or discharge valves, the air being simply forced back and forth under pressure in

Fig. 108. "Electric-Air" Rock-Drill with Alternating-Current Pulsator. Ingersoll-Rand Co., New York City.

a closed circuit. The full advantage of the energy of expansion contained in the air at the end of the stroke, is secured; and, by making the piston of the reciprocating engine relatively large, a powerful blow is secured without the use of high air-pressures. The alternate compression and expansion of the air in the pulsator evidently makes unnecessary the use of any water jacket. These machines are put out by the builder under a guarantee that they will do the work of an equivalent machine of the ordinary pattern, with a saving of at least one-half the power. This higher economy is due to the use of the lower pressures, and to the saving by re-expansion of the compressed air.

Electro-Pneumatic Switch and Signal Systems. In the electro-pneumatic switch-and-signal system for railway service, compressed air from some central plant furnishes the power for throwing the switches and operating the signals, the movement being produced by the action of air on the piston of a cylinder. Electricity in this

Fig. 109. "Electric-Air" Track Channeler.
Ingersoll-Rand Co., New York City.

system is simply the controlling force used for manipulating the valves for admitting and discharging the air operating the heavier mechanism. This subject is a large one, and no attempt will be made at this point to go into the details of construction and operation.

Pneumatic Haulage. Locomotives operated by compressed air are largely used for haulage purposes in mining and industrial work. These machines use air at high pressure, carried in one or more storage tanks, and delivered to the cylinders through a reducing valve so that the actual working pressure is moderate. For haulage work, the air is compressed in multi-stage high-pressure machines, and delivered in pipes to various charging stations along the haulage track. The storage pressure is usually from 900 to 1,200 lbs., which is reduced to from 100 to 150 lbs. as delivered to the cylinders. The reducing valve automatically maintains a constant delivery pressure, whatever the pressure in the storage tanks, the latter of course falling as air is withdrawn.

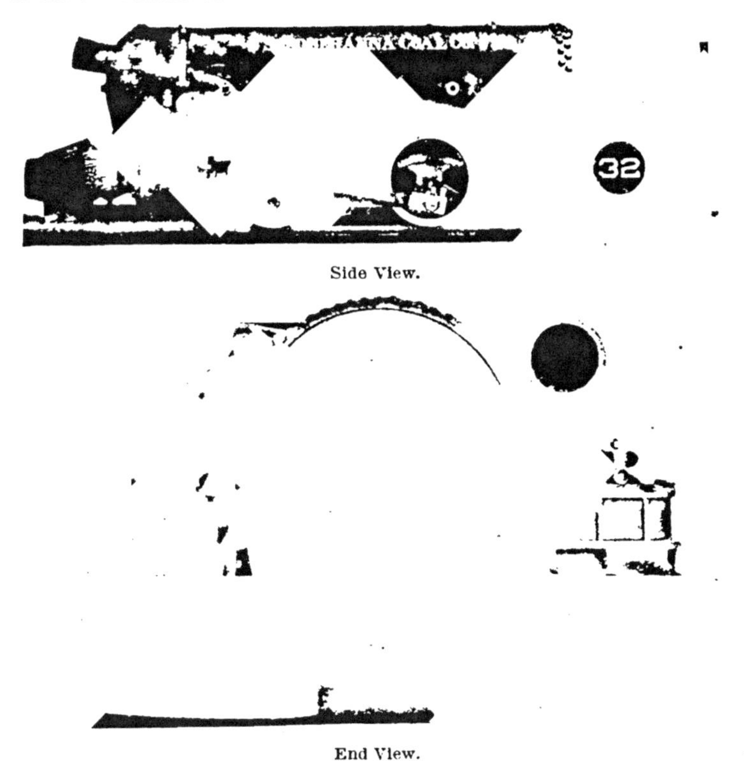

Side View.

End View.

Fig. 110. Side View, and End View (at larger scale), of Compressed-Air Locomotive. Compound type. The air, as it leaves the high-pressure cylinder at a very low temperature, passes through an atmospheric interheater before reaching the low-pressure cylinder. By actual tests, this gives an increase in efficiency over the simple-acting locomotive of 50-65 per cent.
H. K. Porter Co., Pittsburg, Pa.

In small locomotives, the air is ordinarily used in the cylinder without cut-off and expansion. This of course reduces the economy, but results in giving large power from a very compact machine with small cylinders. In locomotives of large size, there is usually provision for adjustable cut-off on the cylinders. Thus no cut-off can be used in starting a train or going over a heavy grade; and the cut-off is shortened when the load is light or the grade easy, when less power is required. It is true that this expansive use of the air produces a low temperature. But the initial storage pressure of the air is so high that, when this air is expanded through the reducing valve and after cut-off, the percentage of moisture in the air is so small that no trouble is caused by freezing. Nevertheless, large, short, direct ports and passages are usually sought in the design. Reheating is sometimes employed, to improve still further the economy. The locomotive cylinders are not covered, but left open to the air so that as much heat as possible may be absorbed from the atmosphere. In some cases the exterior of the cylinder is ribbed, to expose a larger surface for heat absorption.

COMPRESSED-AIR BIBLIOGRAPHY

The following publications are recommended to the student wishing to pursue further the study of compressed air:

Cyclopedia of Compressed Air Information, edited by W. L. SAUNDERS; Compressed Air Magazine Co., New York City, Publishers.

Compressed Air in All Its Applications, by GARDNER D. HISCOX; Norman W. Henley Publishing Co., New York City, Publishers.

Compressed Air, by FRANK RICHARDS; John Wiley & Sons, New York City, Publishers.

Compressed Air Plant for Mines, by ROBERT PEELE; John Wiley & Sons, New York City, Publishers.

Mechanics of Air Machinery, by WEISBACH AND HERMAN; D. Van Nostrand Co., New York City, Publishers.

In addition to these books, the student is referred to *Compressed Air Magazine* (monthly), published by the Compressed Air Magazine Company, New York City.

ENGINE ROOM ST. LOUIS REFRIGERATING AND COLD STORAGE CO.

REFRIGERATION

PART I

Historical. Just as, in the realm of applied electricity, progress has been made, not by accidental discoveries and fortunate inventions, but by careful study and application of principles learned in scientific research, so refrigeration is produced mechanically with apparatus designed and perfected by applying known principles. The history of the refrigerating machine is a history of steady growth and development. As knowledge of physical and chemical laws has become more extensive, particularly as related to the action of gases and heat transfers, the refrigerating machine has been improved, until to-day it is able to make ice that can be sold in competition with the product of Nature's factory. History shows that ice was used for domestic purposes as early as the time of Nero, this monarch having had ice houses built in Rome for storing natural ice. Moreover, it is believed that the ancients had knowledge of means to produce refrigeration by artificial processes, as by the evaporation of water or other liquid in strong air-currents. Also, they are thought to have used freezing mixtures of various kinds, one of the most common being that of salt and ice, as used by every housekeeper of to-day in making frozen creams.

When one considers that all refrigeration is produced by the evaporation of certain liquids, it is seen at once that the ancients used the right principles, whether or not their history—if it could be better known to-day—would show that they had mechanical genius and inventive capacity sufficient to develop a practical machine for producing refrigeration mechanically. Antiquarian researches have shown that in certain particulars the mechanical genius of the ancients was greater than that of our own day; and we may well wonder whether or not they were able to develop a successful refrigerating machine. Whatever knowledge the ancients may have possessed along these lines, was lost in the various upheavals that convulsed the world down to the end of the Dark Ages, at which time modern history begins. Records since that time show that the cooling effect obtained by dissolving certain salts was recognized as a process some three hundred years ago, the method having first been used about 1607, and later, 1762, by Fahrenheit.

Mechanical apparatus for producing cold dates from a much more recent period, the first authentic record of any such invention being the machine of Dr. Cullen for evaporating water under a vacuum. This machine was put in operation about 1755. About this time, also, experiments were made in France by Lavoisier, for using ether in refrigerating machines; but little or nothing came of either of these men's efforts, and it was not until the early part of the nineteenth century that refrigeration by mechanical means began to assume a practical form. About 1810, Leslie experimented with a machine using sulphuric acid and water; and in 1824 a machine was patented by Vallance, in which dry air was circulated over shallow trays of water, the resulting evaporation abstracting a large amount of heat; this latter process for cooling water was used in India from a time prior to the dawn of modern history.

From the beginning of the nineteenth century on, the development of the refrigerating machine was rapid and continuous. Various machines using liquids as the working medium were employed. Until within the last thirty or forty years, however, refrigerating machines met with little success, the inventors in most cases being compelled to stand the heavy cost of development, with little or no returns. An interesting point in connection with the development of mechanical refrigeration, is seen in the fact that physicians were numbered largely in the list of those developing new machines, this presumably being due to the necessity felt in medical circles for cold water and ice to be used in the treatment of diseases. Thus the early physicians who used their substance in an effort to develop a machine to alleviate suffering, were philanthropic heroes of the first type.

Air-Machine. About 1845, Dr. Gorrie invented the *cold-air* refrigerating machine, which was later developed by Windhausen, Bell, Coleman, Haslam, and others. Thus the air-machine invented by a physician was the first successful refrigerating machine used in commercial work; and for a number of years machines constructed on this plan were used exclusively for transporting meats and perishable products over-sea. This was the first large application of mechanical refrigeration. On account of the advantages incident to the absence of any obnoxious gases in the cold-air machine, this type of refrigerating apparatus has been largely used on shipboard even to the present day, notwithstanding the fact that this machine is recognized as having lower mechanical efficiency than other types on the market.

After the invention of Dr. Gorrie, the next important step in development was the invention of the *ammonia absorption* process, by Carré, in 1850; and between 1850 and 1860, independent inventors in Australia and the United States were working to improve the *ether* machine invented by Perkins many years previously. Machines of this type were put in commercial use in Ohio and in England about 1859.

From this time to the present day, there has been no question as to the commercial practicability of machines for producing mechanical refrigeration, and inventors have successively brought out various designs of cold-air compression machines, ammonia and carbon dioxide machines operating on the compression plan, the ammonia absorption system, and the dual system of absorption refrigeration.

Aside from this, there have been at various times a number of special machines in operation; but for commercial purposes on a large scale, the proposition has narrowed itself down to a choice between *ammonia or carbon dioxide compression machines* and *absorption machines using ammonia or a dual liquid.* There is endless discussion among engineers as to which of these two systems is the better, and the matter seems likely not to be settled definitely until some radical change in method of operation or design is made in one or both of the systems. The end of development in designing and constructing refrigerating machinery is not yet; and progressive men confidently expect to see marked changes in the recognized procedure along these lines at no very distant date.

Definitions. Refrigeration may be defined as a process of cooling. It is artifically or mechanically performed by transferring the heat contained in one body to another, thereby producing a condition or state commonly called *cold,* but which is in fact an absence of heat. The terms *hot* and *cold* are entirely relative, and have reference to the manner in which the heat of substances affects the senses. It is quite possible for a substance that feels cold to one person to feel hot to another, so that the terms *hot* and *cold* have no reference whatever to the absolute amount of heat in a given substance. One substance may feel colder to a person than another, and yet contain more heat than the substance that feels warmer. In considering a transfer of heat from one substance to another, as in a refrigerating machine, the student should bear in mind the process by which a steam engine operates, and remember that the working of a refrigerating machine is exactly the reverse of engine operation.

An engine converts heat into mechanical work, while the refrigerating machine converts mechanical work into heat, doing this in such a way as in the end to produce cold. It thus turns out that in order to produce cold in a refrigerating machine, heat must be expended; and this is one of the most puzzling points that the uninitiated man has to ponder over in taking up a study of refrigeration. The explanation is found in the first law of thermodynamics, which states that heat and work, or mechanical energy, are mutually convertible, and that heat cannot be raised from a lower to a higher level of temperature—as in passing from a cold to a hot body—without the expenditure of external energy. This energy, in the case of refrigerating machines, is finally lost in the cooling water that passes from the condensers and other parts of the cooling equipment at a temperature higher than when brought to the plant.

By the expenditure of energy in performing mechanical work, heat is abstracted from one substance and transferred to another at a higher temperature; and the substance from which heat is taken then becomes cold to the senses, and its exact temperature or *sensible heat* is determined by the use of thermometers. In a word, then, a refrigerating machine is a *heat pump*; and in studying such apparatus, it becomes necessary to learn something of *heat* and the units used in measuring it, as well as the method of effecting its measurement in any substance.

Heat. Unfortunately scientists have never been able to determine the exact nature of heat; but it is pretty well agreed among authorities that heat is a form of energy manifesting its presence in a substance by producing a kind of motion among the molecules of which the substance may be considered as made up. The molecules are supposed to be separated by a certain distance at any given temperature of the substance, this distance in the case of a gas being great as compared to the size of the molecules themselves. Molecules are considered as being round and as being in rapid vibratory motion, which motion is much more rapid and intense as the temperature of the substance rises by reason of the addition of heat. So long as the kinetic energy manifesting itself in this motion is not great enough to overcome the force of cohesion—which exists between the molecules of all substances and between all parts of the universe—this substance retains its form; but as soon as the force of cohesion is overcome, there is a change of state, and a solid substance changes to the liquid form, or a liquid to the gaseous state. On the other hand, the reverse changes in state take place when the substance is cooled, for, as the temperature falls, the vibratory motion is less intense, and the distance between molecules becomes less, until the force of cohesion acts with sufficient effect to reduce the gaseous substance first to the liquid, and then, as the temperature continues to fall, to the solid state.

A moment's consideration will show that different substances require widely different changes of temperature to effect a change of state. Thus metals like iron fuse or melt at a temperature which will vaporize other metals like zinc and tin. This temperature, however, is many degrees higher than that required to vaporize liquid substances, such as water, which, although it exists in the solid form, is at ordinary temperature a liquid. In the case of substances in the gaseous form, such as air, a comparatively large drop in temperature must be produced by refrigeration before the force of cohesion acts with sufficient effect to reduce the gas to liquid form. These illustrations of the manner in which the degree of heat affects change in the state of substances, show why it is regarded as a form of energy and motion. That heat is not a material substance, is shown conclusively by the fact that the weight of a substance is unchanged no matter how hot or cold it may be, and that a given quantity of a substance will retain the same weight in the liquid or gaseous state as when existing as a solid.

Units of Heat Measurement. As there is no way to make direct measurement of the motion of molecules, the heat in a given amount of substance at a certain temperature cannot be measured directly. It can be measured only by the effect produced in performing work or in changing the state of some other substance. It is therefore evident that the measurement of energy in the form of heat is an entirely relative or comparative process. When a ball of iron at a certain temperature contains enough heat to melt a given amount of ice, and when heated to another temperature melts twice as much ice, it evidently contains twice as much heat energy in the second case as in the first instance. For convenience in practical operation, scientists have established an arbitrary heat unit by which all amounts of heat may be measured, this unit being the amount of heat necessary to raise unity weight of water, at its maximum density, one degree in temperature. Water has its maximum density at 39.1 degrees F. or 4 degrees C.; and the unit of measurement, therefore, is *the amount of heat that will raise one pound of water from* 39 *to* 40 *degrees F. or from* 4 *to* 5 *degrees C.*, the former being the *British thermal unit*, and the latter the *thermal unit.* For convenience it is customary to use the letters B. T. U. as an abbreviation for the first of these units; and unless otherwise specified, all heat units mentioned in the present treatise will be B. T. U., or British thermal units.

There is a third unit used in France, known as the *calorie*, which is based on the French decimal system of measurement. A calorie is the amount of heat that will raise the temperature of one kilogram of water from 4 to 5 degrees C. Since one degree C. is 9/5 of a degree F. it follows that the thermal unit is 9/5 times as large as the B. T. U. Also, a kilogram is the same as 2.2 pounds of water, so that the calorie is the same as 2.2 thermal units or 3.96 B. T. U. Each of these units of heat measurement has reference to the actual amount of heat—or, in other words, to the total molecular energy in a substance—and has no reference to the sensible heat, or temperature, of the substance.

Sensible heat, or *temperature*, is heat as manifested to our senses by a substance, and is determined accurately by measuring with a *thermometer*. We feel a substance, and say that it is hotter or colder than another body which we feel at the same time, but, owing to the imperfection of our senses, it is not possible to determine accurately just how much hotter or colder it is. Still more is it impossible to ascertain by the sense of touch the temperature of a substance taken alone. It is because of this imperfection of the senses that the instruments known as *thermometers* are used. There are three kinds of these instruments in use; but only one of them is used to any extent in practical work in the United States. All three thermometers are constructed in the same way and operate on the same principle, the difference consisting solely in the method of graduating the scales of the three instruments.

Measurement of temperature or sensible heat is effected in all cases by noting the expansion of certain substances when brought in contact with varying amounts of heat. Substances used for temperature measurement in thermometers may be in any one of the three natural states of matter—namely, solid, liquid, or gaseous; but for all ordinary work, liquids are used, the expansion in the case of solids being too small with ordinary temperatures to be readily measured by the eye on a thermometer scale; while, on the other hand, the expansion of gases is far too large for such measurement. In order that the thermometer may be in condition for use at all times, it is essential that the liquid used be such as will not freeze or change into the gaseous form at the ordinary temperatures for which the instrument is designed to be used. Two materials that have been found to satisfy these requirements are *mercury* and *alcohol*, the former of these being used because it does not boil except at a very high temperature, while alcohol, on the other hand, freezes at such a low temperature (−203F.) that it may be used in the thermometers for refrigeration. Only one of these substances, generally, is used in a single thermometer; but in the case of instruments designed for automatically recording maximum and minimum temperatures, tubes using the two liquids on the same instrument are employed. As alcohol boils at 172.4° F., it cannot be used for high-temperature work, and is suitable only in cold stores and other places where low temperatures are to be recorded.

Much skill is required to make accurate thermometers, and great care must be taken in graduating the scale on the tube. Cheap thermometers having the scale on the frame carrying the mercury tube are of little or no value in careful work, and it is always preferable to use instruments having the graduations on the glass itself. Even then, in making the graduations allowances must be made for the expansion of the glass at different temperatures and for the fact that the bore of the tube cannot be depended on to have a uniform size.

Having sealed the mercury in its tube, the next step in making a thermometer is to determine the *fixed points*, the lower one being that point in the tube at which the mercury stands when placed in melting ice, and the higher fixed point being that at which the mercury stands when placed above water boiling under standard pressure. When these two points have been determined and etched on the tube, the graduation of the space between is a matter of calculation and measurement, taking care that allowances are made for all irregularities of the tube.

There are three methods of graduating thermometer scales, known respectively as the *Fahrenheit*, *Centigrade*, and *Réaumur*, the names being taken from the inventors of the respective methods. It is customary to use the initial letters—F., C., and R.—of these names as abbreviations, as has already been done in this text. Where temperatures below the zero point of a scale are denoted, the subtraction or minus sign of arithmetic is placed before the figure denoting the number of degrees. Thus, − 5 F. would mean five degrees below the zero point on the Fahrenheit scale. On this scale, the freezing point of water is marked 32 degrees, the zero point consequently being 32 degrees below freezing. At sea-level, water under atmospheric pressure boils at 212 degrees F., while the absolute zero point of the scale, *i e.*, the point at which there is no molecular motion of any substance, is 460.6 (approximately 461) degrees below zero, or 492.6 (approximately 493) degrees below the point at which water freezes. Experimenters have never been able to reduce the temperature of any substance to absolute zero (that is, to abstract all the heat energy from any substance); and the figure named is not from positive measurement, but the result of calculations made on the behavior of gases. For one degree F., a perfect gas will expand or contract about 1/493 part of its volume; and for one degree C., the change in volume is about 1/273 part. This, taken in connection with the conception of heat as the manifestation of the energy of molecular motion, directly proportional to the expansion or contraction of a substance, gives −493° F. and 273° C. as the points on the respective scales at which there would be no molecular motion. Some idea of the intense cold at *absolute zero* may be had by considering the fact that melting ice is as much warmer than any substance at the absolute zero point as molten solder is warmer than the ice.

On the Centigrade scale, the freezing point of water is marked "0," and the boiling point "100," the interval being divided into one hundred degrees. This scale is universally employed in scientific work, there being many advantages incident to the use of the decimal system in calculations. The Réaumur scale also has the freezing point at zero, but the boiling point is marked "80." This scale is used in some parts of Europe, particularly in brewery work, but has little to recommend it. It is seen that 180 degrees on the F. scale corresponds to 100 degrees C., so that 1 degree C. is 9/5 degree F. To change a Fahrenheit temperature to Centigrade, subtract 32 degrees, and multiply the result by 5/9. To change C. to F., multiply by 9/5, and add 32 degrees. Table I, page 10, gives a comparison of the three thermometer scales; by reference to this table, the labor of calculation may be avoided.

The *specific heat* of a body or substance is its capacity for absorbing heat, as compared with the capacity of water. Thus, in this case also, the measurement of heat is relative, being referred to the heat-absorbing capacity of water as the unit. As the specific heat of a unit weight of water is taken as unity, the specific heat of other substances is less than unity, being smaller than the unit in almost every case. It has already been shown that the unit of absolute heat measurement (B. T. U.) is the amount of heat that will raise the temperature of 1 pound of water 1 degree F. at the temperature of maximum density; and as the specific heats of other substances are less than that of water, we may define the specific heat of any substance as being that proportion of a B. T. U. which is required to raise the temperature of one pound of the substance one degree.

There is some variation in the amount of heat required to raise the temperature of a substance one degree at different temperatures; but for practical purposes in refrigeration, this may be neglected, the difference being of importance only for those engaged in minute scientific calculations. It is worth noting, however, that this variation must be considered in dealing with the specific heat of gases, as in this case it is of considerable importance, the specific heat of the gas varying by quite appreciable amounts under different con-

TABLE I

Thermometer Scales

Fahr.	Cent.	Reau.	Fahr.	Cent.	Reau.	Fahr.	Cent.	Reau.
212	100	80	120	48.9	39.1	30	-1.1	-0.9
210	98.9	79.1	118	47.8	38.2	28	-2.2	-1.8
208	97.8	78.2	116	46.7	37.3	26	-3.3	-2.7
206	96.7	77.3	114	45.6	36.4	24	-4.4	-3.6
204	95.6	76.4	112	44.4	35.6	22	-5.6	-4.4
202	94.4	75.6	110	43.3	34.7	20	-6.7	-5.3
200	93.3	74.7	108	42.2	33.8	18	-7.8	-6.2
198	92.2	73.8	106	41.1	32.9	16	-8.9	-7.1
196	91.1	72.9	104	40	32	14	-10	-8
194	90	72	102	38.9	31.1	12	-11.1	-8.9
192	88.9	71.1	100	37.8	30.2	10	-12.2	-9.8
190	87.8	70.2	98	36.7	29.3	8	-13.3	-10.7
188	86.7	69.3	96	35.6	28.4	6	-14.4	-11.6
186	85.6	68.4	94	34.4	27.6	4	-15.6	-12.4
184	84.4	67.6	92	33.3	26.7	2	-16.7	-13.3
182	83.3	66.7	90	32.2	25.8	0	-17.8	-14.2
180	82.2	65.8	88	31.1	24.9	-2	-18.9	-15.1
178	81.1	64.9	86	30	24	-4	-20	-16
176	80	64	84	28.9	23.1	-6	-21.1	-16.9
174	78.9	63.1	82	27.8	22.2	-8	-22.2	-17.8
172	77.8	62.2	80	26.7	21.3	-10	-23.3	-18.7
170	76.7	61.3	78	25.6	20.4	-12	-24.4	-19.6
168	75.6	60.4	76	24.4	19.6	-14	-25.6	-20.4
166	74.4	59.6	74	23.3	18.7	-16	-26.7	-21.3
164	73.3	58.7	72	22.2	17.8	-18	-27.8	-22.2
162	72.2	57.8	70	21.1	16.9	-20	-28.9	-23.1
160	71.1	56.9	68	20	16	-22	-30	-24
158	70	56	66	18.9	15.1	-24	-31.1	-24.9
156	68.9	55.1	64	17.8	14.2	-26	-32.2	-25.8
154	67.8	54.2	62	16.7	13.3	-28	-33.3	-26.7
152	66.7	53.3	60	15.6	12.4	-30	-34.4	-27.6
150	65.6	52.4	58	14.4	11.6	-32	-35.6	-28.4
148	64.4	51.6	56	13.3	10.7	-34	-36.7	-29.3
146	63.3	50.7	54	12.2	9.8	-36	-37.8	-30.2
144	62.2	49.8	52	11.1	8.9	-38	-38.9	-31.1
142	61.1	48.9	50	10	8	-40	-40	-32
140	60	48	48	8.9	7.1	-42	-41.1	-32.9
138	58.9	47.1	46	7.8	6.2	-44	-42.2	-33.8
136	57.8	46.2	44	6.7	5.3	-46	-43.3	-34.7
134	56.7	45.3	42	5.6	4.4	-48	-44.4	-35.6
132	55.6	44.4	40	4.4	3.6	-50	-45.6	-36.4
130	54.4	43.6	38	3.3	2.7	-52	-46.7	-37.3
128	53.3	42.7	36	2.2	1.8	-54	-47.8	-38.2
126	52.2	41.8	34	1.1	0.9	-56	-48.9	-39.1
124	51.1	40.9	32	0	0	-58	-50	-40
122	50	40						

ditions of pressure and volume as brought about by varying temperature. As the capacity of a substance for absorbing heat is determined directly by its specific heat, and as all refrigerating work is done to dispose of the heat absorbed by substances, it is evident at once that a knowledge of the specific heats of various substances is of first importance to those concerned with refrigerating work.

Aside from leakage and conduction losses, the refrigeration that must be performed in any case depends directly on the specific heat of the substance to be cooled; and it is on account of this fact that scientists and practical refrigerating men have devoted a great deal of time and attention to the accurate determination of the specific heat of all substances and materials ordinarily handled in refrigerating establishments.

Owing to the inherent difficulty in determining the specific heat of a subtance, values obtained by independent investigators have been found at times to differ, but Table II gives the figures commonly accepted for a number of the more important substances, while Table III gives the specific heat and the specific gravity of beer wort. By the use of these tables, it is possible to calculate in any given case the amount of refrigeration that will be required to handle a known quantity of goods, the leakage of heat through the insulating material of the cold stores being previously determined by experiment.

Latent heat is a term used to designate the quantity of heat absorbed or given up by a substance to effect change of state without a change of temperature. Thus one pound of ice at 32 degrees, on being melted into water at the same temperature, absorbs 142.65 heat units; and one pound of water at 212 degrees, when evaporated into steam at the same temperature, absorbs 966.6 units. It is this absorption of heat on change of state that causes cooling by the evaporation of water and other liquids. For each pound of water evaporated at atmospheric pressure, about 966 B. T. U. are absorbed, and become latent in the vapor passing off from the water. This heat must be taken up from the surrounding objects, which are thereby cooled. One pound of water evaporated, then, is sufficient to cool 966 pounds of water one degree. This is the principle used in all practical refrigerating machines of the present time; but it is impracticableto use water for the evaporating liquid, on account of the fact that it boils at a comparatively high temperature, the temperature at atmospheric pressure being 212 degrees F., while that of liquid ammonia, for example, at the same pressure is – 28 degrees. This is the main reason why ammonia is used so largely in refrigerating work, in spite

TABLE II

Specific Heat of Various Substances Under Constant Pressure

SOLIDS

Substance	Specific Heat	Substance	Specific Heat
Copper	0.0951	Cast Iron	0.1298
Gold	0.0324	Lead	0.0314
Wrought Iron	0.1138	Platinum	0.0324
Steel (soft)	0.1165	Silver	0.0570
Steel (hard)	0.1175	Tin	0.0562
Zinc	0.0956	Ice	0.5040
Brass	0.0939	Sulphur	0.2026
Glass	0.1937	Charcoal	0.2410
Oak	0.570	Brickwork	0.200
Pine	0.650	Stone	0.270
Cast Iron	0.130	Marble	0.209 to 0.215
Coal	0.241	Fat Beef	0.60
Coke	0.203	Lean Beef	0.77
Fish	0.70	Fat Pork	0.51
Chicken	0.80	Veal	0.70
Eggs	0.76	Fruit and Vegetables	0.50 to 0.93

LIQUIDS

Substance	Specific Heat	Substance	Specific Heat
Water	1.0000	Milk	0.90
Alcohol	0.7000	Lead (melted)	0.0402
Mercury	0.0333	Sulphur (melted)	0.2340
Benzine	0.4500	Tin (melted)	0.0637
Glycerine	0.5550	Sulphuric Acid	0.3350
Strong Brine	0.700	Oil of Turpentine	0.4260
Vinegar	0.920	Anhydrous Ammonia	1.020
Cream	0.680	Carbonic Acid	0.980

GASES

	Constant Pressure	Constant Volume
Air	0.23751	0.16847
Oxygen	0.21751	0.15507
Nitrogen	0.24380	0.17273
Hydrogen	3.40900	2.41226
Superheated Steam	0.48050	0.34600
Carbonic Oxide	0.24790	0.17580
Carbonic Acid	0.21700	0.15350
Ammonia	0.393	0.508

of the fact that its latent heat is but little more than half that of water (or 573 B. T. U.) when expanded at atmospheric pressure.

The temperature at which a solid changes to the liquid form is known as its *temperature of fusion*; and that at which it passes into the form of vapor, is known as the *temperature of vaporization*. Similarly, we have the terms *latent heat of fusion* and *latent heat of vaporization*, or the heat required in each case to effect the change in state of the substance without changing its temperature.

Table IV gives the temperatures and latent heats of fusion and vaporization for a number of substances for which this data has been determined by experiment. It will be noticed from the blank spaces in the table, that considerable is yet to be learned on this subject in regard to some substances. Thus, for example, the temperature at which alcohol may be changed to the solid form has never been

TABLE III

Specific Heat and Specific Gravity of Beer Wort

Strength of Wort in % after Balling	Corresponding Specific Gravity	Corresponding Specific Heat	Strength of Wort in % after Balling	Corresponding Specific Gravity	Corresponding Specific Heat
8	1.0320	.944	15	1.0614	.895
9	1.0363	.937	16	1.0657	.888
10	1.0404	.930	17	1.0700	.881
11	1.0446	.923	18	1.0744	.874
12	1.0488	.916	19	1.0788	.867
13	1.0530	.909	20	1.0832	.861
14	1.0572	.902			

determined, as no process has ever been devised for freezing this liquid. Then, again, the vaporization of tin and lead takes place at such high temperatures as to make accurate measurement impossible by any known means of recording temperatures.

Unit of Plant Capacity. Ordinarily the capacity of a refrigerating machine or plant is stated in *tons*—that is, one ton is the heat equivalent of a 2,000-pound ton of ice at 32 degrees F., melted into water at the same temperature; or, conversely, the amount of heat that must be abstracted from 2,000 pounds of water at 32 degrees to change it into ice at the same temperature. Since the latent heat of ice is about 142 B. T. U., the ton of refrigerating capacity used in rating apparatus is equivalent to 142 × 2,000 = 284,000 B. T. U.

Thermodynamics, as the name implies, is the science treating of heat as a form of energy and of its relation to other forms of energy, particularly its relation to and transformation into mechanical energy

TABLE IV

Fusion and Vaporization Data of Substances

Substance	Temperature of Fusion F.	Temperature of Vaporization F.	Latent Heat of Fusion	Latent Heat of Vaporization
Water	32°	212°	142.65	966.6
Mercury	-37.8°	662°	5.09	157
Sulphur	228.3°	824°	13.26	
Tin	446°	.....	25.65	
Lead	626°	.....	9.67	
Zinc	680°	1,900°	50.63	493
Alcohol	Unknown	173°	......	372
Oil of Turpentine	14°	313°	......	124
Linseed Oil	....	600°		
Aluminum	1,400°			
Copper	2,100°			
Cast Iron	2,192°	3,300°		
Wrought Iron	2,912°	5,000°		
Steel	2,520°			
Platinum	3,632°			
Iridium	4,892°			

or work. It is not possible in brief space to enter into a discussion of thermodynamics in detail; but brief mention must be made of the fundamental principles of this science that have to do with the operation of refrigerating apparatus.

Some reference has already been made to the *first law*, which is a special case of the general law expressing the mutual convertibility of all forms of energy. According to this law, as already mentioned, heat is equivalent to work or mechanical energy, each unit of heat being equivalent to 778 foot-pounds of work, or the amount of work that must be performed to raise 778 pounds a vertical distance of one foot against the action of gravity. This first law of thermodynamics must be qualified to some extent, for, although heat and work when convertible are theoretically equivalent to each other, the actual conversion of one into the other is not, in every case practicable—being, in fact, practicable in every case only so far as the conversion of work into heat is concerned. In other words, while it is always possible to change a given amount of work into heat energy, it is not possible in every case to convert heat into work, for there is always a certain amount of *unavailable heat* which it is found impracticable, with devices at present in use, to convert into work.

The *second law* of thermodynamics states, therefore, that it is impossible to transfer heat from a body of low temperature to one of higher temperature without the application of some external form of energy.

Thus in every case, for a given temperature, there is a certain well-defined portion of the total heat in the substance that can be converted into work, the remaining heat being *unavailable for conversion.* If this were not true, it would be possible to reduce the temperature to absolute zero, when the substance would have no volume. Thus matter would be annihilated. As this is impossible, however, it is plain that absolute zero temperature can never be attained. In every case where heat is converted into work, there is a lowering of the temperature of the body from which the heat is taken, and the fall in temperature is an index of the amount of heat energy converted into work. The energy existing as heat at low temperature is unavailable and must be dissipated, it being absorbed on discharge from the heat engine by a still colder body, which in the case of steam engines is the atmosphere or condensing water. Much depends on the character of the heat engine, as to just what amount of energy in the form of heat will be transformed into useful work. A conspicuous example of this is the case of low-pressure turbines, which, within the last few months, have been so perfected as to reclaim as much energy from the *exhaust* steam of reciprocating engines as the engines themselves utilize in the first place.

As already pointed out, the molecular activity of a substance is increased by heating or raising its temperature; and it is this molecular activity that gives an index to the energy present and available in the form of heat. There are many ways of converting this energy into work; but, in that most commonly used, advantage is taken of the pressure produced by the molecular activity of a gas that has been heated. The gas in its heated state is placed behind a piston; and the molecules in their efforts to get away from each other, or to occupy a larger space, exert a pressure on the piston that moves it forward. As the gas expands, its heat energy is expended, and its temperature is lowered until the force of cohesion acting between the molecules is sufficient to prevent the performance of further external work. The one condition of work being performed is, that there shall be resistance to movement of the piston—there being no work, with resulting fall of temperature, where a gas has free room to expand, as in a vacuum.

This leads to a consideration of the way in which gases may be compressed and expanded; and it is well for the student to give attention to this subject which lies at the bottom of all efficient work in refrigerating machines. When a gas is expanded or compressed without addition or subtraction of heat, the process is said to be *adiabatic,* and the temperature of the gas will rise with compression and fall with expansion. If it is desired to maintain the temperature of the gas constant, heat must be abstracted as the gas is compressed, or supplied as it is expanded, except in the case of free expansion, in which case there is practically no change in temperature. In this discussion, however, it is assumed that work is expended in compressing the gas, and that the gas performs work in expanding.

As the heat abstracted from a gas in a heat engine is the equivalent of the work performed, it should be possible theoretically to restore the temperature of the gas to its original state by reversing the operation of the heat engine, there being no loss of heat in the process. In practical work, a certain part of the work obtained from heat in an engine is dissipated at once in overcoming the frictional resistance of the moving parts of the machine, and appears as heat in the bearings, etc., so that the theoretical conditions of a complete reversible cycle cannot be carried out. If it were possible to convert heat into mechanical work by direct process without the intervention of heat engines or other mechanism, there would be much greater efficiency than at present; but so far no method of doing this has been found. Where a gas is allowed to expand freely without performing work, its energy is dissipated, and there is no way to restore it to its original condition without the expenditure of some external form of energy. It is possible, therefore, for the entire heat energy of the world to be dissipated, as the only way in which waste heat is reclaimed is in its indirect effect on growing vegetable matter which can be used as fuel.

In the operation of a refrigerating machine, it is necessary to have a continuous conversion of heat into work. This presupposes dissipation of a certain amount of heat energy, as it is impossible to carry on continuous conversion without the loss of the unavailable heat that must be rejected by the machine. Thus the heat pump of a refrigerating establishment is a machine designed to abstract a certain amount of heat from the body to be cooled by changing the form of the surrounding medium used in the abstraction so that its temperature becomes lower than that of the body itself. It is therefore evident that, with the possible exception of the vacuum process, in which the evaporation of a portion of a body of water absorbs sufficient heat to freeze the remaining part of the water, it is impossible for a machine, however designed, to refrigerate the body to be cooled without using some working medium. Several kinds of mediums are used, and will be discussed in detail later.

In practically all cases, water is used for the cooling body to absorb the waste heat, the water being circulated over the condensers and cooling coils of the machine, so as to take up the heat that has been concentrated in a comparatively small volume of the working medium by the action of the refrigerating machine. It is therefore in order to see how the machine effecting this result is constructed; but before doing this, it is well to see in what way heat transfers are made.

Heat transfers may be made in three ways—by *radiation*, by *convection*, and by *conduction*. An illustration of the meaning of these terms is had in the case of an iron bar thrust into the fire until one end becomes hot. When the bar is withdrawn from the fire, it loses heat in each of the three ways mentioned, part of the heat being radiated into space, part of it being conducted along the length of the bar, and part being carried off by air currents which circulate around the bar, the heated air rising. At first the temperature of the iron will fall rapidly; but as it becomes cooler, the transfer of heat to the atmosphere and surrounding objects becomes less rapid. This brings us to the law of Newton, which holds good for moderate temperatures such as those used in refrigerating work—namely:

"The rate of cooling of a body is proportional to the difference between the temperature of its surface and that of its surroundings."

In the case of liquids, we have the similar law, also evolved by Newton, in the words:

"The amount of heat lost in a given interval of time by a vessel filled with liquid is proportional to the mean difference of temperature between the liquid and its surroundings."

Radiation. Taking again the illustration of the hot bar of iron, if the hand is placed at a certain distance above the bar, a greater intensity of heat is felt than if it is placed at the same distance below the bar, this being due to the fact that in the first case the hand feels heat emitted from the bar both by radiation and convection, while in the second case with the hand underneath the bar, the heat or radiation alone is felt. Another illustration of radiant heat is the common experiment performed with the optical lens by means of which the rays are focused on a given point, when the heat becomes so intense as to burn the flesh or ignite a dry substance. Radiant heat rays pass readily through glass, but are reflected by smooth polished surfaces; while, on the other hand, a rough surface covered with lamp black will absorb the rays. The rays pass readily through air, but are absorbed to some extent by carbonic acid gas, and still more so by ammonia gas.

A warm body exposed to the air will lose a certain amount of its heat by radiation; but when placed in a closed chamber the walls of which are at the same temperature as itself, will lose no heat. This does not mean, however, that radiation of heat from the body stops under such circumstances, but simply that as much heat is radiated from the walls of the containing chamber to the body as is radiated from the body to the walls, so that the temperature of walls and body remains constant. This is known as *Prevost's theory of radiant heat exchanges;* and in the case cited, the radiation of heat is equal from the body and from the walls. Where bodies are at different temperatures, the exchange of radiant heat goes on in the same manner; but the amount radiated from the warmer body is greater than that from the cooler body, so that there is a tendency to equalize the temperatures of the two bodies.

Where two bodies covered with lamp-black are enclosed in a chamber the walls of which are at the same temperature as the bodies, the temperatures throughout will remain constant; but as the black surface of one body will absorb heat readily and this heat must be supplied from external sources, it is evident that the other body must emit heat rapidly. In other words, the exchange process of radiant heat goes on much more rapidly between bodies covered with lamp-black—and, in general, between all dark bodies—than between bodies of lighter color and those that are polished. Thus good absorbers of heat are also good radiators; and surfaces designed to absorb or radiate heat should preferably be of a dark color, while those designed to prevent radiation should be smooth and of as light color as practicable Thus a polished copper steam pipe radiates much less heat than a similar black pipe, and a plain cast-iron radiator is better than the same radiator covered with one of the bright metallic paints so frequently used. For this reason, also, the walls and sides of a cold storage house should be whitewashed or constructed of white enamel brick to reflect radiant heat that otherwise would be absorbed by the walls and conducted through the insulating material of the rooms to the cold stores from which it would have to be absorbed by the expenditure of considerable work in refrigerating.

Convection. It is by this process that heat is readily diffused through liquids and gases; for, when one portion of the liquid is heated, its density is decreased and it is displaced by the heavier and colder portions of the liquid, which in turn are themselves displaced as the heating process continues, until finally the temperature is practically uniform throughout. The currents set up in this process of heating are known as *convection currents.*

The same thing takes place in the case of heat applied to a body of gas. Thus the air in a room, for example, being heated by a stove or other means, rises to the ceiling of the room and displaces the colder air, which, being heavier, falls to the floor to be heated in its turn. In this case we have convection air currents; and it is owing to currents of this character that great care must be taken in constructing insulating walls, where air spaces are used, in such a manner that the spaces will be comparatively small, thus not allowing room enough for the setting up of such currents, which, if formed, would be a means of transferring heat to the cold stores instead of acting as an insulation.

Neglect of this matter has frequently resulted in disappointment with insulation where *dead* air spaces have been depended on to a considerable extent. The air in contact with the warm wall on the outside of the chamber becomes heated and rises to the top of the space, whence, as it is gradually cooled, it falls down along the comparatively cold inner wall, imparting its heat to this wall during passage. By the time the air has passed down this inner wall to the bottom of the space, it is comparatively cold, and then comes in contact a second time with the outer wall. In this way a continuous current is set up, and acts as a conveyor of heat from the outer to the inner wall of the building. Having these facts in view, insulating men have agreed that the smaller the air space can be made, the *deader* it is; and in modern work, this is reduced to a nicety where there are no air spaces larger than the minute cells in the structure of cork, which is used for insulating purposes in the best work of the present time.

Conduction. This term has reference to the manner in which heat is propagated through a substance, or from one substance to another where the two substances are in contact. Taking again the case of the iron bar heated at one end, the molecules at the heated end may be considered as being in a state of violent agitation so that each possesses a definite amount of kinetic (active or moving) energy. In the cooler portion of the bar, the molecules will be agitated to a less extent; but those molecules in contact with the similar molecules in the hotter portion of the bar are gradually affected by the impact of these latter molecules, by which means their rate of motion among themselves is gradually increased, they receiving in the contact a portion of the energy of the more violently agitated molecules. In this way the heat energy of the molecules in the cooler portion of the bar is considerably increased; and the heat gradually passes thus from molecule to molecule toward the cooler end of the bar, until finally the temperature has been made uniform throughout the entire length of the bar by the process of conduction.

PRODUCTION OF COLD

Production of cold is in general effected by transfer of heat from one body or substance to another at lower temperature. Something has already been stated as to the manner in which heat transfers take place and the effects produced by such transfers. It remains to be seen, then, how the transfers are brought about in practice in such a way as to give the desired results.

There are three ways in which heat transfers may be made to produce cold, the first of these being by chemical action as exemplified in the so-called *freezing mixtures.* It has already been seen that when a solid changes to the liquid form, the heat becomes latent and the temperature correspondingly lowered, the change being effected by separation of the molecules of the substance in melting. It is equally true that the latent heat is absorbed when the change of state to the liquid form is made otherwise than by melting, as in the case where a solid is dissolved in water. Heat, then, may become latent with change of state by the process of dissolving as well as by that of melting, the fall in temperature in either case being brought about by the expenditure or exchange to the latent form of the heat energy necessary to separate the molecules of the solid substance so that it assumes the liquid form.

To illustrate the lowering of temperature produced by solution, take a glass of water and place in it a thermometer. On dissolving sugar or salt in this water, the temperature will be seen to fall, the effect being much more marked if the disolving process is hastened

TABLE V

Composition of Freezing Mixtures	Red. of Temp. in Deg. F.		Amt. of Fall in Deg. F.
	From	To	
Snow 4 parts; Muriate of lime 5 parts	32	−40	72
" 1 " Common salt 1 part	32	0	32
" 2 " Muriate of lime crys. 3 parts	32	−50	82
" 3 " Dil. sulphuric acid 2 parts	32	−23	55
" 3 " Hydrochloric acid 5 parts	32	−27	59
" 7 " Dil. nitric acid 4 parts	32	−30	62
" 8 " Chloride of calcium 5 parts	82	−40	72
" 2 " " crystallized 3 parts	32	−50	82
" 3 " Potassium 4 parts	32	−51	83

by stirring, so that the heat will not be absorbed from the surrounding subjects before the reduction of tempeyature occurs. One part nitrate of ammonia mixed with one part of water at 50° F. gives a reduction of 46 degrees, or to 4° F. Where two solids are mixed, one of them being at the freezing point, the cooling action is still more marked. Two parts of snow mixed with one part of common salt gives a reduction of 50 degrees; while four parts of potash mixed with three parts of fine snow or crushed ice gives a drop from 32° to −51°, or a total of 83 degrees. Table V gives the reduction in temperature for a number of other mixtures, and is of considerable value to manufacturers of ice cream, in enabling them to determine what materials may be used with greatest economy in freezing or packing cream.

It is in work of this kind that freezing mixtures have their chief value, the cooling produced being too slight in proportion to the amount of material used to be of any value in producing refrigeration on a large scale in commercial work.

A more practical and at the same time a rather expensive method of producing refrigeration for commercial purposes, is that in which a non-condensable gas is expanded adiabatically, or without the addition or subtraction of heat. The gas, after being compressed and cooled, is allowed to expand while doing work against a piston, with the result that its temperature is lowered. In this machine the gas is never condensed to the liquid form, but merely compressed to greater density than its natural condition. Air is used in all practical machines employing the principle of adiabatic expansion; but in no case is the air reduced to liquid form, as liquid air has far too low a temperature to be of any practical use for refrigeration under normal conditions. It is this difference in handling the working medium that distinguishes the compressed-air machine from other compression machines in which the liquid is compressed and then condensed to the liquid form by cooling.

The third and most important method of refrigeration is that in which a volatile liquid is vaporized to absorb heat, as represented by the latent heat of the medium used. The heat of vaporization is absorbed from objects surrounding the working medium; and these objects are cooled to the temperature desired, the material used for the cold body in most refrigerating plants being a strong brine solution. Thus the liquid expanding in the cooling coils absorbs heat from the brine in the tank surrounding these coils, and the cold brine is used to freeze ice or is circulated through the cold storage rooms, the application of the cold produced by the expansion of the working medium varying according to the circumstances and requirements of each case.

In the special case of vaporization or *latent-heat* machines operating on what is known as the *vacuum system*, water is at once the working medium and the cold body, the cooling being done by evaporating part of a body of water under a vacuum so that the latent heat taken up in evaporation reduces the temperature until the portion of water remaining in the apparatus is frozen. Owing to the fact that the latent heat of water is large as compared with other liquids used, the freezing is very rapid, so that the ice produced is usually opaque, there not being time for separation of the air from the water. It should be noted that this system depends for its operation on a vacuum being produced, as otherwise the boiling point of water is at 212°, which is altogether too high a temperature for refrigeration work. The chief difficulty, then, with the vacuum process, is the necessity of maintaining the vacuum, for which complicated apparatus is required.

In the vacuum process, external energy is expended to drive the vacuum pumps and other machinery connected therewith; and a moment's consideration will show that in every system external energy is utilized at some point in the cycle, thus obeying the thermo-dynamic laws. It is seen at once that pressure and temperature tell the whole story in refrigerating work, the whole object of such work being the reduction of temperature, which reduction depends on the pressures and corresponding temperatures in the different parts of the system. As already pointed out, water cannot be used except in a vacuum on account of its high temperature of vaporization at ordinary pressures. Since it is not desirable to operate with a vacuum in all cases, other working mediums or *refrigerants* than water must be chosen, and thus it comes about that the temperature at which a substance will vaporize at a given pressure is of first importance.

For any given substance in the form of vapor—that is, a fully expanded gas containing no moisture—there is a certain temperature above which it is impossible to liquefy the substance no matter how great the pressure. This is the *critical temperature.* The pressure that will cause liquefaction at the critical temperature is known as the *critical pressure.* These two critical points of temperature and pressure determine largely whether or not a given substance is suitable as the working medium in refrigerating machines. Aside from its latent heat, which should preferably be high, the substance should have such *critical data* as to make it possible to work it in the refrigerating machine at ordinary pressures and temperatures, for otherwise the special apparatus required to manipulate it will be too expensive to be practical. Table VI gives the critical pressure and temperature, with the corresponding density, for a number of substances. It is seen that *ammonia, carbon dioxide,* and *sulphur dioxide* are the only three substances that have the critical points at anything like normal conditions of temperature and pressure. Hence the choice of a refrigerant from among the many volatile liquids known to chemistry is narrowed down to these three substances.

There is considerable discussion and expression of opinion among engineers as to which of these three is best. Generally speaking, the choice of any substance from an engineering standpoint depends on the latent heat of the liquid per pound; the boiling point at ordinary pressures; the number of cubic feet that must be compressed

TABLE VI

Critical Data

Substance	Boiling Point at Atmospheric Pressure in Degrees F.	Freezing Point at Atmospheric Pressure in Degrees F.	Critical Temp. in Degrees Fahrenheit	Critical Pressure in Atmospheres 1 Atmos.=14.7 Lbs.	Critical Volume in Cu. Ft. of 1 Lb.
Water .... H_2O	+212	+ 32	+657	205	0.037
Alcohol .... C_2H_6O	+172	−148	+423	67	0.114
Sulphur dioxide .... SO_2	− 14	−105	+313	81	.....
Ammonia .... NH_3	− 27.4	−106.6	+266	115	0.048
Carbonic acid (carbon dioxide)... CO_2	−110	−110	+ 88	75	0.035
Oxygen .... O	−296	−269	−180	52	.....
Atmospheric air	−312	.....	−220	39	.....
Nitrogen .... N	−317	−353	−231	36	.....
Hydrogen .... H	−400	.....	−382	21	0.037
Air	.....	....	−231.9	45	.....
Nitrous oxide	.....	.....	+ 96	75	.....

to produce a certain refrigerating effect (or, in other words, the size of the compressor necessary); the pressure required to produce liquefaction of the gas at certain temperatures; and the specific heat of the liquid. Table VII gives the boiling point and latent heat of a number of substances at 14.7 pounds, and also gives the specific heat of the liquids used in refrigerating work.

Under atmospheric pressure, carbon dioxide boils at −110° F., or far below the temperatures required in ordinary refrigerating work. By reference to Table VI, it is seen that this refrigerant must be liquefied under about 900 pounds pressure, and in view of this it would seem advisable to carry a higher pressure on the suction line to the compressor than is carried with the machines using the other refrigerants, which are liquefied at lower condensing pressures. With a pressure of 342 pounds a square inch, carbon dioxide boils with a temperature of 5° F., its latent heat under these conditions being 121.5 B. T. U. This temperature is about as low as is usually required in refrigerating work, and gives, therefore, an index of the suction pressure that may be carried.

Passing by, then, those refrigerating agents that have been tried and found wanting, such as the various forms of ether and its com-

TABLE VII

Boiling Point and Latent Heat of Substances

Substance	Temperature of Boiling Point	Latent Heat B. T. U.	Specific Heat of Liquid
Nitric acid	248° F.	....	....
Saturated brine	226° F.	....	....
Water	212° F.	966	1.0000
Alcohol	173° F.	....	....
Chloroform	140° F.	....	....
Ether, sulphurous	95° F.	170	.5299
Ether, methyl	-10° F.	....	....
Sulphur dioxide	14° F.	168.7	.4100
Anhydrous ammonia	-28.5° F.	573	1.0058
Carbon dioxide	-110° F.	141	.9550

binations with sulphur dioxide; and also such agents as cryogene, acetylene, naphtha, and gasoline, it will be sufficient to give in some detail the properties of sulphur dioxide, carbon dioxide, and ammonia, which are in common use as refrigerants. Table VIII gives the qualities of these three refrigerants, and should be given careful study, as it shows up the good and bad points of each refrigerant in the clearest manner. It will be seen in the column next to the last, that the size of the sulphur dioxide compressor is about twenty times that of the carbon dioxide machine, or three times as large as the ammonia machine, which itself is something like five times the size of the carbon dioxide machine. The size of machine, of course, determines to a large extent the amount of friction losses. Other things being equal, the smaller the machine, the better.

The great disadvantages of the carbon dioxide machine are the high pressures required and the comparatively high specific heat of the liquid, which means that considerable of the cooling effect pro-

TABLE VIII

Qualities of Principal Refrigerants

Substance	Absolute Pressure in Lbs. per Sq. Inch at 0° F.	Heat of Vaporization per Pound at 0° F. (Old Data)	Heat of Vaporization per Pound at 0° F. (Lorenz)	Volume in Cubic Feet per Pound at 0° F. (Old Data)	Volume in Cu. Ft. per Lb. at 0° F. (Lorenz, Stetefeld)	Specific Heat of the Liquid (Old Data)	Specific Heat of the Liquid (Mollier, Lorenz, Stetefeld)	Heat of Vaporization per Cu. Ft. (Old Data)	Heat of Vaporization per Cu. Ft. at 14° F. (Stetefeld)	Loss in % Due to Cooling the Liquid (Lorenz, Stetefeld)	Useful Heat of Vaporiz. per Cu. Ft. Compr. Volume (Lorenz, Stetefeld)	Volume in Cu. Ft. per Lb. at 14° F. (Stetefeld)	Loss in % Due to Clearance, Friction, and Leakage (Stet.)	Relative Volume of Compressor for Equal Refrigeration	Loss Due to Cooling Liquid, per cent.
Carbon dioxide .... CO_2	310	123.2	115	0.27	0.3	1.00	0.79	444.7	483.5	17	400	0.44	19	3.24	0 81
Sulphur dioxide .... SO_2	10	171.2	161	7.35	7.4	.41	0.277	23.3	32	5	28	5.3	43	61.70	0.24
Ammonia .... NH_3	30	555.5	588	9.10	9.5	1.02	1.153	61.7	83.7	4.5	80	7.0	42	23.30	0.18

Tank Room of Pittsburg Ice Company.

duced by evaporation will be absorbed in reducing the temperature of the liquid from that of the condenser to that in the expansion coils or cooler. This is shown up clearly in the last column of the table, where it is seen that the loss due to cooling the liquid—as shown in percentage for every degree difference of temperature between the condenser and cooler—is less for ammonia than for any other liquid, the loss being high with carbonic acid. The chief point in favor of sulphur dioxide or sulphuric acid is the low pressure of its vapor, but the large size of machine required for this refrigerant has prevented its coming into general use.

Table IX gives the comparative refrigerating values of the refrigerants most in use, and shows at a glance the standing of each on the principal scores of value.

TABLE IX

Comparative Values of Three Refrigerants

Substance	Theoretical Compr. Vol. Required at 14° F. per Ton per Min. in Cu. Ft.	Loss Due to Radiation in the Refrigerator (Brine System)	Loss Due to Clearance in %	Loss Due to Friction in the Suction Pipe and Valves in %	Loss Due to Leakage of Piston and Valves in %	Total Loss in %	Actual Required Compr. Vol. at 14° F. per Ton per Min. in Cu. Ft.	Compressor Vol. Usually Allowed in Practice per Ton per Min. in Cu. Ft.	Theoretical Efficiency per I. H. P. in B. T. U.	Practical Efficiency per I. H. P. in B. T. U.
Carbon dioxide..... CO_2	0.49	6	5	4	9	24	0.60	0.61	14,855	13,370
Sulphur dioxide.... SO_2	7.04	6	6	14	23	49	10.556	10.6	16,673	13,300
Ammonia.......... NH_3	2.4	6	6	11	25	48	3.56	4.00	16,673	13,300

In case of the carbon dioxide machine, particular attention should be given to the temperature of the cooling water, as the critical temperature of this refrigerant is not much above the ordinary summer temperature of river water from natural sources of supply. Where the initial temperature at the condensers is 70° or more, it is advisable to increase the supply of cooling water so as to maintain an average condenser temperature of 75°. With temperatures greater than this, the efficiency of the carbon dioxide machine falls off as compared with the other two systems, but even with water at 90 degrees, the machine will develop about 70 per cent of its normal capacity. Theoretically the efficiency of the carbon dioxide machine is about 12 per cent less than that of the ammonia and sulphur dioxide machines, but practical compensating features enable the machine to make up for this. Stetfeld has found that the losses resulting from radiation in the clearance spaces of the refrigerator, the resistance of the gases on their way from the refrigerator to the compressor and in passing the suction valve, and friction and valve leakage, all together, average 49 per cent of the losses in the ammonia and sulphur dioxide systems, and not more than 25 per cent when carbon dioxide is used.

With the carbon dioxide machine, for example, the piston leakage averages about 9 per cent, as against 25 per cent in the ammonia and sulphur dioxide machines. In carbon dioxide machines, the great density of the gas permits making the valves, passages, and suction pipes large enough to materially reduce frictional losses, and yet not so large as is necessary in the other machines. In view of these facts, engineers have generally concluded that the practical efficiency of the three refrigerants mainly employed is about equal when each is used to best advantage. This is shown in the last column of Table IX. The choice therefore depends on circumstances and the local conditions in any case. To determine the fitness of the refrigerant for any set of conditions, its natural characteristics must be taken into consideration.

While ammonia and sulphur dioxide have a sharp penetrating odor, carbon dioxide is odorless; and if leaks are to be detected readily, the charge must be made odoriferous by adding a small amount of alcohol impregnated with camphor. At high temperatures, ammonia dissociates into its constituent gases and loses its value as a refrigerant, while the gases formed have a detrimental effect on the working of the machine. It is not definitely settled as to the exact temperature at which this dissociation takes place; but it is certain that above 900° F., ammonia gas is gradually decomposed, until at about 1,600°, complete dissociation takes place. It is believed that the action goes on to some extent at lower temperatures, but just under what conditions and to what extent are not definitely known.

Carbon dioxide does not decompose under any conditions, and is a fire extinguisher, while ammonia mixed with lubricating materials, etc., may support combustion in case of an explosion. Ammonia to some extent, and sulphur dioxide particularly, have a corrosive effect on metals, and the machines must be designed with this in view, a special close-grained steel cylinder being used in ammonia compressors of the best manufacture. Carbon dioxide is entirely neutral in its action on metals, and in the event of accidents resulting in large leaks, it has a marked advantage, as 8 per cent of the gas in the air can be inhaled with safety while less than 1 per cent of ammonia gas is dangerous to life. Large losses have frequently resulted by damage to goods in store where ammonia has escaped, but this cannot happen where carbon dioxide machines are employed, as the gas does no damage.

Sulphur dioxide was used as a refrigerant in the early stages of modern machine development after the ether machine had its day. Owing to the high cost of ether, and other disadvantages connected with its use—principally its inflammability—investigators took up sulphur dioxide and studied its properties as a refrigerant. It was found to require a higher condensing pressure than ether, but did not need to be evaporated under a vacuum, so that the compressor could be made smaller for a given capacity. On account of the higher condensing pressure, it was necessary to build the compressor stronger than had formerly been done, and more attention was given to the elimination of clearance spaces. Even though the machine for use with this refrigerant is smaller than that formerly used with ether as a refrigerant, it is still much larger than ammonia and carbon dioxide machines, as has been shown in the tables. Table X gives the properties of sulphur dioxide.

Carbon dioxide, although not used extensively until within a comparatively recent period, is coming into favor, for it is made as a by-product in certain industries and can be obtained cheaply. The gas is not readily absorbed by water or by lubricating materials; and, not being easily dissociated, the system using it remains free of non-condensable gases and in efficient condition. Table XI gives the properties of this refrigerant.

Ammonia, the most widely used of all refrigerants, is composed of one part of nitrogen in combination with three of hydrogen, this

TABLE X

Properties of Saturated Sulphur Dioxide

Temperature of Ebullition in Deg. F.	Absolute Pressure in Lbs. per Sq. In.	Total Heat Reckoned from 32° Fahr.	Heat of Liquid Reckoned from 32° Fahr.	Latent Heat of Vaporization	Density of Vapor, or Weight of 1 Cubic Ft.
Deg. F.	Lbs.	B. T. U.	B. T. U.	B. T. U.	Lbs.
-40	3.16	155.22	-17.76	172.98	.048
-31	4.23	156.39	-16.55	172.94	.062
-22	5.56	157.55	-15.05	172.60	.079
-13	7.23	158.69	-13.26	171.95	.009
- 4	9.27	159.82	-11.18	171.00	.124
5	11.76	160.93	- 8.82	169.75	.154
14	14.75	162.02	- 6.17	168.19	.190
23	18.31	163.10	- 3.23	166.33	.232
32	22.53	164.16	0.00	164.16	.282
41	27.48	165.24	3.52	161.69	.341
50	33.26	166.24	7.32	158.92	.410
59	39.93	167.25	11.41	155.84	.491
68	47.62	168.25	15.79	152.46	.584
77	56.39	169.23	20.45	148.78	.692
86	66.37	170.20	25.41	144.79	.819
95	77.64	171.15	30.65	140.50	.965
104	90.32	172.08	36.18	135.90	1.131

TABLE XI

Properties of Saturated Carbon Dioxide

Temperature of Ebullition in Deg. F.	Absolute Pressure in Lbs. per Sq. In.	Total Heat from 32° F.	Heat of Liquid from 32° F.	Latent Heat of Vaporization	Density of Vapor, or Weight of 1 Cu. Ft.
-22	210	98.35	-37.80	136.15	2.321
-13	249	99.14	-32.51	131.65	2.759
- 4	292	99.88	-26.91	126.79	3.265
5	342	100.58	-20.92	121.50	3.853
14	396	101.21	-14.49	115.70	4.535
23	457	101.81	- 7.56	109.37	5.331
32	525	102.35	0.60	102.35	6.265
41	599	102.84	8.32	94.52	7.374
50	680	103.24	17.60	85.64	8.708
59	768	103.59	28.22	75.37	10.356
68	864	103.84	40.86	62.98	12.480
77	968	103.95	57.06	46.89	15.475
86	1,080	103.72	84.44	19.28	21.519

being the only proportion in which these two gases combine. *Anhydrous ammonia* thus formed, when dissolved in water, gives the *aqua ammonia* of commerce, used in absorption machines. When heat is applied to this aqua ammonia in the generator of the absorption

TABLE XII

Properties of Saturated Ammonia Gas

DE VOLSON WOOD AND GEO. DAVIDSON

Gauge Pressure Pounds Per Square Inch	Absolute Pressure Pounds Per Square Inch	Temperature Degrees F.	Absolute Temperature Degrees F.	Latent Heat of Evaporation in Thermal Units	Volume of 1 Pound Vapor in Cubic Feet	Weight of 1 Cubic Foot of Vapor in Pounds	Volume of 1 Pound of Liquid in Cubic Feet	Weight of 1 Cubic Foot of Liquid in Pounds
-4.01	10.69	-40	420.66	579.67	24.38	.0410	.0234	42.589
-2.39	12.31	-35	425.66	576.68	21.32	.0469	.0236	42.337
-0.57	14.13	-30	430.66	573.69	18.69	.0535	.0237	42.123
1.47	16.17	-25	435.66	570.68	16.44	.0608	.0238	41.858
3.75	18.45	-20	440.66	567.67	14.51	.0690	.0240	41.615
6.29	20.90	-15	445.66	564.64	12.83	.0779	.0241	41.374
9.10	23.80	-10	450.66	561.61	11.38	.0878	.0243	41.135
12.22	26.92	- 5	455.66	558.56	10.12	.0988	.0244	40.900
15.67	30.37	0	460.66	555.50	9.03	.1107	.0246	40.650
19.46	34.16	5	465.66	552.43	8.07	.1240	.0247	40.404
23.64	38.34	10	470.66	549.35	7.23	.1383	.0249	40.160
28.24	42.94	15	475.66	546.26	6.49	.1541	.0250	39.920
33.25	47.95	20	480.66	543.15	5.84	.1711	.0252	39.682
38.73	53.43	25	485.66	540.03	5.27	.1897	.0253	39.432
44.72	59.42	30	490.66	536.91	4.76	.2099	.0255	39.200
51.22	65.92	35	495.66	533.78	4.31	.2318	.0256	38.940
58.29	72.99	40	500.66	530.63	3.91	.2554	.0258	38.684
65.96	80.66	45	505.66	527.47	3.56	.2809	.0260	38.461
74.26	88.96	50	510.66	524.30	3.24	.3084	.0261	38.226
83.22	97.92	55	515.66	521.12	2.96	.3380	.0263	37.994
92.89	107.59	60	520.66	517.93	2.70	.3697	.0265	37.736
103.33	118.03	65	525.66	514.73	2.48	.4039	.0266	37.481
114.49	129.19	70	530 66	511.52	2.27	.4401	.0268	37.230
126.52	141.22	75	535.66	508.29	2.09	.4791	.0270	36.995
139.40	154.10	80	540.66	505.05	1.92	.5205	.0272	36.751
153.18	167.88	85	545.66	501.81	1.77	.5649	.0273	36.509
167.92	182.62	90	550.66	498.55	1.64	.6120	.0275	36.258
183.65	198.35	95	555.66	495.29	1.51	.6622	.0277	36.023
200.42	215.12	100	560.66	492.01	1.39	.7153	.0279	35.778
218.28	232.98	105	565.66	488.72	1.289	.7757	.0281	
237.27	251.97	110	570.66	485.42	1.203	.8312	.0283	
258.7	272.14	115	575.66	482.41	1.121	.8912	.0285	
275.79	293.49	120	580.66	478.79	1.041	.9608	.0287	
301.46	316.16	125	585.66	475.45	.9699	1.0310	.0289	
325.72	340.42	130	590.66	472.11	.9051	1.1048	.0291	
350.46	365.16	135	595.66	468.75	.8457	1.1824	.0293	
377.52	392.22	140	600.66	465.39	.7910	1.2642	.0295	
405.79	420.49	145	605.66	462.01	.7408	1.3497	.0297	
435.5	450.20	150	610.66	458.62	.6946	1.4396	.0299	
466.84	481.54	155	615.66	455.22	.6511	1.5358	.0302	
499.70	514.50	160	620.66	451.81	.6128	1.6318	.0304	
534.34	549.04	165	625.66	448.39	.5765	1.7344	.0306	

One atmosphere in this table is equal to a pressure of a column of mercury 29.9 in. high
Specific heat of ammonia gas and vapor at constant pressure. = 0.508
The same at constant volume.................................. = 0.3913
Weight of 1 cubic foot liquid ammonia at 32 degrees F..... = 39.108 pounds
Volume of 1 pound liquid ammonia at 32 degrees F........ = 0.02557 cubic feet
Specific heat of liquid ammonia.................. = 1.01235 + 0.008378 t^2.

machine, the gas is distilled off, because it evaporates at a lower temperature than the water; and, when freed from all moisture in the analyzer and rectifier, the gas is again in the anhydrous form. This anhydrous gas, when liquefied by cooling under pressure and allowed to evaporate at atmospheric pressure, has a temperature of − 28.5°. When subjected to a temperature of − 94° F., the liquid anhydrous ammonia freezes solid. Table XII gives some of the properties of ammonia.

Aqua ammonia, or *ammonia liquor*, is the ordinary ammonia known to commerce, as distinguished from anhydrous ammonia either in the form of gas or as liquid. It is nothing more than a solution of the anhydrous gas in water. At 32° F. and atmospheric pressure, water absorbs 1,140 times its volume of ammonia gas, and the amount of the gas that can be absorbed under other conditions is governed by the temperature of the water and the pressure of the gas, as shown in Table XIII.

TABLE XIII

Solubility of Ammonia in Water

SIMS

Absolute Pressure in Lbs. per Sq. In.	32° F.		68° F.		104° F.		212° F.	
	Lbs.	Vols.	Lbs.	Vols.	Lbs.	Vols.	Grams	Vols.
14.67	0.899	1.180	0.518	.683	0.338	.443	0.074	.970
15.44	0.937	1.231	0.535	.703	0.349	.458	0.078	.102
16.41	0.980	1.287	0.556	.730	0.363	.476	0.083	.109
17.37	1.029	1.351	0.574	.754	0.378	.496	0.088	.115
18.34	1.077	1.414	0.594	.781	0.391	.513	0.092	.120
19.30	1.126	1.478	0.613	.805	0.404	.531	0.096	.126
20.27	1.177	1.546	0.632	.830	0.414	.543	0.101	.132
21.23	1.236	1.615	0.651	.855	0.425	.558	0.106	.139
22.19	1.283	1.685	0.669	.878	0.434	.570	0.110	.140
22.16	1.336	1.754	0.685	.894	0.445	.584	0.115	.151
24.13	1.388	1.823	0.704	.924	0.454	.596	0.120	.157
25.09	1.442	1.894	0.722	.948	0.463	.609	0.125	.164
26.06	1.496	1.965	0.741	.973	0.472	.619	0.130	.170
27.02	1.549	2.034	0.761	.999	0.479	.629	0.135	.177
27.99	1.603	2.105	0.780	1.023	0.486	.638	.....	....
28.95	1.656	2.175	0.801	1.052	0.493	.647	.....	....
30.88	1.758	2.309	0.842	1.106	0.511	.671	.....	....
32.81	1.861	2.444	0.881	1.157	0.530	.696	.....	....
34.74	1.966	2.582	0.919	1.207	0.547	.718	.....	....
36.67	2.020	2.718	0.955	1.254	0.565	.742	.....	....
38.60	.....	.....	0.992	1.302	0.579	.764	.....	....
40.53	.....	.....	.....	.....	0.594	.780	.....	....

As gas is absorbed in the water, its density changes, the solution being lighter than water or of less specific gravity, and it is this fact that affords means of measuring the density of aqua ammonia. Such measurements are made by an instrument called a *hydrometer*, which consists of a mercury tube having a bulb near the middle and a second bulb at the lower end, the first bulb serving to give the instrument buoyancy, while the one at the bottom, which is partially filled with mercury, gives balance and holds the tube vertical. Fig. 1 shows the instrument in use in a vessel of liquid the density of which it is desired to ascertain.

Fig. 1. Hydrometer in Use.

As seen in the illustration, the upper part of the tube is graduated, the ordinary method of graduating being that devised by Baumé, in which the point on the tube at the surface of a liquid composed of 10 parts salt and 90 parts water is marked "0," and the point to which the tube sinks when put in pure distilled water is marked "10" degrees. The space between the two fixed points thus determined is divided into ten parts, and the graduations are continued to the top of the tube or stem. In another method of graduating, the reading for distilled water is marked "0" instead of "10" degrees, but this instrument is not used extensively. To avoid calculation in using the hydrometer, it is convenient to use the data given in Table XIV, which shows the number of parts of ammonia gas in 100 parts of the solution, and the specific gravity of the liquid, as well as the hydrometer reading corresponding thereto.

Tests of Refrigerants. Sulphuric acid or sulphur dioxide is not used to any extent at the present time; but where used, the only tests to be made are for chemical purity, and these are best made by a chemist. Carbon dioxide is ordinarily contaminated by air, carbon bisulphide, hydrocarbons, water, and oil or grease, all of which should be eliminated as far as possible. The chief trouble caused by water is due to its tendency to freeze and clog the system with icicles that interfere with the circulation of the gas. Its presence may be detected by a test made with a piece of filter paper which is prepared by soaking in a solution of copper sulphate and drying thoroughly, when it has a slightly greenish color. The test is performed by holding the paper in a blast of the carbon dioxide allowed to escape from a drum by opening a valve, this being done before the refrigerant is charged into the system. If the color of the paper changes to blue, it is evidence of too much moisture in the gas; and to get rid of the water, the drum is turned upside down, when the water will settle to the bottom of the drum, underneath the carbon dioxide, and may be drained off by opening the valve.

TABLE XIV

Strength of Aqua Ammonia

Percentage of Ammonia by Weight	Specific Gravity	Degrees Baumé	
		Water 10°	Water 0°
0	1.000	10.0	0
1	.993	11.0	1.0
2	.986	12.0	2.0
4	.979	13.0	3.0
6	.972	14.0	4.0
8	.966	15.0	5.0
10	.960	16.0	6.0
12	.953	17.1	7.0
14	.945	18.3	8.2
16	.938	19.5	9.2
18	.931	20.7	10.3
20	.925	21.7	11.2
22	.919	22.8	12.3
24	.913	23.9	13.2
26	.907	24.8	14.3
28	.902	25.7	15.2
30	.897	26.6	16.2
32	.892	27.5	17.3
34	.888	28.4	18.2
36	.884	29.3	19.1
38	.880	30.2	20.0

Where water is present in the piping and connections of a machine using carbon dioxide, it will gradually work into the suction line at the compressor; and when the machine is shut down, the water may be drawn off, care being taken to disconnect the expansion coils and the suction line before opening the drain-cock on the line at the compressor. As a rule, there is not enough oil in carbon dioxide to cause trouble, but care should be taken not to allow the lubricating material to foul the pipes and connections.

Non-condensable gases give the most trouble, and should never be allowed in the system in excess of 3 per cent. A test for such gases is performed by drawing a given quantity of the refrigerant into a clean glass tube, and absorbing the carbon dioxide gas with caustic potash shaken up in the tube while the end is closed with the finger. As absorption goes on, more alkali is added by dipping the mouth of the tube into the potash solution, and when absorption ceases, the permanent gases remaining may be compared with the quantity of the refrigerant drawn off. If air and other gases are shown in excess of 3 per cent by this test, the drum containing carbon dioxide should be set upright and the gas vented off until a second sample taken meets the test.

Ammonia Test. In discussing aqua ammonia, the method of testing its density by the hydrometer has been explained in detail. So long as such ammonia is of the desired density and is composed of pure anhydrous ammonia absorbed in reasonably pure water, there is nothing further to be desired. It is important, however, that the anhydrous gas used in making the solution of aqua ammonia be pure; and still more important that there be no doubt about the purity of such anhydrous ammonia when it is to be used in compression machines. As in the case of carbon dioxide, only a small percentage of non-condensable or so-called *permanent* gases should be tolerated.

Some of the impurities found in ammonia damage the rubber gaskets and packing used in the plant, and oil in the system coats the inside surfaces of the pipes, etc., so as to interfere with efficient transfer of heat. This oil is removed by blowing out the coils with steam and air after the ammonia charge has been removed. Each part of the plant may be taken in its turn, the charge being pumped over into another part of the system *ad interim* by means of the by-passes with which all modern plants are provided. The purity of anhydrous ammonia is entirely a matter of keeping the system clean in this way, purging off the gases at the condenser, and distilling the ammonia at periodical intervals. At overhauling time, the charge of a plant should be withdrawn and distilled, or sent to the ammonia factory to be *re-worked* in case the plant has no distilling facilities. All large plants should be provided with such apparatus. In absorption plants, where impurities are present, the ammonia generator may be used to distill the charge until practically all the anhydrous ammonia is removed from the liquor. This weak aqua ammonia is then discarded, and the solution made up by adding clean water to the system. A practical engineer judges the purity of his charge by the manner in which it acts, and has no time to perform tests which are entirely up to the chemist and his laboratory.

SYSTEMS OF REFRIGERATION

There are two kinds of refrigerating machines in common use —namely, *first,* those machines producing refrigeration by means of the adiabatic *expansion* of a non-condensable gas performing work, as the cold-air machine; and *second,* all those machines of various types which operate by the vaporization of a volatile liquid—in other words, *latent-heat* machines. There are three main types of these latter machines, known respectively as the *vacuum,* the *absorption,* and the *compression* processes or systems. In some cases the compression and absorption systems are used in combination in the same plant; and in other cases absorption machines using a dual or combination liquid refrigerant are employed.

THE COLD-AIR MACHINE

This type of refrigerating machine, of which there are a number of constructions built principally in England and continental Europe, was used extensively in the early applications of mechanical refrigeration on shipboard; but the machine is inherently of low efficiency, owing to the fact that the heat capacity of air is only about 0.2377 B. T. U. per pound, or 0.034 B. T. U. per cubic foot for each degree F. of temperature. On account of this fact, the lower limit of temperature must be made very low, and this calls for a comparatively large amount of work. Also, large quantities of air must be handled so that the air machines are of great size for the refrigerating duty produced, as compared with machines using volatile liquids. This of course means large frictional losses, heavy wear and tear, and increased maintenance cost; so that, with heavy first cost and maintenance cost standing against the compressed-air machine, it is not surprising that this type of apparatus has been in little favor since latent-heat machines have been perfected.

Aside from this, there is the difficulty with moisture in the air, which freezes and clogs up the system. As all air contains a certain amount of moisture, which is precipitated to a greater or less extent as the temperature is reduced in the refrigerating machine, the difficulty with clogged passages is unavoidable except where special apparatus for drying the air is added to the system. This of course means additional complications and increased cost, so that, all things considered, the compressed-air machine has been in little favor in the United States, where the design and construction of latent-heat machines have been so highly perfected. Even in the Old Country, where air machines were formerly in high favor, they are being gradually replaced by the machines using volatile liquids.

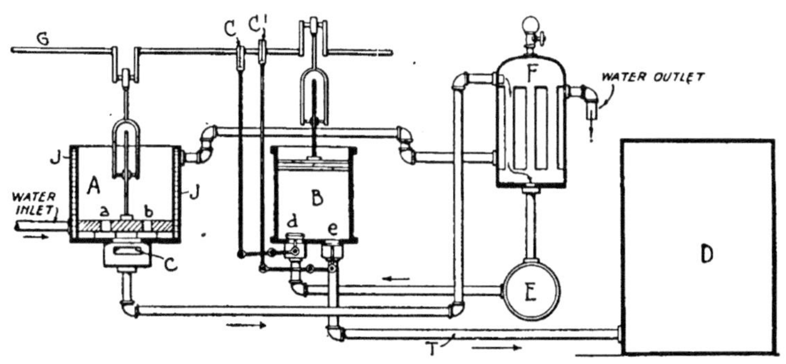

Fig. 2. General Arrangement for Air-Ice Machine.

Arrangement of Air System. Fig. 2 shows the general arrangement of the compressed-air machine, the equipment consisting of a single-acting compression cylinder *A*; an expansion cylinder *B*, which is also single-acting; and a condenser *F*, through which water circulates and cools the compressed air, without, however, condensing it. At *D* is shown a cold-storage room or refrigerator kept cool by the machine. In cylinder *A* the piston is provided with suction valves *a* and *b*, which open inward, and a discharge valve *c*, opening outward, as shown. Around the cylinder is the water jacket *J*, which removes part of the heat generated by compressing the gas. The diameter of cylinder *B* is a little less than that of cylinder *A*, and its piston is solid. In the cylinder-head, two valves are provided, designated in the figure by *d* and *e*, and operated by the eccentrics *C* and C^1, one of these being a suction and the other a discharge valve.

Connections are made from the receiver *E* to the condenser and to the inlet valve *d* of the expansion cylinder.

In operation, air at atmospheric pressure is taken into the cylinder *A* through the valves *a* and *b*, and on being compressed is discharged through the valve *c* to the condenser *F*, where it is cooled. At the same time, the valve *d* in the expansion cylinder is kept open by the eccentric *C*, so that air passes from the receiver *E* into the cylinder until it is filled, when the valve closes. The eccentrics are so arranged that the valve *d* closes at the beginning of the compression stroke of the cylinder *A*, so that the air in cylinder *B*, by its expansion, does work and helps the steam cylinder (not shown in the figure) to drive the compressor cylinder. Cold air from the expansion cylinder cools the refrigerator *D*. In early applications of the compressed-air machine, the cooled air coming from the expansion cylinder was discharged directly into the refrigerator; but it was found that the moisture in the air, as well as the oil from the machine, by which the air was contaminated, affected the goods in storage so as to make them unpalatable. Owing to this fact, the modern air machines are arranged to return the air to the machine after passing it through the cooling coils in the refrigerators or cold stores so that the air is used in a continuous cycle.

Commercial Form of Air Machine. Figs. 3 and 4 give general views of the Allen dense-air ice machine made by H. B. Roelker of New York City. There are three main cylinders having slide valves not unlike those used on steam engines, *A* representing a steam cylinder arranged to drive the cylinder *B* in which air at 14.7 pounds pressure is compressed to 150 pounds. The steam cylinder is operated by a single *D*-valve and controlled by a throttling governor suitable for use on shipboard, where these machines are chiefly employed, they being in fact the standard for the vessels of the United States Navy. Power from the steam cylinder is transmitted by connecting rod and disc crank, through the shaft *H*, to a center crank arranged to drive the compressor cylinder *B*. On the opposite end of the crank-shaft is a third crank disc to which the connecting rod of the expansion cylinder is attached. In this cylinder the compressed air, after passing through the cooling coil *C*, is expanded until its temperature is 40 to 65 degrees F. below zero. *F* is a plunger piston pump for circulating cooling water around the coils of the tank *C*, and *G* is a priming air-pump.

In the *location of the cylinder cranks*, the crank driving the air-

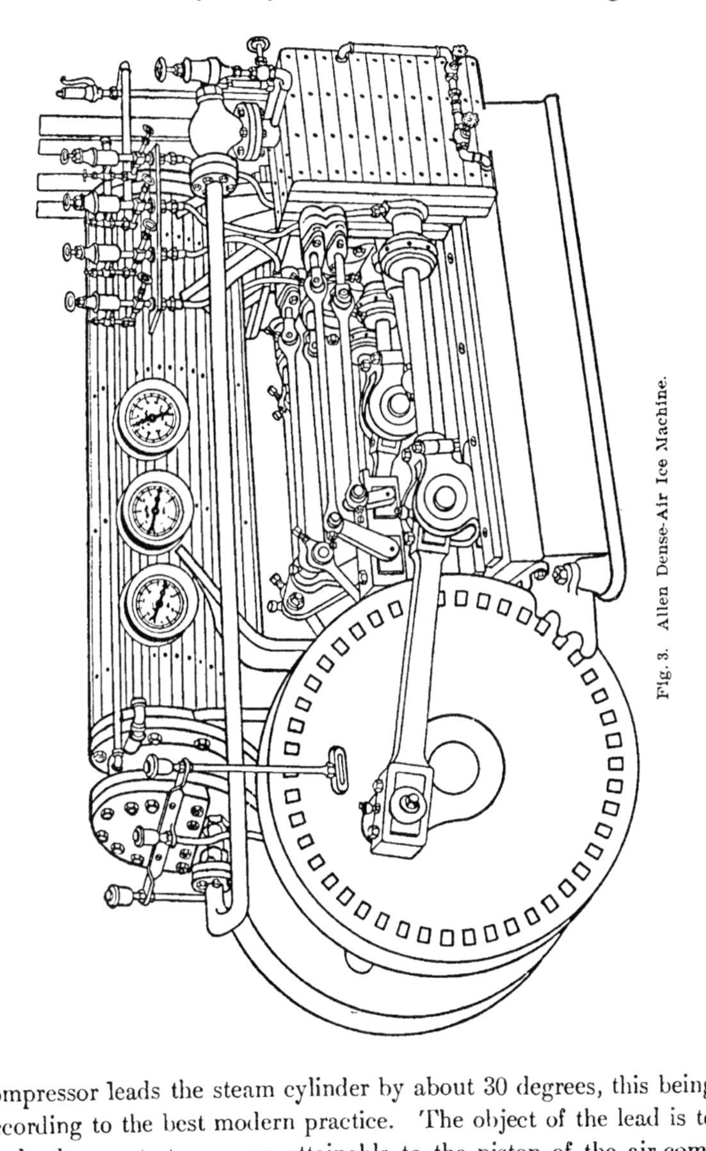

Fig. 3. Allen Dense-Air Ice Machine.

compressor leads the steam cylinder by about 30 degrees, this being according to the best modern practice. The object of the lead is to apply the greatest pressure attainable to the piston of the air-com-

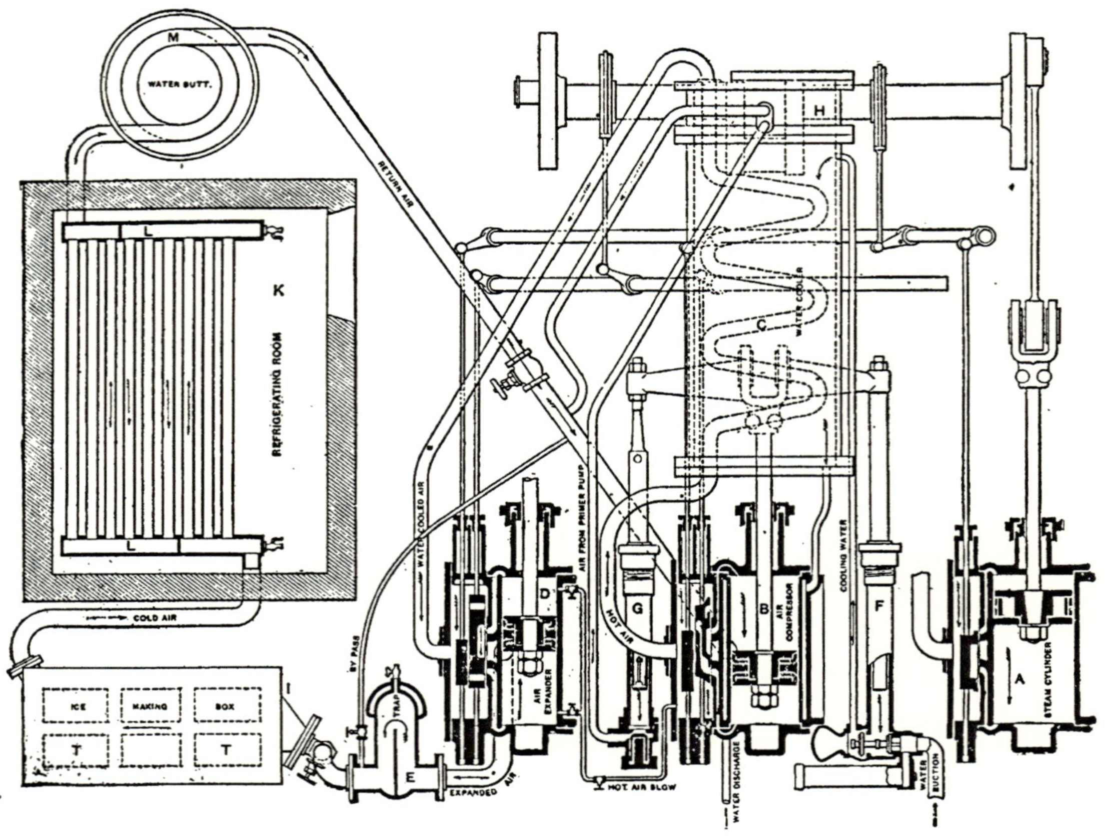

Diagram of Construction and Connection of Allen Machine

pressor at the time it is completing its stroke, when the angle of the crank-pin nears the position exerting the greatest effort on the crank and previous to the time of cut-off on the steam cylinders. By this method a much lighter fly-wheel than otherwise can be used, since the power developed by the engine is applied directly through the shaft, and not transmitted to the fly-wheel to be given off when the compressor cylinder is taking the maximum amount of power and the steam cylinder is approaching the end of its stroke.

Compressor Cylinder. Suction is through an opening in the bottom of the valve chest, located in the same way as the exhaust on a slide-valve engine, while discharge is through the face of the valve chest by a passage similar to that used for steam supply in such an engine. All valves are of the slide-valve type. The advantages of this type over the poppet valves used on ammonia and other compressors are—comparatively noiseless action, no hammering of seats or face of valves in closing, and absence of unbalanced pressure on the valve to be overcome by the engine.

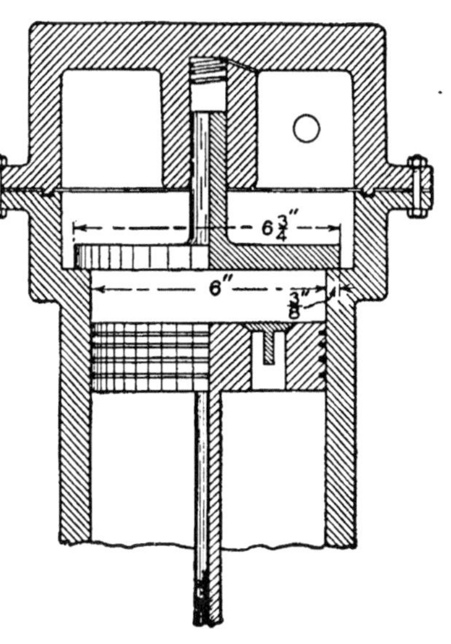

Fig. 5. Diagram Illustrating Unbalanced Valve Pressure.

An illustration of unbalanced pressure is had in a study of Fig. 5. In this diagram it is assumed that the design calls for a 6-inch compressor cylinder with a valve seat of $\frac{3}{8}$-inch face, giving a bearing surface on top of the valve $6\frac{3}{4}$ inches in diameter. The area of the cylinder is 28.274 square inches, while the area of the top of the valve, including the seat, is 33.183 square inches. With a working pressure of 150 pounds, the total pressure to open the valve is 150×33.183, or 4,977.45 pounds. The power exerted on the under side of the valve when the pressures are equal, is 28.274×150, or 4,241.1 pounds; so that in order to open the valve, there must be $4{,}977.45 - 4{,}241.1$

= 636.35 pounds additional pressure applied. This means 28.53 pounds a square inch in excess, or a total pressure of 178.53 pounds a square inch on the piston. This pressure is overcome by placing springs or compressing chambers on top of the valve, the latter being preferred. The loss due to this unbalanced pressure may be as much as 20 per cent of all the energy required to compress the gas, besides the wear and tear on the valves.

The valves of the Allen machine are operated by two rocker shafts which are controlled by eccentrics in the manner already mentioned. The rocker nearest the crankshaft operates the valve admitting steam to the cylinder *A*, and also the rider valve on each of the air cylinders. The other rocker shaft operates the main valves of the air cylinders; and an arm extending horizontally from the cross-head of the compressor cylinder operates the charging air-pump and the water-circulating pump. The action in the expansion cylinder *D* is the same as in a steam cylinder having slide-valves. With an initial pressure of 260 pounds and final pressure of 60 pounds, the tempertaure of the cold air will be from 70 to 90 degrees below zero, depending on the condition of the machine. Should the pressure in the expanding cylinder become greater than that in the discharge chamber due to the distortion of rods or slipping of eccentrics, the valves will spring back from their seats and relieve the pressure. Oil and water traps and blow-off cocks are arranged in different parts of the system as shown in Fig. 4. The cold air is utilized first in an ice box, filled with calcium brine, and then in a refrigerator, the air passing finally through a coil in a water-cooler before it returns to the suction of the compressor cylinder.

Operation of the Allen machine, according to the rules used in the United States Navy, should be as follows:

On starting the machine, have the blow-valves of the expansion cylinder and the pet-cocks of the various traps open until no more grease or water discharges. The two 1½- or 2-inch valves of the main pipes must be open, and the 1-inch by-pass pipe closed; also the ½-inch hot-air valves from the compressor to the expander cylinder must be closed.

Be sure that the circulating water is in motion. The full pressure is 60 to 65 lbs. low pressure, and 210 to 225 lbs. high pressure.

During the running, open the pet-cocks of the water-trap in order to take the water out of the air from the primer pump frequently enough so that it will never be more than half-filled. If the water should be allowed to enter the main pipes, it is liable to freeze and clog at the valves. By keeping all

TANGYE FRAME HORIZONTAL LINDE COMPRESSOR.
Fred W. Wolf Company.

stuffing-boxes well lubricated by the lubricator cups, the pressures are easily maintained with but little screwing-up of the packing. If the low-air pressure is not maintained, the fault is almost always due to leaks at the stuffing-boxes. Under all circumstances it is due to some leak into the atmosphere, as the primer pump valves have never yet been found to be at fault.

The packing of valve-stems and piston-rods consists of a few inner rings of Katzenstein's soft metal packing, then a hollow greasing ring, then soft fibrous packing (Garlock packing).

The sight-feed lubricators of the compressor and expander should use only a light, pure, mineral machine oil from which the paraffine has been removed by freezing—usually three drops per minute in the compressor, and one or two in the expander.

The pistons of the compressor and expander cylinders are packed with cup leathers, which commonly last about one or two months of steady work. When these leathers give out, the high pressure decreases in relation to the low pressure, and the apparatus shows a loss of cold. A leak at any other point of high pressure into low pressure will have the same effect. These packing leathers are made of thick kip leather, or of white oak-tanned leather of somewhat less than ⅛-inch thickness. They are cut ⅝-inch larger in diameter than the cylinders. The leathers must be kept soaked with castor oil and must be well soaked in that before using; and a tin box containing spare leathers and castor oil must be kept on hand.

Once, or sometimes twice a day, it is necessary to clean the machine by heating it up and blowing out all the oil and ice deposits. This is done as follows: The 1-inch valve of the by-pass is opened. Then the two 1½- or 2-inch valves in the main pipes are closed; then the two ½-inch valves in the hot-air pipe from the compressor chest to the expander are opened, and the 1¼-inch valve of the expander inlet is partially closed; then the live steam is let slowly into the jacket of the oil-trap, keeping the outlet from the steam jacket open enough to drain the condensed steam.

Run in this manner for about one-half hour; during this time, frequently blow out the bottom valve of the oil trap, also the blow-off from the expander, until everything appears clean. Then shut off the steam, and drain connections of the jacket of the trap and the hot-air pipe from the compressor to the expander. Then open the two 1½-inch valves in main pipes. Then close the 1-inch by-pass pipe and all pet-cocks, and run as usual.

Whenever opportunity offers to blow out the manifolds of the meat-room and the ice-making box (that is, whenever they are thawed), this should be done. If it is suspected that a considerable quantity of oil and water has got into the pipe system and is clogging the areas and coating the surfaces, the pipes can be cleaned by running hot air through them as is done during the daily cleaning of the machine. The oil and water are then drawn off at the bottom of the ice-making box and the manifolds of the refrigerating coils.

The clearance of the two air pistons and of the primer plunger is only ⅛-inch; therefore not much change of piston-rods and connecting rods is permissible, and when the piston nuts are unscrewed to change the piston leathers, the rod should be watched that it does not unscrew from the cross-head. Whenever it is noticed that the brine freezes, more chloride of calcium should be added, and should be well stirred into the brine.

VACUUM PROCESS

In machines working on the vacuum system, the essential feature, as the name implies, is production of a vacuum in the vessel containing the liquid to be cooled. Probably the first machine constructed on this principle was that invented by Dr. Cullen in 1755. His apparatus has served as a pattern for all later types. In some aspects, the vacuum machine, as to its principle of operation, resembles the latent-heat machines using volatile liquids, in all of which evaporation is had under comparatively low pressures, the difference being that, in the case of the vacuum process, the pressure is very much lower than with the other latent-heat machines. Thus, with an ammonia compression machine, the back-pressure in the suction coils may be from 15 to 25 pounds, or possibly more in some cases; whereas, with the vacuum process, it is necessary to have the vacuum as good as possible with the best pumps made, there being only about 0.1 pound pressure per square inch, this being necessary where water is to be frozen by evaporation of water, as its temperature of evaporation is too high to get good freezing for pressures any higher than this.

In ordinary practice, the discharge of the vacuum pump is connected to a condenser in which the pressure is about 1.5 pounds. About one-fifth the water put into the vacuum chamber is evaporated, so that for each five tons of water used there is a net return of about four tons of ice. Under the best conditions less water will be evaporated with a better return. About 340 gallons of condensing water is required per ton of refrigeration, assuming a temperature range of 30 degrees. The vapor cylinder must be about 150 times as large as the cylinder of an ammonia latent-heat machine of equal capacity. This enormous size of cylinder places the vacuum machine out of competition with other apparatus on the market; and to get around this difficulty inventors have developed various modifications of the machine, using various substances to absorb the vapor chemically, the most practical apparatus of this kind being that employing sulphuric acid as an absorbent.

Fig. 6 is a diagram showing the arrangement of parts in the sulphuric acid vacuum machine, the air-pump P being employed to produce a vacuum in the chamber A, so that the water in this chamber begins to evaporate. Vessel B contains sulphuric acid, which is allowed to fall into vessel C in the form of spray. Since the acid has a great affinity for water, it will absorb the vapor in the chamber A, and, after being thus diluted, will flow to the vessel D. An injector F is arranged to supply fresh water to the vessel A, and at the same time to draw the brine from the coils E, brine being used in A as it is not desired to do the freezing direct in this vessel. Brine from the vessel A is circulated through the coil E in the refrigerator, in a closed cycle. A pump (not shown in the illustration) is used to return the acid from the reconcentrator to vessel B to be used over again. The chief objection to this type of machine is the fact that the pipes must be made of lead or lined with lead in order to prevent corrosion.

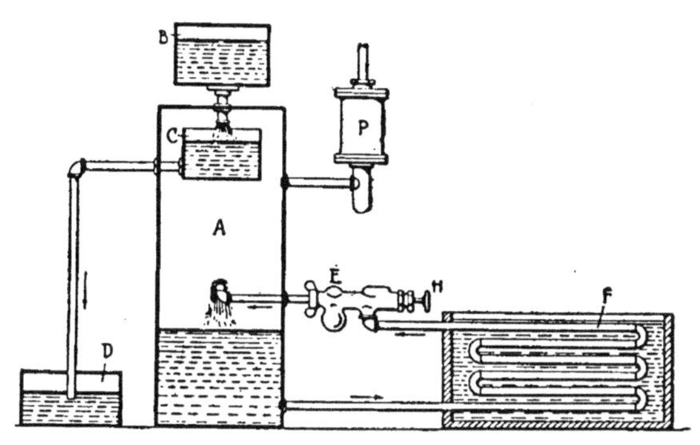

Fig. 6. Diagram of Sulphuric Acid Vacuum Machine.

Aside from the size of the apparatus and the difficulty in using the acid, the vacuum system involves considerable complication in the pumping apparatus, as it is difficult to keep the packing glands and joints of the vacuum pumps tight for the low pressure required. Ice produced by this system is not of good quality, it being porous or opaque unless frozen in specially prepared moulds previously chilled, which process is too expensive for commercial use. Owing to the necessity for circulating the acid and cooling water, about 10 or 12 tons of liquid must be handled for each ton of ice produced; and in doing this about 180 pounds of coal are burned, this being the fuel necessary to supply steam to the pumps and heat the coils of the acid evaporator used in reconcentrating.

Vacuum Pump. Many pumps have been developed for use in the vacuum process of refrigeration; but one typical of others is that of Lange, which is illustrated in Fig. 7. In this pump three pistons are employed, as indicated in the illustration by the letters A, B, and C, the arrangement being such that the pistons are in vertical line one above the other, with each working in its own cylinder. The three cylinders are connected by valves as shown, the valves being sealed with oil and so arranged that each of the pistons exhausts the contents of the cylinder below it. On leaving the top cylinder, the mixed oil and air is discharged into the separator *D*, from which the air is allowed to escape into the atmosphere, while the oil passes to the receptacle below. From this receptacle the oil is returned to the chamber in the lower end of the pump as needed.

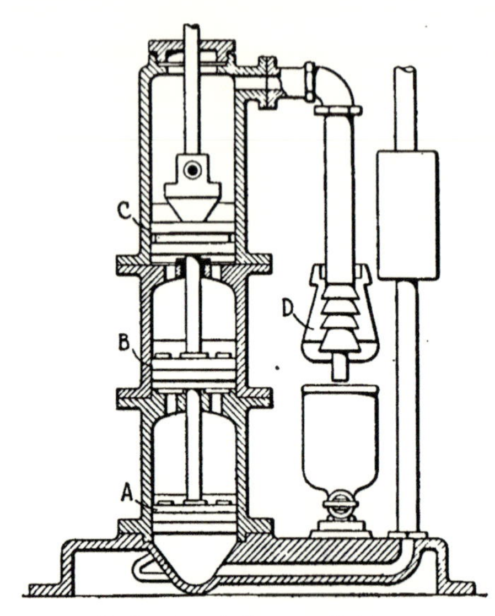

Fig. 7. Lange's Vacuum Air Pump.

About the only vacuum plant of any importance in the United States is the 100-ton establishment of the Patten Vacuum Ice Machine Co., Baltimore, Md. This plant has been in operation for some years.

ABSORPTION SYSTEM

This depends for its operation on the affinity of certain liquids for each other and on the process of *fractional distillation*—that is, the distilling of one or more liquids from a solution under such conditions of temperature and pressure that the other liquids do not vaporize and are left behind. When a liquid can be vaporized readily and the resulting vapor is easily absorbed by another liquid, we have the complete set of conditions for the absorption system. In some respects this system is like the vacuum process, while in other respects it is similar to the compression system. The difference from the one is the absorption of the gas by water instead of by acid; from the other, the compression of gas by direct application of heat in the generator instead of by mechanical action, as in the compressor. Mention has already been made of the great affinity of ammonia for water, on account of which practically all absorption machines use aqua ammonia for the working medium.

For the invention of the absorption process, the world is indebted to Ferdinand Carré, brother of Edmund Carré, inventor of the sulphuric acid machine, who made a study of the phenomena of evaporation and, about the year 1850, evolved the primitive absorption apparatus, shown in Fig. 8, consisting of two strong iron jars connected together by an iron pipe. Jar *A* was used as the ammonia still or generator and had placed in it a quantity of strong ammonia solution. A spirit lamp placed under the jar supplied the heat necessary for distillation, and an air cock *I* afforded the means of venting off the air as soon as the jar became heated. With continued application of heat, the pressure increased in all parts of the system. Jar *C* was placed in a tank and surrounded by cold water *D*, the supply being changed constantly so as to maintain the original temperature at about 60° F. Ammonia vaporized in the jar *A* passed to the jar *C* and was liquefied by the cooling of this jar at about 120 pounds pressure. At this pressure the boiling temperature of water is 230° F., so that vaporization of water was impossible under the conditions prevailing.

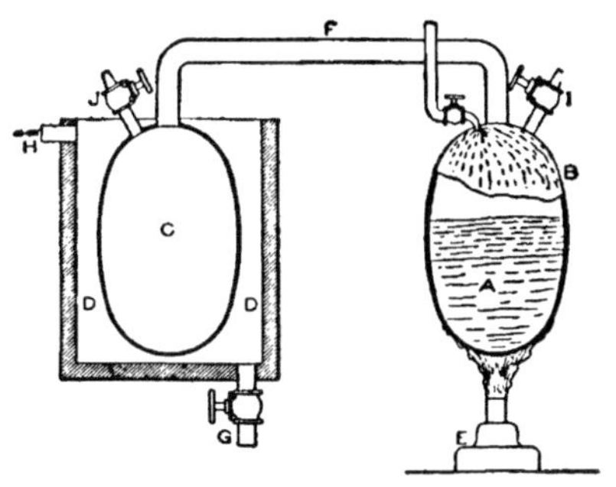

Fig. 8. Carré's Experimental Absorption Machine.

When the process of distilling ammonia was completed, the lamp was removed from jar *A*, the water drawn off from around jar *C*, and the tank *D* filled with whatever substance it was desired to freeze. A water pipe allowed water to flow over jar *A* and cool its contents, with the result, that the pressure was removed from both the jars, and the liquid anhydrous ammonia in jar *C* began to vaporize and return to jar *A*, where it was absorbed by the water from which it had been distilled. By the change of anhydrous ammonia from the liquid into the gaseous form, the heat was taken from the liquid in the tank *D*, and held latent by the gas, so that this liquid was cooled. This same heat was given up as the gas was reabsorbed in the jar *A*, and was transferred to the cooling water passing over the jar.

The intermittent action of Carré's machine makes it impracticable for anything more than experimental uses; and for the decade following its invention, great efforts were made by inventors to improve the apparatus in such a way as to get continuous operation. This was finally accomplished about 1858. The apparatus then developed, similar in a general way to that employed at the present time, consisted of three distinct sets of appliances: the first, for distilling and liquefying the ammonia; the second, for producing cold by means of an evaporator and an absorber; and the third, for pumping the rich liquor from the absorber to the generator, from which the cycle is started afresh. These three operations are distinct, but the apparatus in each part of the plant is dependent on all the other parts, so that all must operate continuously when in use in order to form a complete closed cycle. Fig. 9, taken from the *Southern Engineer*, is a good diagram illustration of an absorption plant, consisting of an ammonia boiler or generator *J*, an analyzer *E*, exchanger *B*, condensing coils *F*, cooler *C*, absorber *D*, receiver *G*, and the ammonia pump *I*.

The Generator. The generator is a cylindrical drum containing coils of pipe through which steam from the boiler is circulated for distilling the ammonia. This part of the apparatus, which is also known as the *ammonia still*, is constructed in several forms, both horizontal and vertical. The horizontal generator is the more expensive of the two forms, but has the advantage of supplying drier gas. In the vertical generator, the evaporating surface is comparatively small and the boiling so rapid, that it is difficult to keep the steam coils covered with liquor. These coils, when uncovered, become pitted by the combined action of the gas and the ammonia liquor. Generators of the vertical type are also likely to boil over, when forced, and the height is inconveniently great when properly installed in connection with the analyzer.

Analyzer. This apparatus is set directly above the generator and connects thereto, as shown at *E* in the illustration. It consists ordinarily of a cylindrical drum containing a series of pans or deflectors over which the hot distilled gas must pass for the purpose of separating any water that may have been distilled or carried over with the ammonia gas. The water thus separated is drained back into the generator. Rich ammonia liquor, being returned from the

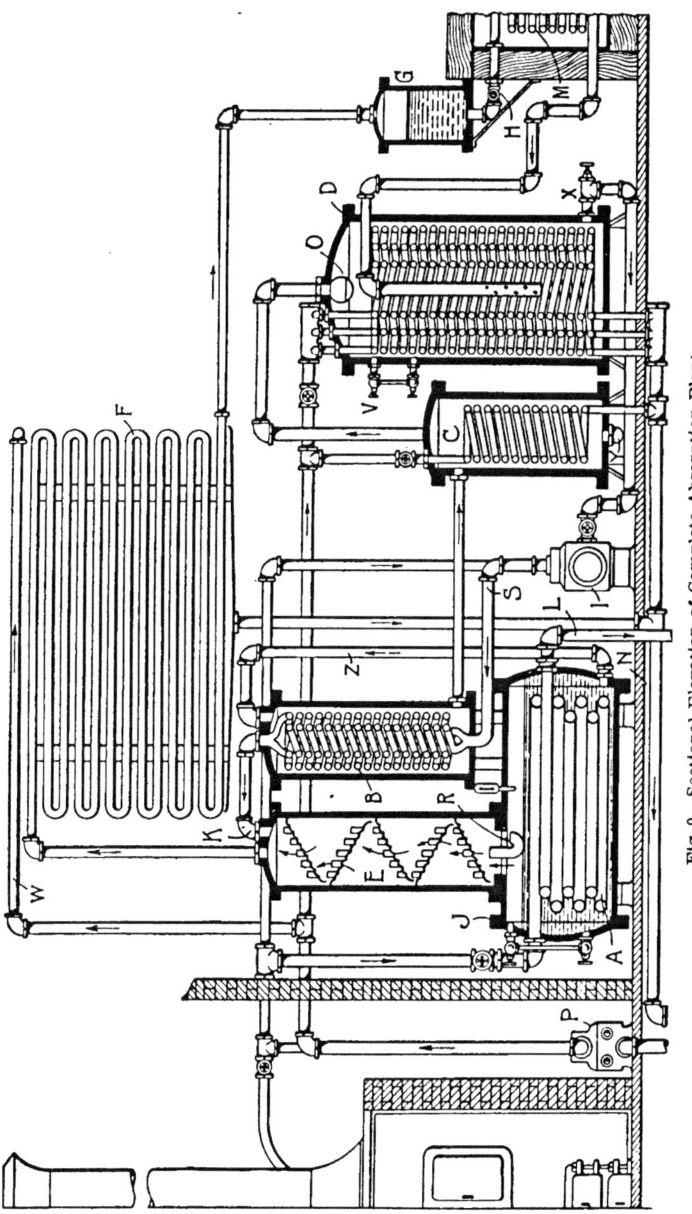

Fig. 9. Sectional Elevation of Complete Absorption Plant.

absorber to the generator, is discharged from the ammonia pump into the top of the analyzer. It flows over the pans in such a way as to absorb heat from the gas. This results in a saving of steam required to heat the generator, as well as in the amount of water required at the condenser to liquefy the ammonia gas.

The Condenser. The condenser F is a series of pipes over which water flows, cooling the dry gas, which liquefies under the pressure created in the generator. The liquid ammonia flows to the receiver G. It is then ready to be passed to the cooling coils through the expansion valves, which are adjusted to supply the amount of liquid required to accomplish the desired cooling.

Rectifier. This is an apparatus added to the absorption plant in recent years with a view to giving increased efficiency. It is not shown in the illustration but is usually mounted on top of the generator in connection with the analyzer. The construction varies with different makes of machines, the object in all cases being, to remove moisture from the gas coming to the analyzer, as moisture is very detrimental to the operation of the absorption machine. The most usual construction consists of a nest of tubes enclosed in a cylinder having, at the bottom, a chamber for collecting the moisture separated from the gas. The rich liquor from the absorber is taken through these tubes, either before or after passing through the heat exchanger, and goes thence to the top of the analyzer. The gas, on its way to the condenser, passes through this cylinder, surrounds the tubes and is thus cooled sfficiently to cause any moisture that it may contain to be precipitated. In some cases the rectifier is in the form of a pipe coil similar to the ordinary atmospheric or double-pipe condenser, in which case care must be taken to avoid supplying cooling water sufficient to cause condensation of the ammonia in the apparatus.

The Equalizer. The equalizer or heat exchanger B, as shown in the illustration, is a drum in which are several coils of pipe through which the liquor from the pump I is forced on its way to the generator J. The pipes in the drum are surrounded by the hot weak liquor entering at the top through the pipe U from the generator. Thus the hot liquor is made to give up heat to the rich liquor in the coils, on its return passage to the generator. In horizontal machines properly proportioned, at least 6 square feet of exchange surface should be allowed for each ton of refrigerating capacity. This cannot be done with machines using vertical generators, owing to the increased danger of boiling over. In such machines it is not practicable to allow much more than 2 square feet of exchange surface per ton. This, however, is not sufficient to reduce the temperature of the weak liquor much below 135° F., so a special weak liquor cooler *C* is employed.

Even with the horizontal generator, this cooler is sometimes employed to advantage, as shown in Fig. 9. The cooler in this, as in most instances, is of drum shape similar to the exchanger, and contains a coil of pipe through which water is circulated for cooling the liquor coming from the exchanger on its way to the absorber. The liquor flows from the cooler to the absorber through the pipe *V*, a regulating valve being placed on this pipe at the absorber, in the best practice. Where the equalizer and rectifier are both used, the rich liquor passing through them returns to the generator at a comparatively high temperature. It must not be forgotten that—the higher the temperature, the less the moisture removed by the liquor in the analyzer. On this account rectifiers are frequently made to use cooling water, and as the water so used can be passed over the ammonia condensers after leaving the rectifier, there is not much added expense on account of water. Rectifiers of this type are made in pipe coils, as above mentioned.

Absorber. This part of the apparatus *D*, performs the reverse operation to that performed in the generator. Heat supplied in the generator separates the ammonia from the solution, while in the absorber the weak aqua ammonia absorbs the gas at low temperature produced by cooling water. The absorber shown in the illustration is similar in general form to the weak liquor cooler, which consists of a number of cooling coils in a cylindrical shell, water being circulated through the coils to give the desired reduction of temperature. The construction shown is known as the *empty* form of absorber, which is about the most satisfactory design yet evolved. The weak liquor is sprayed into the shell of the absorber at the top, passing down over the cooling coils and coming in contact with the gas, which is admitted in the form of spray near the bottom of the chamber. Thus absorption of gas is rapid and continuous, the rich liquor being drawn off from the bottom of the absorber as shown.

In another form of apparatus known as the *submerged* or *tank* absorber, a series of coils is so arranged in the containing shell that the gas enters them at the top, together with a spray of the weak liquor, which absorbs the gas in descending through the coils. A receiver at the bottom of the tank stores the rich liquor, and cooling water entering the shell near the bottom fills the space surrounding the coils and flows off at the top, the drum being full of water. With this type of apparatus about 35 square feet of surface should be allowed per ton of capacity. In the *full* absorber, coils are nested in a cylindrical shell into which the gas and weak liquor are admitted at the bottom, so that the gas is compelled to pass through the entire body of ammonia liquor in the shell. This gives complete absorption, but there is the disadvantage incident to a loss of pressure equal to the head of liquid between the top level of the liquor and the gas inlet.

Ammonia Pump. As it is desirable to have means for testing the piping of a plant to 350 or 400 pounds pressure, the ammonia pump should be selected with this in view, notwithstanding the fact that in its ordinary work there is only the pressure of the generator to work against. Care should be taken in selecting the materials and constructing the pump in order to make it proof, as far as possible, against the action of ammonia. The stuffing-box gland should be particularly designed to prevent leakage. It should have good depth so that, with good ammonia packing, the gland need not be set up tight enough to prevent leakage, thereby causing loss of work in friction and perhaps damaging the piston rod so that it will have to be renewed or turned down. As duplex pumps have more moving parts and more glands requiring attention than single-acting pumps, such pumps are not suitable for use in pumping aqua ammonia, so that single-acting pumps are invariably used in both the vertical and horizontal types, arranged for direct steam drive or drive by power. As the ammonia pump is vital to the operation of the plant it should be of the very best construction, and in large plants duplicate pumps should be installed. Some manufacturers provide this duplicating feature also for the small plants, by specially constructing the pumps with duplicate cylinders, that can be driven by power or steam, only one of the cylinders being required in ordinary operation.

Ammonia Regulator. As in the absorption system, regulation of the working depends principally on the strength of the rich liquor supplied to the still, it is desirable that the density of this liquor be as constant as possible at the given figure for which the machine is designed to operate, this ordinarily being about 26 degrees on the Baumé scale. With the full absorbers and other types where quantities of the weak liquor come in contact with the gas to be absorbed, the regulation of density in the rich liquor is simple; but with the empty absorber, which is preferable on other accounts, it is necessary to regulate the supply of weak liquor carefully, if the density is to be maintained constant. For this purpose ammonia regulators are used on the absorber, governing the flow of weak liquor, the density of which is 16° to 18° Baumé.

Operation. As steam is turned into the generator, the rich liquor is heated and ammonia gas distilled. This gas rises into the analyzer *E*, and, coming in contact with the pans or deflecting surfaces and the rich liquor at lower temperature, is cooled so that what steam it contains is condensed, leaving the gas free of moisture as it leaves the analyzer for the condenser, or for the rectifier when this apparatus is used to ensure absolute dryness. The hot dry gas enters the condensing coils at the top as shown, flowing downward; and the cooling water flows over the coils *F* from the pipe *W*, so as to cool the gas until it condenses under the pressure of the system. The liquid anhydrous ammonia thus formed flows into the receiver *G* and thence to the expansion valve *H* from which it passes to the cooling coils *M*. The condensing pressure is from 150 to 200 pounds. Modifications may be found in some plants causing them to vary from the plan shown, but an understanding of the plant illustrated in Fig. 9 will enable the student to handle any equipment, the prinicples being the same as here set forth.

Power for Absorption Plant. In computing the size of boiler required for an absorption machine, it is first necessary to find the heat required in the generator which the boiler is to supply. The weight of steam corresponding to this amount of heat added to the weight of steam required for the operation of the auxiliary machines gives the total weight of steam necessary, and from this data the size boiler which will deliver the required amount of steam continuously is computed.

The heat required in the generator is equal to the heat lost in the condensing coils added to the heat lost in the absorber, minus the heat gained in the cooling coils, minus the heat gained by the work of the ammonia pump in raising the rich liquor from the pressure in the absorber to the pressure in the generator.

The heat lost in the condensing coils is equal to the temperature of the entering gas, minus the temperature of the liquid ammonia, plus the latent heat of ammonia. Latent heat of ammonia is found by the rule, 555.5−(0.613 × temp.)−(0.000219 × sq. of temp).

TABLE XV

Heat Generated by Absorbing Ammonia

Ammonia in Poor Liquor, Per Cent	Ammonia in Rich Liquor, Per Cent	Heat of Absorption by One Pound of Ammonia in Units	Pounds of Rich Liquor for Each Pound of Active Ammonia
a	c	H_a	P_r
10	25.	812	6.0
10	36	828	3.45
12	35.5	828	3.74
14	25	854	7.8
15	35	811	4.25
17	28.75	840	7.0
20	25	840	16.0
20	33	819	6.1
20	40	795	4.0

The heat lost in the absorber is equal to the heat produced by the absorption of one pound of ammonia by the poor liquor, plus the heat brought into the absorber by a corresponding amount of poor liquor (per cent ammonia in rich liquor ÷ per cent of ammonia in poor liquor) and the negative heat produced by one pound of gas from the cooling coils. Table XV gives the heat generated in the absorber per pound of gas under various conditions.

The heat brought into the absorber by the poor liquor for each pound of active ammonia gas is equal to the number of pounds of rich liquor for each pound of active ammonia minus the difference in temperature between the incoming poor liquor and the outgoing rich liquor. The specific heat of the poor liquor may be taken as 1, and disregarded.

The negative heat brought into the absorber is equal to the difference in temperature between the outgoing rich liquor and the gas in the cooling coils, multiplied by the constant 0.5.

The heat absorbed by one pound of the refrigerant in the cooling coils while doing work is equal to the latent heat of ammonia, plus the difference in the temperatures of the ammonia liquid and the refrigerator, multiplied by the specific heat of ammonia.

The heat produced by the pump is that generated by raising the rich liquor from the pressure in the absorber to that in the generator, and in approximate calculations may be disregarded. It may be found by the following rule:

Heat produced is equal to the weight of rich liquor to be pumped multiplied by the difference between the height in feet of a column of water one square inch in area which will give the pressure in the generator, and the height of a similar column of water giving the pressure in the absorber. This amount is to be divided by the specific gravity of the rich liquor multiplied by the constant 778.

The weight in pounds of rich liquor to be circulated for each pound of liquid ammonia obtained in *G* is equal to 100 minus the percentage of ammonia in the poor liquor, divided by the difference between the percentage of ammonia in the poor liquor and the percentage of ammonia in the rich liquor. A column of water one square inch in area and 2.3 feet high weighs one pound.

After finding the heat required per hour by the generator for each pound of rich liquor entering, the total heat required per hour will be equal to the heat in one pound multiplied by the number of pounds of rich liquor circulated in an hour. The weight of rich liquor circulated in an hour may be found by multiplying the area of the pump cylinder in square inches by the length of stroke in inches, and by the number of strokes an hour, and dividing this amount by 27.7. This result should be multiplied by the specific gravity of the rich liquor.

We now have the number of British thermal units required per hour in the generator. This number divided by the latent heat of steam corresponding to the generator pressure gives the pounds of steam required.

The weight of steam required for the auxiliaries may be calculated as follows

Weight of steam per hour for each steam cylinder equals 0.0348 × the square of the diameter of the cylinder × the length of stroke, both in inches, × the number of strokes per minute × the density or weight of one cubic foot of the steam at the initial pressure.

The sum of the weight of steam required by the generator and that required by all the auxiliaries gives the total amount of steam required per hour to distill the ammonia from the rich liquor and to operate all the machinery about the plant. The total weight of steam divided by 13.8 gives the number of square feet of effective heating surface required in the boiler.

Binary Systems. In the development of the absorption machine, many experiments have been made with various refrigerants and with special machines using combination or dual working mediums. In case of the dual liquid, one of the substances should be capable of liquefaction at a comparatively low pressure, at the same time taking the other substance into solution by absorption. In some machines of this type, the refrigerating agent is liquefied partly by absorption and partly by mechanical compression. A machine developed by Johnson and Whitelaw uses bisulphide of carbon which is first vaporized and then, together with air introduced by a force pump, is passed through chambers charged with oil where the bulk of the moisture in the gas is taken up or absorbed, provision being made for extracting the moisture from the air by passing it over chloride of calcium on its way to the pump. Pictet's refrigerating agent is a combination of carbon dioxide and sulphur dioxide, the latter constituting 97 per cent of the mixture. This fluid gives a boiling point 14 degrees F. lower than is had with pure sulphide dioxide, and its latent heat is practically the same as that of this material.

In an apparatus designed to work on the vacuum principle, by De Motay and Rossi, the refrigerating agent is a mixture made up of common ether and sulphur dioxide, known as *ethylo-sulphurous oxide*. At ordinary temperatures, liquid ether has the power of taking up, or absorbing sulphur dioxide, the absorption in some cases being as much as 300 times its own bulk, while the tension of the vapor given off from the compound or dual liquid is below that of the atmosphere for a temperature of 60° F. This dual liquid is placed in the refrigerator and evaporated by reducing the pressure with air pumps, as in the vacuum system, so that the pressure is no greater than required to cause liquefaction of ether. Owing to the absorption of sulphur dioxide by ether, the pump need not have as large capacity as if ether alone were used, but must necessarily have greater capacity than if pure sulphur dioxide were used. Neither the vacuum system nor the binary system with dual liquid refrigerant have come into general use. Practically all refrigeration is done with simple refrigerants.

Care and Management. To be successful in handling an absorption plant, the man in charge must be regular in his habits and methods and, above all, must know his plant from top to bottom, having the location of every valve and connection in mind. Regular inspections should be made to see if all parts are in good condition and if the apparatus is performing its full duty. Gauges and thermometers with test cocks enable the engineer to ascertain this latter point. There should be a pressure gauge on the generator and one on the low-pressure side. Some engineers prefer also to have such a gauge on the absorber. These gauges should have pipe connection with shut-off valves, so that they can be inspected and repaired if necessary. It is convenient to run the pipe to a gauge board located where readily seen, so that the condition of all parts as to pressure may be seen at a glance. Owing to the danger of bursting, some engineers prefer to work their plants without glass liquid-level gauges, especially on the high-pressure side. In some cases this difficulty is overcome by using special casings for the glass tube and special shut-off valves that close automatically in case of a break. In other cases the gauge glasses on the high-pressure side are kept shut off except when a reading is desired. Where used, such gauges are placed on the generator, the absorber, and the ammonia condenser or liquid receiver. Test cocks are placed on the generator and, in some cases, on the absorber.

Thermometers should be used freely on all principal pipe lines. One should be placed on the ammonia liquid line near the expansion valve; another, in the poor liquor pipe near the absorber; while a third is connected direct to the absorber, a fourth, to the manifolds of the bath coils, and a fifth, to the cooling water pipe near the condenser. Instruments can also be used to advantage on the rectifier, analyzer, and exchanger. Portable thermometers, and hydrometers for measuring the temperature and density of brine and other liquids, should be provided; while the engineer should have indicating apparatus for the ammonia and steam cylinders, as well as scales and measuring vessels for performing rough tests in measuring coal and water used, ice turned out, ammonia supplied to system, etc. Two forms of connections for thermometers placed on the pipe lines are shown in Fig. 10.

Preliminary to charging and putting a plant in operation it must be tested under pressure for leaks. This is done by connecting the suction of the ammonia pump to a water supply and pumping water into the system until full and a pressure of 150 pounds is had. Inspection is then made for leaks and, if everything proves tight, the pressure is increased gradually to 300 pounds or something more. This pressure is maintained for at least half an hour while careful inspection of all joints is made. While the pressure is on, the valves on the gauge piping connections are closed and the gauges disconnected and cleaned by blowing out. Drains are then opened and the water allowed to flow out of all parts of the system except from the generator, which is left half full.

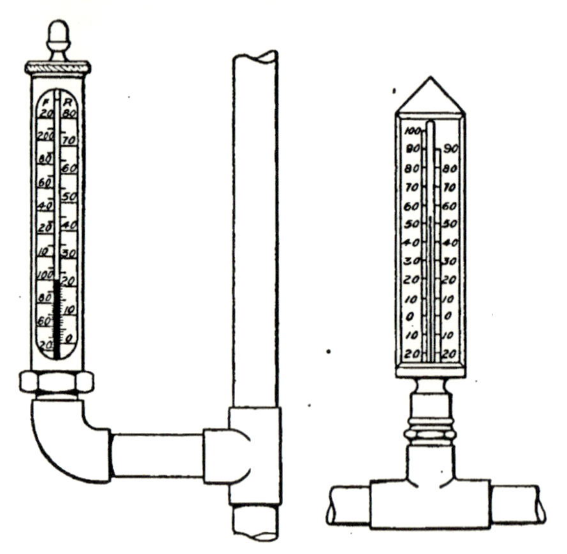

Fig. 10. Thermometer Attachments.

Steam pressure of 50 pounds is then turned on the generator coil, which results in a pressure of 30 pounds in the generator as soon as the water is heated and sufficient steam made to fill the high pressure side. At this stage the valves on the weak liquor pipe to the absorber are opened and the pressure forces the water from generator to absorber. Suction of ammonia pump is now connected to absorber, and the

15 TO 20 TON REFRIGERATING MACHINE
Frick Company.

water and steam circulated through the system, while the vent cocks are left open so that all parts are thoroughly cleaned and blown out. After this has gone on for an hour or so, the pump is stopped, the steam shut off the generator coil, and the system allowed to cool. As soon as steam stops issuing from vent cocks, they are closed, and the stop valves separating the various parts of the system are closed so that, as vacuum is formed in different parts by condensing steam, a leak in one part need not affect the rest of the plant.

Charging. With the system under a vacuum, as after steaming out, aqua ammonia may be forced into the generator and absorber by atmospheric pressure. Otherwise, the ammonia pump must be used to fill the two parts of the system—one after the other. An auxiliary suction makes connection to the pump, and the end of the suction pipe is inserted in the $1\frac{1}{2}$-inch bung hole of an aqua ammonia drum to within 1 inch of the bottom. All valves on the discharge side of the pump are opened and the ammonia is pumped into the system, care being taken to keep the drum as cool as possible. The generator is filled to the working level, as shown by the gauge cocks, and the absorber is filled to the top of the gauge glass. The ammonia pump is now connected with its suction to the absorber; all valves are adjusted for regular working conditions with steam admitted to the generator coils; the ammonia gas generated is allowed to pass over to the condenser, over which cooling water is circulated. When the pressure is about 120 pounds on the high side, the ammonia pump is started slowly, and the liquor level in the absorber is gradually reduced to the normal about mid-height of the gauge glass. With cooling water circulating through the absorber coils and the weak liquor cooler, the plant is in full operation.

At least once a day the coils of the absorber should be examined to make sure that they are clear. In case the pressure gets above 12 pounds, attention should be given the coils and the expansion valves. The temperature at this pressure should be between 86 and 90 degrees, and if it increases to more than this the supply of cooling water through the coils should be increased. Every two or three days the foul gas should be burned off the absorber at the purge cock. Where the weak liquor cooler is employed, care should be taken to see that the coils are in good condition—this, in fact, applies to all coils in the plant—so that the liquor reaches the absorber as cool as possible.

Once or twice a week, and oftener, if necessary, the rich liquor should be tested for density and should show at least 26° Baumé. If run at 28°, the ammonia pump may be slowed down. In some plants this greater density can be used to advantage.

When the quantity of liquor in the system gets low, as shown by the gauges on the generator and absorber, aqua ammonia is charged into the system by connecting to the absorber; while, if the liquor gets weak, anhydrous ammonia is added by connecting the drum between the ammonia receiver and the expansion valves. It is best to place the drum on scales and to charge only a part of the ammonia in it at one time. As soon as the effect of the amount charged is seen, more can be added if thought advisable. With proper working, the density of the weak liquor leaving the generator should not be more than 18 degrees, and if found to be more than this, the cause must be removed. It may be due to running the ammonia pump too fast; but if not this, the trouble will be found in the analyzer where one or more pans will probably be found broken.

To stop the machine, close the steam valve to the generator, allowing the ammonia pump to run until the pressure is raised to 150 pounds, when it is stopped and the expansion valves closed. Thus the machine is left to cool of its own accord, and this is much better than cooling it quickly. When, however, circumstances require that the temperature be lowered rapidly, the steam valve to the generator is closed and the pipe disconnected so as to admit water to the coil, through which the circulation of water is kept up until there is no apparent odor of ammonia.

Efficiency Tests. It is not the intention in the brief space here available to give a complete code of rules for conducting tests. Such a course is impossible. For the rules now recognized as standard in this country, the student must have reference to the Proceedings of the American Society of Mechanical Engineers. The rules adopted by this society for testing refrigerating machines is very complete and elaborate. Briefly, it may be said, that a complete test must involve every part of the plant from the coal pile to the ice dump, careful and accurate measurements being made of the fuel and water used, as well as of the temperatures in the different parts of the system and the corresponding pressures. A test should be run for at least one week, or long enough to get all average conditions.

Economy of Absorption Machine. Dry gas governs the economy of the machine more than any other factor. It was the difficulty in getting dry gas with the early machines that put them out of the race with compression apparatus to such an extent that they have never regained the lost ground, the compression system being used in the great majority of plants to-day. For hot climates the absorption machine has some advantages; and it is highly efficient for low temperatures, as required, for example, in fish freezing plants. Practically all moisture should be eliminated and in no case should more than 5 per cent be tolerated. Owing to presence of moisture, many of the machines in operation must carry a low back pressure with corresponding loss of economy. Twelve pounds back pressure is the lower limit. With perfectly dry gas the machine should be able to carry considerably more than this with increased economy. One should strive to have the back pressure as high as practicable as, the higher this pressure the stronger the rich liquor and the less pumping required per ton of refrigeration produced. The relative economy of the absorption and compression systems is an open question and depends largely on the local conditions of any given case.

COMPRESSION SYSTEM

This system dates from the invention by Jacob Perkins, made about 1834, of a machine operating on the compression plan and using a volatile liquid derived from the destructive distillation of caoutchouc. Perkins' machine was little more than an experiment and was never used commercially. About 1850 Twining made improvements in the apparatus, using the same principles as Perkins, and developed his machine so that it came into practical use in one or two cases, particularly in Cleveland, O., where a machine designed for 2,000 pounds of ice per 24 hours, actually turned out 1,600 pounds, and was used off and on for three years. None of the early machines were much in use until the adoption of ether as a refrigerant. The successful application of the compression machine may be said to date from the invention of the ether machine by James Harrison about 1856. Sulphuric ether as used in Harrison's machine is obtained from the action of sulphuric acid on vinous alcohol. It has a specific gravity of 0.72, with latent heat of vaporization 165, and boiling point 96° F. at atmospheric pressure. Various forms of the ether machine were in use for a few years following the invention of Harrison. Owing, however, to the disadvantages already mentioned, it was soon succeeded by the sulphuric acid apparatus, and this, in turn, by the ammonia and carbon dioxide compression machines, as soon as the absorption machine had shown itself inefficient and the methods of iron manufacture had been developed sufficiently to construct machines of strength necessary in handling ammonia. Since 1860 to 1865, the ammonia compression machine has practically held the field, though in the last decade it is meeting considerable competition by the absorption systems as improved to operate with dry gas and the compression machines using carbon dioxide.

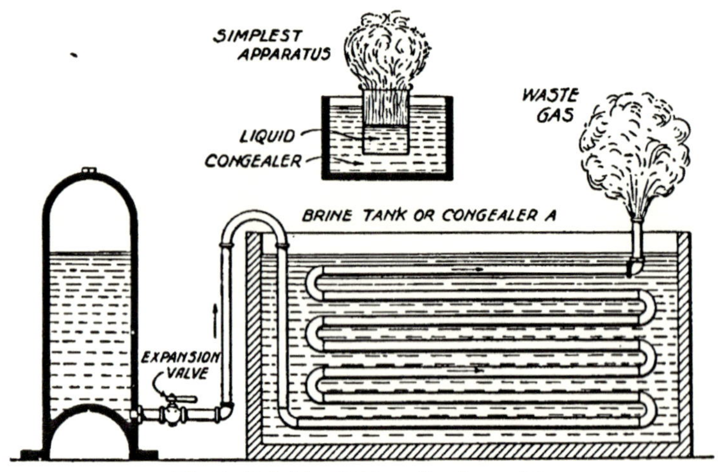

Fig. 11. Primitive Refrigerating Apparatus.

Operating Principle. As already shown, production of refrigeration in latent-heat machines depends on the principle of heat absorption or change to latent form when a volatile liquid evaporates. This is shown in Fig. 11, where the small vessel, at the top of the illustration, containing a volatile liquid, and set in a second vessel containing a liquid to be frozen, represents the simplest possible form of refrigerating apparatus. If volatile liquids could be obtained at nominal cost, no other refrigerating apparatus than that shown would be necessary, though it would be convenient to connect the supply drum of the liquid to the evaporating vessel by pipe so as to give a continuous flow of the volatile liquid to take the place of that evaporated and dissipated in the atmosphere. The simplest system would thus be made up of an evaporator arranged to cool any substance desired and a receptacle for the refrigerant with a connecting pipe between having an expansion valve to regulate the flow of the liquid to the evaporator.

As the cost of ammonia, for example, is about $200 per ton of refrigerating capacity, where wasted, as in Fig. 11, it is evident that some means of conserving the refrigerant and using it over must be adopted. To do this, the heat absorbed by it in evaporating must be removed; and it is this removal that makes necessary the use of machinery in a refrigerating plant. In such an establishment there are three processes—the gas is first reduced in volume by a compressor in which expended work takes the form of heat and raises the temperature of the gas; this hot gas is then passed to a condenser where the heat is given up to cooling water and the gas restored to the liquid form, in which it passes to the third stage of the process, to be evaporated in such a manner as to absorb additional heat.

Fig. 12 shows the complete compression plant in its simplest form. The supply drum is connected to the evaporator, which in turn is connected to the suction of the compressor that discharges to the condenser, from which the liquefied refrigerant passes to the supply drum. From the discharge pipe E of the compressor, the ammonia passes to the oil separator F and goes thence to the condenser C, made up of a series of pipes over which water flows to take up the heat in the gas. Liquid anhydrous ammonia from the bottom of the condenser passes through the pipe H to the drum I, which is known as the *ammonia receiver*. As there is some loss of heat in cooling the liquid ammonia from the temperature of the condenser or ammonia receiver to that of the evaporator, it is seen that the three stages of *compression*, *condensation*, and *expansion* as applied in commercial refrigerating machines, although making a closed operating cycle, do not make a complete reversible cycle. The cost of machinery necessary to utilize work by expansion in cooling the liquid ammonia being more than the value of the work performed, this cooling is done at the expense of the refrigeration of the system. It is then in order to see how the apparatus for carrying out the three essential processes is constructed and to study the proportion and combination of the different parts in making up the refrigerating plant.

Compressors. Compressors may be divided into two principal classes, single and double-acting; and each class subdivided into the vertical and horizontal. They are of the vertical type if single-acting, and of both vertical and horizontal if double-acting, although the

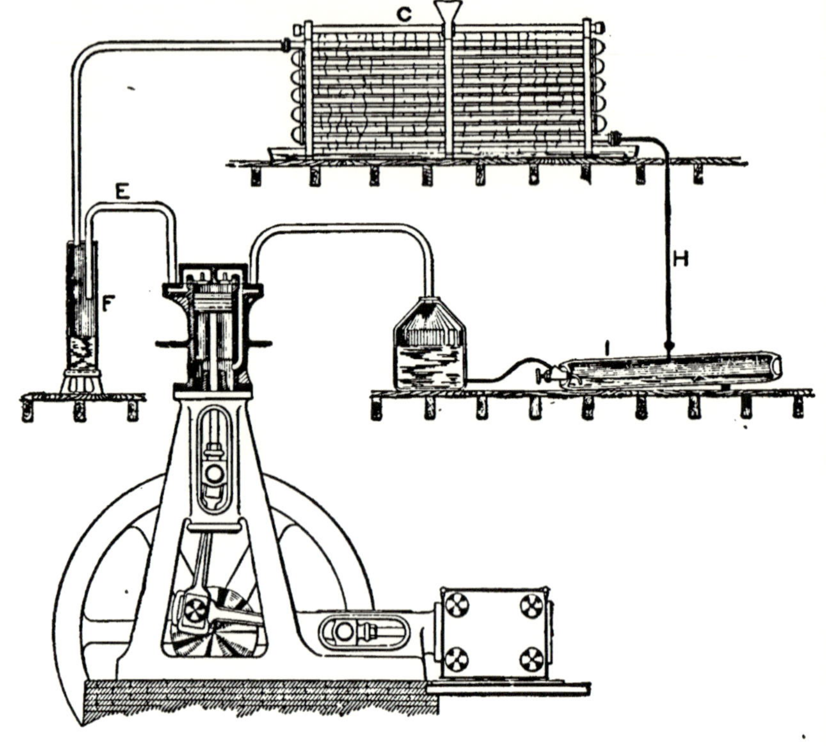

Fig. 12. Diagram of a Complete Compression Plant.

majority of the double-acting are of the horizontal type. Machines may also be classed according to the mode of driving, whether by engine connected direct or by belt, or by motor drive with geared or belted connection. The engine may be used vertical or horizontal, and within this classification comes almost any machine of modern build. In all except the smallest machines, and in some designs for extremely large units, the engines are direct-connected to the same crank-shaft as the compressors. Either simple or compound engines may be used, and if there is enough cooling water the engines may be run condensing. In some cases the connecting rod of the engine is attached to the same crank pin as that of the compressor and in others, to a separate crank on the same shaft. One form of the horizontal machine has the engine connected to a crank at one end of the shaft, while the other end has a crank arranged to drive two single or double-acting horizontal compressors. Another arrangement has the engine connected to the middle of the crank-shaft so as to drive two compressors of either the horizontal or vertical, single- or double-acting type. All of the various arrangements use fly-wheels to give steady working, these being placed in various ways according to the disposal of the other parts of the machine. Where compound engines are employed, each cylinder usually drives its own compressor cylinder, the connecting rods of steam and compressor cylinders being attached to a common crank pin, with the fly-wheel at the middle of the shaft. In the tandem machine, having the ammonia and steam cylinders in line on a common center with their pistons on the same rod, a connecting rod drives a crank-shaft on which are two fly-wheels placed at the ends.

In comparing the single and double-acting compressors, it is seen at once that the stuffing-box—a vital part—can be kept tight easily in the single-acting machine, where it is subjected only to the comparatively low pressure of the suction gas instead of the pressure of the condenser which ranges from 125 pounds, upward. On the other hand, a double-acting compressor is more economical because, at each revolution of the crank-shaft, it deals with almost twice as much gas as a single-acting machine of the same diameter and stroke. With the exception of the extra friction resulting from the necessarily tighter stuffing-box gland of the double-acting machine, the friction of the two machines is the same. With improved stuffing-box construction, the friction is much reduced, and in a box properly adjusted may be little, if any more, than for the single-acting machine, particularly taking into account the friction of the two boxes necessary for a machine of this type to have the same capacity as the double-acting equipment.

Altogether it is estimated that, as against a machine having two single-acting gas compressors, the double-acting machine will save about one-eighth the total power necessary to compress a given amount of gas. Also less material is required for the single cylinder used in the double-acting machine, but this is partly offset by the extra care and expense necessary to construct the latter type of machine. It is more difficult to adjust the piston for clearance in the double-acting machine but with proper care it can be done correctly. An advantage of two single-acting compressors is the fact, that in case of accident to one of the cylinders, the other may be kept going at increased speed to keep the temperature of the coolers down so that the pipes will not drip. For this and other reasons it is customary to use two cylinders where the single-acting machine is employed and the cranks are set opposite or at 180 degrees to one another, so that one compressor is filling, and one compressing and charging, at each half revolution of the crank-shaft.

There is some difference of opinion as to the relative merits of the vertical and horizontal types of machines, but in respect of uniform wear of the moving parts it is plain that the vertical machine has the advantage. As little oil as possible should be used in refrigerating machines and prevention of undue wear is an important consideration. Vertical machines are not subject to bottom wear of the pistons as are horizontal compressors in which the weight of the piston is supported by the lower half of the cylinder wall. When the piston is near the crank end of the stroke, part of the weight is supported by the stuffing-box glands, resulting in unequal wear. In the vertical compressor the valves and piston work up and down, so that the wear is equal in all directions and there is little friction. Owing to the fact that the vertical machine is comparatively expensive to build and care for properly, there are greater fixed charges and more depreciation with this type of machine. As to space occupied by the two forms of machine, there is little difference, the choice depending principally on whether vertical or horizontal space is more valuable.

Essentials in Compressors. Owing to the extreme tenuity of gases used in refrigerating, considerable care must be taken in designing and operating compressors if the maximum possible efficiency is to be had. Handling ammonia, for example, is quite another matter from handling steam, and a joint that will hold steam at high pressure may be a perfect sieve for leakage when it comes to ammonia. This is even more true with carbon dioxide with which the pressures are much higher than in machines using ammonia. Provision should be made to remove the heat of compression, and the clearance should be made as small as possible, consistent with safe working. Any gas left in the clearance spaces expands on the return stroke of the piston and prevents the cylinder filling properly with the low-pressure gas, so that there is a large loss in efficiency. Stuffing-boxes, valves, and pistons should be kept tight to prevent leakage of the gas and yet should not be so tight as to cause undue friction. This calls for great care in adjusting and lubricating the machine, methods of doing which will be discussed more in detail later.

The piston and valves are the heart of the compressor and practically all troubles in operation may be traced to one or the other of these parts, when it will be found that lubrication is imperfect or that they are out of adjustment for the prevailing conditions of pressure and speed of operation. In most designs of ammonia compressors, the suction and discharge valves operate in cages, the valves on one end of a cylinder being usually mounted in a common casing which acts as the cylinder head and may be removed bodily. In the double-acting machines, there being no provision for water jacket on the ends of the cylinder, the valves are accessible singly if desired, and in some designs, as the Triumph for example, the construction is such that the adjustment for spring tension in the valves may be made while the machine is in operation.

Compressor Valves. There must of necessity be a gas-tight joint between the valve cage and the cylinder head, which may be made in a number of ways. In one method, a square recess or shoulder is cut in the head and a corresponding shoulder is cut on the face of the valve cage. A lead gasket about $\frac{1}{8}$ inch in thickness is placed in the shoulder of the head and the cage adjusted with set screws provided for the purpose until in proper position. This makes a durable joint except where the openings at the facing surfaces of the shoulders are such that the lead is pressed out—a disadvantage of lead as a gasket material. When this happens, the gasket is gone and a leak develops before the engineer is aware that trouble is brewing. Another objection to a lead gasket is that it compresses and *knits* into the interstices between the cage and the cylinder head, so that it is often impossible to remove a valve or the cage without the aid of a chain block and tackle.

Another form of gasket for the valve cage, but one not popular on account of lack of confidence in its permanency, is common lamp or candle wicking saturated with oil and wound tightly or smoothly in the corner against the shoulder on the cage. It is put in place and fully compressed by pulling down on the valve cap until the cage is in the desired position. Recently engineers have come to the conclusion that any form of gasket joint is a nuisance for joining the cylinder head or valve cage on an ammonia compressor cylinder, so that the ground joint has come into use, the facing surfaces being ground to an absolute smooth surface, before the parts are joined. This is the type of joint that has been employed for some years, with gratifying results, in pipe work on vessels where highly superheated steam is used. The joint made is perfectly tight and there is no factor of depreciation as by gasket becoming worn and ineffective, but great care must be taken in overhauling not to scar the surfaces, when they would have to be reground at considerable expense.

Assuming that the joint between the valve cage and the cylinder head has been made gas-tight, it is next of importance to see that the valve makes a perfect joint in closing against the seat in the cage. This is the more difficult of the two propositions for the valve, being in constant motion under severe conditions of changing pressure, is likely to act erratically and, in some cases, pounds itself and the seat to such an extent that the surfaces have to be reground to a joint. From this it is evident that the best material obtainable should be used in the valves and seats and in the valve stems, which should be immune from breakage as far as possible. The valve stem must fit the guides in the cage as closely as possible and still allow free movement, and the seat between the valve disk and cage (preferably made at an angle of 45 degrees) must be machined and ground to accurate dimensions and surface. Having made it impossible for the gas to pass the cage and valves, means must be provided for closing the valves at the proper time to prevent loss, as a valve that is slow in reaching its seat causes double damage in loss of efficiency and in irregular action on the other parts of the machine. Springs are ordinarily used to operate the suction and discharge valves (as already mentioned in discussing unbalanced valve action), these being placed between the cage and a washer on the end of the stem in the case of the suction valves, and between the casing and a similar washer placed near the middle of the stem in the case of the discharge valves. Nuts on the end of the stems hold the springs and washers in position and afford means of adjusting the spring tension, there being two nuts on each stem so that they may be locked at any point desired. Buffer springs are used to give steady action in some valves but it is considered better practice to use a dashpot chamber on the valve stem for this purpose, as the working of a valve having such chamber is more regular and noiseless, there being little or no hammering action on the seats.

Suction and discharge valves should be constructed to give ample

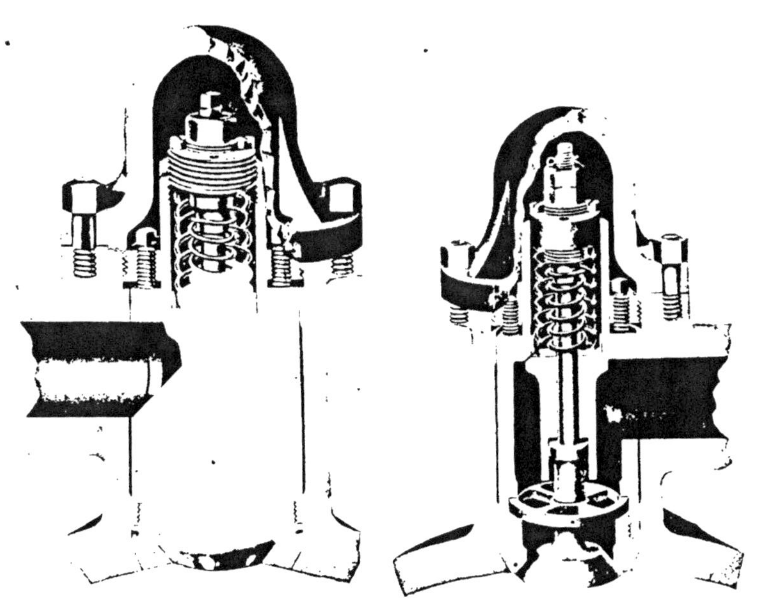

Fig. 13*a*. Triumph Suction Valve. Fig. 13*b*. Triumph Discharge Valve.

area of opening, with lightness of moving parts, quick seating, and noiseless action. As compressors must be constructed with little clearance, it is important that the valves be so made that they cannot fall into the cylinder in case of accidental breakage. Space does not permit discussion of the various forms of compressor valves on the market, but a good idea of construction for double-acting compressors may be had by reference to Fig. 13, which shows the suction and discharge valves as constructed by the Triumph Ice Machine Co. The suction valve is shown at the left, where it is seen that the cross-section of the stem is smallest above the safety collar so that any possible break will occur above this collar and the valve be prevented from falling into the cylinder. This collar, it is noted, also works in a dashpot that steadies the operation of the valve. The tension and buffer springs with the lock nuts for adjustment are seen at the upper end of the valve stems, and the valve cages as a whole are held in position against ground joints at top and bottom, by set screws, so

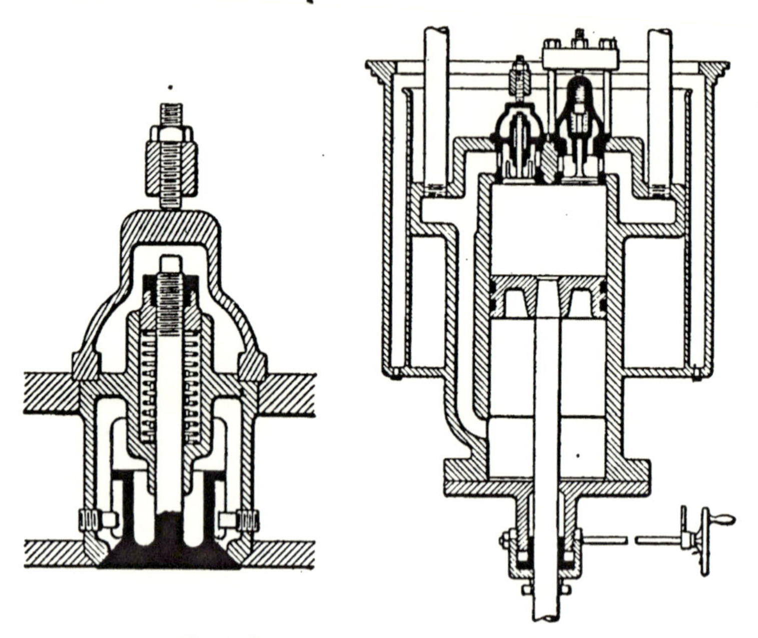

Fig. 14. Featherstone Balanced Suction Valves.

that gas cannot escape to the inside of the outer casing, which is used for protection only and may be removed freely for adjustment of the valve. Fig. 14 shows at the left the construction of valve made by the Featherstone Foundry and Machine Co. for use in single-acting machines. The manner in which the valve is set in position in the safety head of the compressor is shown at the right of the figure. Various constructions and arrangements of valves are adopted by the different manufacturers; but after understanding the valves illustrated in Figs. 13 and 14, the student will be able to handle any special valve with which he comes in contact.

Valve Operation. With any type of valve, it is required that the gas be admitted to the cylinder as the piston recedes from the end of the cylinder in which the valve is located. The valve must close as the piston reaches the opposite end of the stroke at the instant the crank is passing over the center. If the spring is stronger than necessary, the gas must exert a certain pressure to overcome its action and the cylinder will not be filled to the limit of its capacity. Also in closing, the strong spring drives the valve against its seat with considerable force, making a disagreeable noise that means excessive wear. Considerable skill and experience are required to obtain the best results with compressor valves, but with good judgment and a few trials most of the imperfections in adjustment can be overcome. Generally the closing spring of the suction valve should not be stronger than necessary to close the valve when held in the hand in the position —either vertical or at an angle—it naturally occupies in the cage of the machine. By taking the valves and cages in the hands and pressing down on the top of the stem with one of the fingers, it will readily be seen when the springs are of proper strength to effect closure.

Discharge valves operate in the reverse direction to that of the suction valves. As the piston advances on the compression stroke, the pressure of the gas in the cylinder increases until it is as high as that in the ammonia condenser plus the pressure required to overcome the tension of the springs on the discharge valve. At this pressure the valve opens and the gas is discharged into the pipe leading to the condenser, the valve remaining open until the crank passes the center corresponding to the end of the cylinder from which the gas is being discharged. At this instant, the valve should close if it is of proper proportions and the tension of the spring is what it should be. This tension should be no more than necessary to make the valve close promptly and easily as the crank passes over the center, any greater tension increasing the loss by unbalanced pressure on the faces of the valve. The suction valve should open at the same instant that the discharge valve closes, that is—as the crank is passing over the center nearest the end of the cylinder in which the valve is placed.

It will be noticed that the suction valve opens and closes while the piston is practically without motion, but the discharge valve opens while the piston is nearly at its maximum, and closes while the piston is at its minimum speed. From this it will be apparent that the thrust or effort on the discharge valve is much greater than on the inlet, that is, in an upward or outward direction; hence the device for arresting the upward motion of the valve must be more effective than the other. Also the valve must close in a shorter space of time because the great pressure above the valve would cause it to fall with considerable force were it not to reach its seat while the piston was still at the top of its travel, and the pressure above and below equal. Should the piston begin its return or downward stroke before the valve closes it would have the condensing pressure above and the inlet pressure

TABLE XVI

Cubic Feet of Gas to be Pumped per Ton

Temperature of Gas in Degrees F.	Corresponding Suction Pressure, Lbs. per Sq. In.	Temperature of the Gas in Degrees F.								
		65°	70°	75°	80°	85°	90°	95°	100°	105°
		Corresponding Condenser Pressure (Gauge) Lbs. per Sq. In.								
		103	115	127	139	153	168	184	200	218
	G. Pres.									
-27°	1	7.22	7.3	7.37	7.46	7.54	7.62	7.70	7.79	7.88
-20°	4	5.84	5.9	5.96	6.03	6.09	6.16	6.23	6.30	6.43
-15°	6	5.35	5.4	5.46	5.52	5.58	5.64	5.70	5.77	5.83
-10°	9	4.66	4.73	4.76	4.81	4.86	4.91	4.97	5.05	5.08
- 5°	13	4.09	4.12	4.17	4.21	4.25	4.30	4.35	4.40	4.44
0°	16	3.59	3.63	3.66	3.70	3.74	3.78	3.83	3.87	3.91
5°	20 ✓	3.20	3.24	3.27	3.30	3.34	3.38	3.41✓	3.45	8.49
10°	24	2.87	2.9	2.93	2.96	2.99	3.02	3.06	3.09	3.12
15°	28	2.59	2.61	2.65	2.68	2.71	2.73	2.76	2.80	2.82
20°	33	2.31	2.34	2.36	2.38	2.41	2.44	2.46	2.49	2.51
25°	39	2.06	2.08	2.10	2.12	2.15	2.17	2.20	2.22	2.24
30°	45	1.85	1.89	1.89	1.91	1.93	1.95	1.97	2.00	2.01
35°	51	1.70	1.72	1.74	1.76	1.77	1.79	1.81	1.83	1.85

below, a difference of from 150 to 175 pounds. This pressure would cause it to seat with an excessive blow which would soon cause its destruction, and also cause the escape of a portion of the compressed gas into the compressor resulting in great loss in efficiency of the machine.

Valve Proportions. Knowing the amount of gas that must be compressed and passed through the valves as well as the velocity at which it is advisable to force the gas through the machine, it is possible to determine the size of valve opening. Table XVI gives the number of cubic feet of gas that must be pumped per minute at different suction and condenser pressures to give 1 ton of refrigeration in 24 hours. It is customary to base the area of the suction valve on a velocity of 4,000 feet a minute and that of the discharge valve on 6,000 feet. Taking, then, the case of a 10-ton compressor with back pressure of 13 pounds and condenser pressure of 184 pounds, it is seen, from the table, that 4.35 × 10 cubic feet of gas must be pumped per minute. Since the amount of gas pumped is found by multiplying the area of the valve opening by the velocity of the gas, it is evident that the area can be found by dividing the quantity that must be pumped by the velocity, the units of measurement being the same, *i. e. feet,* throughout. Thus 43.5 ÷ 4,000 = 0.010875 sq. ft. or, 1.566 sq. in. corresponding to a diameter of 1.47197 in. For the discharge valve, the area is 43.5 ÷ 6,000 = 0.00725 sq. ft. or, 1.044 sq. in. This means a diameter of 1.15253 inches.

Valves proportioned in this way have proved satisfactory at piston speeds of from 270 to 400 feet a minute and from 40 to 100 revolutions. The York Manufacturing Co. has made tests to determine the effect of velocity of flow through the valves on the efficiency of the machine and the power required to compress a given amount of gas. As a result, this company has found that the area of the discharge valve opening may be made somewhat less than that corresponding to 6,000 feet a minute velocity of gas, without loss of efficiency. In fact the experiments show that up to a velocity of about 9,000 feet a minute, the efficiency increases, more gas being pumped without increase in power in proportion to work done. The same experiments show that it is better to have a lower velocity through the suction valve. The velocity for best results is about 3,000 feet, which calls for a considerably larger valve opening on the suction than figured on a basis of 4,000 feet a minute.

Compressor Piston. Having provided the inlet and outlet valves with proper opening and closing devices and made them capable of retaining the gas passing them, a piston for compressing the gas and discharging it from the compressor must be provided. For the vertical type of single-acting compressor in which both inlet valves are in the upper compressor head, the piston is best made as a ribbed disk with a hub at the center for the piston rod and a periphery of sufficient width to be grooved for the necessary snap rings. Three to five of these rings are generally used.

In double-acting machines it is customary to use the spherical form of piston as will be seen in discussing this form of compressor. Fig. 15 illustrates the simplest form of this type of piston, *A* being the cast head, *B* the snap rings, and *C* the piston rod. The surface is faced square with the bore of the hub, and the rod forced in and riveted over, filling the small counterbore provided for this purpose. The cast head and rod having been previously roughed out, the piston is now finished in its assembled condition. The rod is made parallel

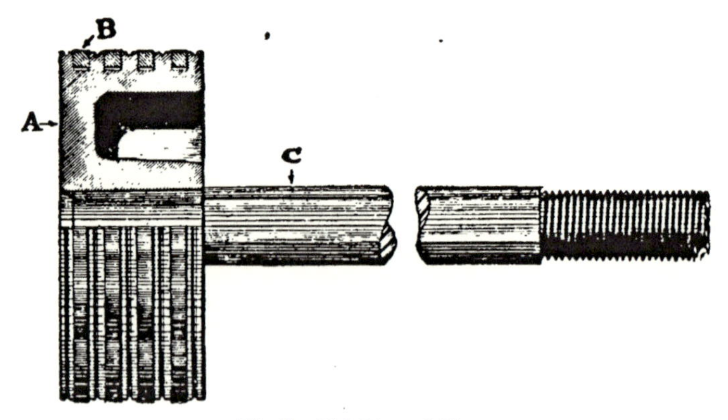

Fig. 15. Disk Type of Piston.

and true to gauge and threaded to fit the crosshead. The grooves are turned for the snap rings, which are made slightly larger than the bore of the compressor. A diagonal cut is made through one side and enough of the ring is cut out to allow it to slip into the cylinder without binding. It is then scraped on its sides until it fits accurately the groove in the piston. It is also well to turn a small half-round oil groove in the outer face of the piston between each ring, which gathers and retains a portion of the oil used for lubrication, thus increasing the efficiency of the piston and collecting dust or scale and lessening the liability of cutting of the cylinder due to any of the usual causes.

The *piston rod* requires special care both in workmanship and material. In order to be effective it must be true from end to end;

SKATING RINK BEFORE FLOODING, SHOWING BRINE PIPES

and to be lasting under the variety of conditions which it operates, it should be of a good grade of tool steel. The end which is usually made to screw into the crosshead is turned somewhat smaller, usually from $\frac{1}{8}$ to $\frac{1}{4}$ inch in diameter, than the portion passing through the packing or stuffing box, principally to allow of re-turning or truing up the rod when it becomes worn, and also to allow it to pass through the stuffing box. After the rod is screwed into the crosshead it is secured and locked with a nut to prevent turning. The nut also allows the position of the piston to be changed to compensate for wear on the different parts of the machine by simply loosening the lock nut and turning the piston and the rod in or outof the crosshead.

Stuffing Box. The stuffing box of the compressor is one of the most difficult parts to keep in proper order. This is owing principally

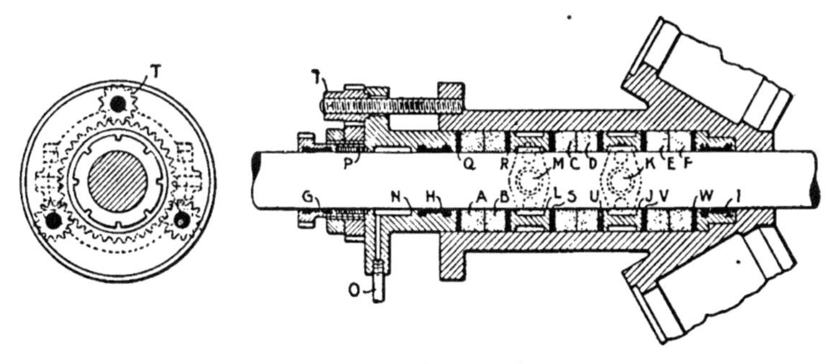

Fig. 16. Type of Double-Acting Stuffing-Box.

to one of two causes: not being in line with the crosshead guide or bore of the compressor, or, the great difference of temperature to which it is subjected owing to the possible changes taking place in the evaporator. However, with the compressor crosshead guides and stuffing box in perfect alignment, and a constant pressure or temperature on the evaporating side, it is a simple matter to keep the stuffing box tight and in good condition. If, however, either of these conditions is changed, it becomes practically impossible to do so. With single-acting compressors the stuffing box is a simple gland with any good ammonia packing.

One form of box used in the double-acting compressor is shown in Fig. 16, this being the construction employed by the Featherstone

Foundry and Machine Co. A, B, C, D, E, and F, indicate composition split packing rings, while Q, R, S, U, V, and W denote pure tin rings of an inside diameter $\frac{1}{16}$ inch larger than that of the piston rod. These tin rings should not be split. At J is a lantern forming an oil reservoir, the oil being supplied by pipe at the connection marked K. This pipe is connected to the oil trap and, as the passage is always open, oil is forced into the stuffing box by the high pressure of the gas in the trap, so that any little oil carried into the cylinder by the rod is instantly replaced and the lantern kept full. A second lantern L has connection to the suction line at M, so that any gas that may have escaped the rings C, D, E, and F is drawn back. With this arrangement packing rings A and B have to withstand only the suction pressure. The stuffing-box gland N has a chamber that is supplied with oil through the connection O by a small rotary oil pump operated from the main shaft. A secondary gland P retains the oil in the gland and should be adjusted just tight enough for this purpose. G, H, and I are points of contact with the rod and are babbitted to an exact fit. In case the rod should have to be turned down or when this babbitt is worn, for any cause, so that it does not fit exact, it should be renewed. The general principles involved in this stuffing box are used in more or less modified form by all the leading manufacturers of double-acting machines.

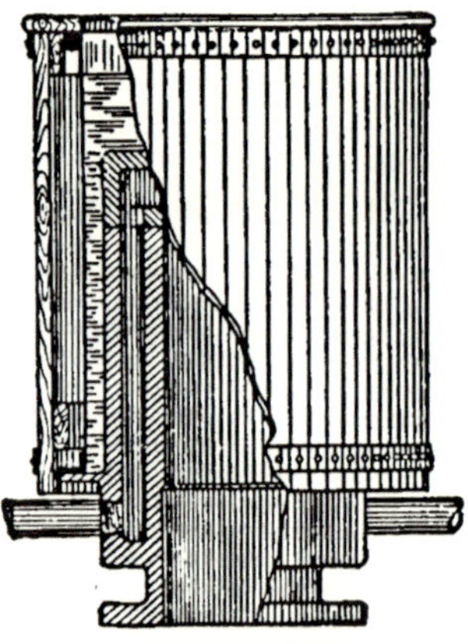

Fig. 17. Jacket of Single-Acting Compressor.

Water Jacket. In the vertical single-acting type of compressor it is usual to provide a water jacket, which may be cast in combination with the compressor cylinder or made of some sheet metal secured to an angle, which is bolted to a flange cast on the cylinder. It is usual to have this water jacket start at about the middle of the compressor (or a little below, as shown in Fig. 17) and extend enough above to cover the compressor heads, valves, and bonnets with water; the principal object of which is to keep these parts at a normal temperature and thereby improve the operation as well as protect the joints against the excessive heat which would be generated by the continued compression. It is also an advantage in the operation of the plant, since by reducing the temperature in the compressor and adjacent parts, the compressor is filled with gas of a greater density. Also the heat extracted or taken up by the water at this point is a certain portion of the work performed in the condenser and therefore not a waste. Double-acting compressors may or may not use the water jacket, but in those where it is used the water surrounds the cylinder only, there being no means of circulating water around the heads, where the valves are placed in their casings. In such machines the cylinder is made of special close-grained steel and is forced into the frame of the machine under hydraulic pressure, thus forming an annular space between it and the frame in which the cooling water is circulated, being admitted near the bottom of the space at one side and drawn off at the top. In the wet compression machines and in those using a liquid oil base, it is not considered necessary to use a jacket, but the latter of these types of machines is now going out of use, owing to the complications incident to keeping the oil out of the piping of a refrigerating plant where it interferes with the proper transfer of heat.

In the operation of the plant it is well to have plenty of water flow through the jackets, as the cooler the compressors are kept the better; but in plants in which water is scarce the quantity may be reduced correspondingly until the overflow is upwards of 100 degrees F. In extreme cases of shortage of water the overflow water from the ammonia condenser is sometimes used on the jackets—that is, the entire amount of available water is delivered to the condenser, and a supply from the catch pan (if it be an atmospheric type) is taken for the water jackets, in which case a greater quantity may be used but at a higher temperature. In the vertical single-acting machine it is customary to admit the water through the flange forming the bottom of the water jacket and to connect the overflow near the top into a stand pipe which is connected at its lower end through the flange to a system of pipes to take it away. To prevent condensation on the outer surface of the jacket, and to present a more pleasing appearance, it is frequently lagged with hardwood strips and bound with finished brass or nickel-plated bands. It is well to have a washout connection from each jacket.

Lubrication. Excessive lubrication is an objection, owing to the insulating effect upon the surfaces of the condensing and evaporating system. Therefore it is well to feed to the compressors as little as is consistent with the operation of the machinery. A proper separating device should be located in the discharge pipe from the compressor to the condenser. To properly admit, or feed the lubricant to the compressors, sight feed lubricators should be provided, by which the amount may be determined and regulated. These may be of

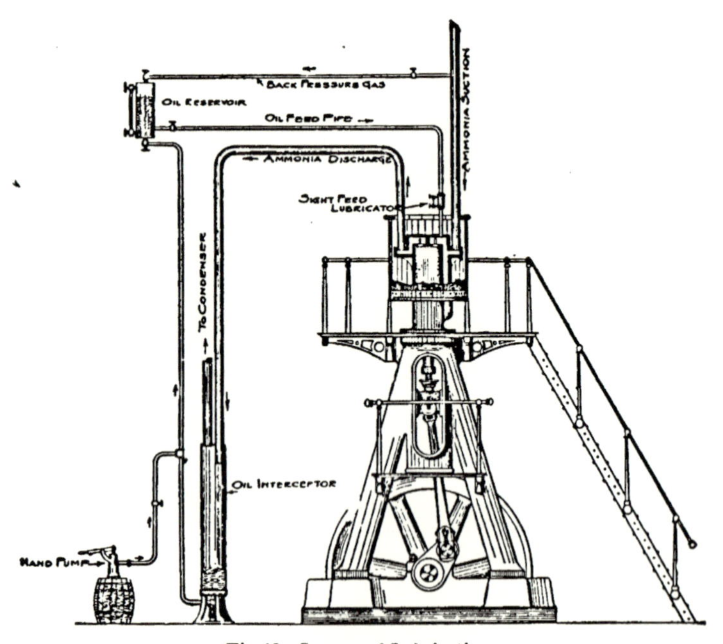

Fig. 18. System of Lubrication.

the reservoir type, or better still the droppers, fed from a large reservoir through a pipe, and which may be filled by a hand pump when necessary, Fig. 18. Owing to the action of ammonia on animal or vegetable oils, other than these must be used as lubricants for the compressor. The principal oil for this purpose (and when obtained pure, a very good one) is the West Virginia Natural Lubricating Oil or Mount Farm, which is a dark-colored oil not affected by the action of the ammonia or the low temperature of the evaporator.

Of late years the oil refining companies have put on the market a light-colored oil which appears to give good results for the purpose. Care should be used, however, in the selection; and oil should not be used unless it is of the proper grade, as serious results follow the use of inferior oils. The usual result is the gumming of the compressors and valves or the saponifying under the action of the ammonia through the system.

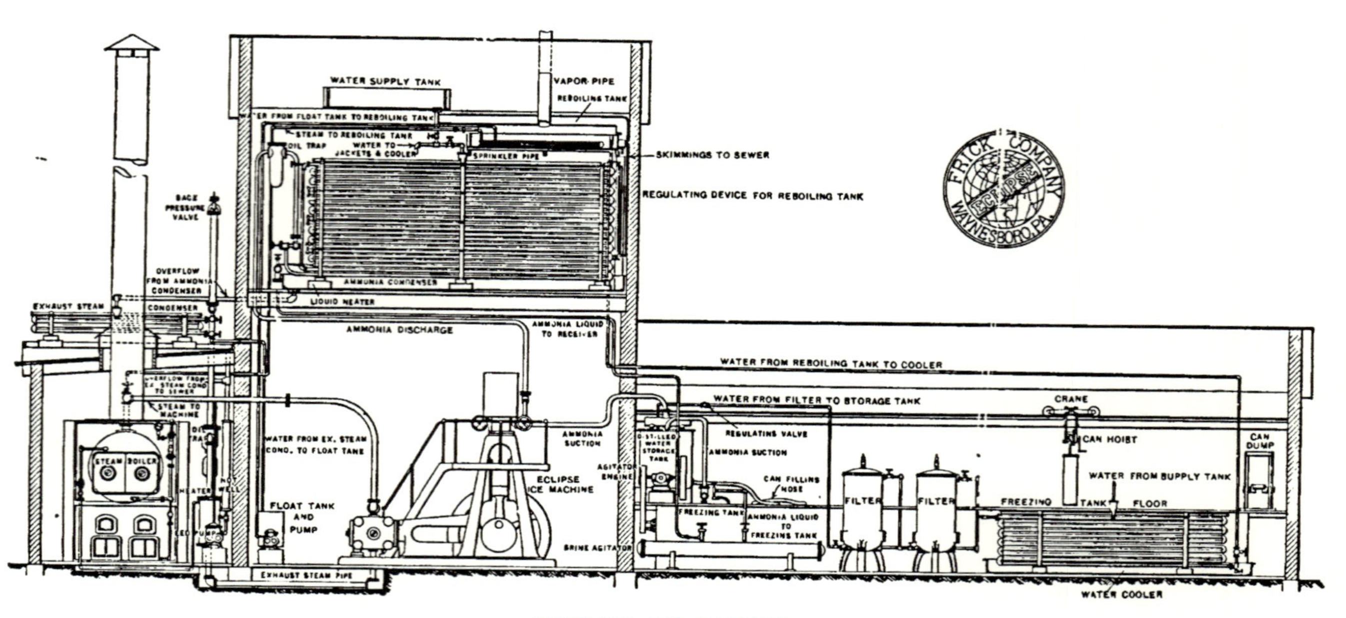

COMPLETE ICE FACTORY.
Frick Company.

REFRIGERATION

PART II

COMMERCIAL MACHINES

No attempt is made here to catalogue all the many refrigerating machines on the market, but some of the leading points of typical machines of the principal makes will be pointed out briefly. Thus the student can get an insight into the general principles of design and arrangement as applied to refrigerating machines, and can supplement the descriptions here given by detail study of manufacturers' catalogues, by which means a detailed knowledge of any particular machine or group of machines may be had.

Horizontal Double-acting. *Triumph.* Fig. 19 shows in a striking manner the simple and massive construction of this machine. One continuous bed plate gives a firm support for the compressor cylinder, the crosshead, and the crankshaft. The crosshead is of compact form and is supported on shoes. These have large bearing surfaces with an arrangement to take up wear by wedge and screw adjustment. The crosshead pin is a ground taper fit, and is held in position by a bolted disk. Owing to the piston being screwed into the crosshead and fastened with a lock nut, it is a simple matter to adjust the clearance for both ends of the cylinder. The connecting rod is made from a heavy steel forging and is provided with a bronze box at the crosshead end, while a babbitt-lined brass box is used on the crank pin, both boxes being fitted with wedge and screw adjustments.

Something has already been said of the valve adjustment which is made by means of a nut on the stem, the tension of the cushion spring being regulated by turning the nut after the lock nuts have been loosened. On the collar under the adjusting nut is a secondary collar with which the working spring is adjusted, and these two collars are held in their correct positions by keepers. This adjustment feature of the valves has an important bearing on the economy of the compressor, as it is evident that the same pressures cannot be used under varying conditions with maximum economy at all times.

Particular attention is directed to the stuffing box of this machine, which is divided into three parts separated by two cages, which are of spider frame construction as shown in the figure. One of the cages forms a relief chamber from which any gas that may leak past the first packing is returned to the suction manifold, while the other serves

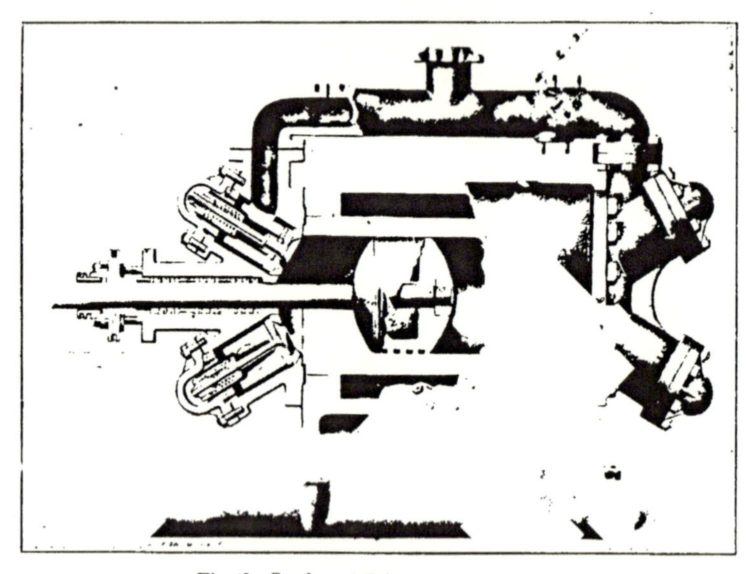

Fig. 19. Section of Triumph Compressor.

as an oil reservoir that keeps the rod and packing well lubricated. Oil is circulated through the gland by means of a small power pump driven from the shaft of the machine. The oil is drawn from a chamber provided in the base of the machine under the cylinder. Thus there is a continuous circulation of oil and it is necessary to pump against only the suction pressure. Fig. 20 shows an outside view of a 70-ton Triumph unit, in which the stuffing-box connections and lubricating apparatus are shown more in detail. It is noted that all the bolts of the gland are connected by inside gear so that in turning one nut, uniform adjustment is given to all. Thus there can be no trouble from *cocking* the gland by unequal adjustment.

Linde. This machine is similar in many respects to the Triumph, and is manufactured by the Linde Refrigerating Co., New York, with the usual heavy-duty type of frame and bored guides for the cross-head. The piston is of the spherical form used in all the best horizontal machines of the double-acting type, and enables the valves to be set close in the head so as to reduce clearance to a minimum. To facilitate handling, each valve is made so as to form virtually a single

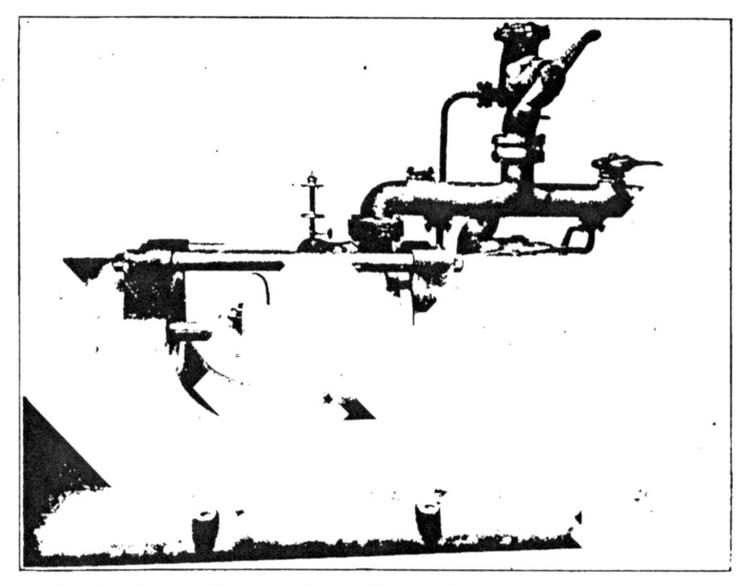

Fig. 20. Exterior Triumph Machine Showing Stuffing-Box Arrangement.

part with its seat, enabling the valve and seat to be removed together with the same labor that would be necessary to remove either separately. The valve-seat casting rests on the cylinder head and is held in place by the valve-stem guide which is secured in position by the cap or bonnet. With the bonnet off, the valve and all its parts can be removed practically as one piece and without disturbing any part of the machine.

Two self-expanding rings make the piston tight, and strength is given by heavy reinforcing ribs. The stuffing-box packing may be seen readily in Fig. 21, which is a sectional view of the cylinder of this machine. It will be seen that the arrangement is similar to the other double-acting stuffing boxes already described, the outer packing having to withstand only the pressure of the suction gas. Particular

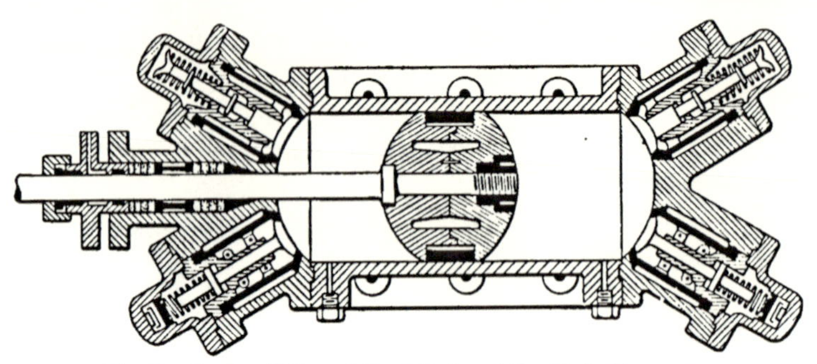

Fig. 21. Sectional View of the Cylinder of the Linde Compressor.

note should be made of the fact that there is no water jacket on the cylinder of the Linde machine. The distinctive feature of this machine is its operation on the *wet system,* where the cylinder is kept cool

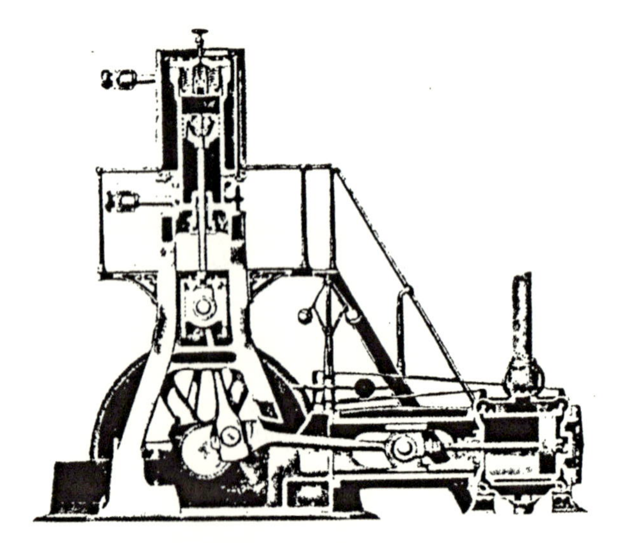

Fig. 22. Longitudinal Sectional Elevation of York Machine.

by the injection of a small amount of liquid ammonia at the beginning of the compression stroke; or, arranging the system so that a small part of the liquid is not evaporated and goes back to the compressor.

Vertical Compressors. Although some manufacturers make a vertical double-acting machine, the most notable being that of the De La Vergne Machine Co., the great majority of such machines are single-acting. The discussion will therefore be confined to this type of apparatus.

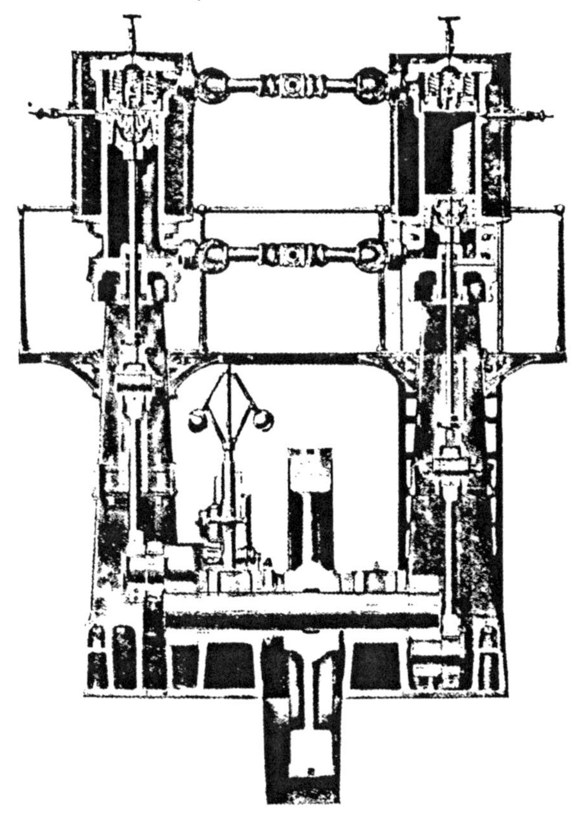

Fig. 23. Cross-Sectional Elevation of the York Machine.

York. Figs. 22 and 23 show longitudinal and cross-sectional views of the York machine direct-connected to its steam engine. It consists of two single-acting compressor cylinders mounted on vertical A-frames and driven from a Corliss engine of the horizontal type. A fly-wheel is mounted on the middle of the crank-shaft, and a crank on each end of the shaft drives the compressor cylinders, the cranks being set 180 degrees apart for reasons already mentioned. The connecting rod of the engine is attached to one of the cranks, as shown in the illustrations. Gas enters the cylinders through valves at the bottom underneath the piston and, on the down stroke of the piston, is forced through the valve in the piston to the space above so as to fill the cylinder. On the return or compression stroke of the piston the gas is compressed and, at the end of the stroke, is forced out through the valve in the upper head, going thence through the pipe connection to the condenser. The upper head itself, being of the safety type, may be considered as one huge valve, as in case anything should get in the cylinder, or the clearance become too small for any reason, the piston may strike the head without doing damage. The effect is similar to lifting the head against the action of heavy buffer springs, shown in the illustration, and allowing the charge in the cylinder to pass over to the condenser by the regular connections.

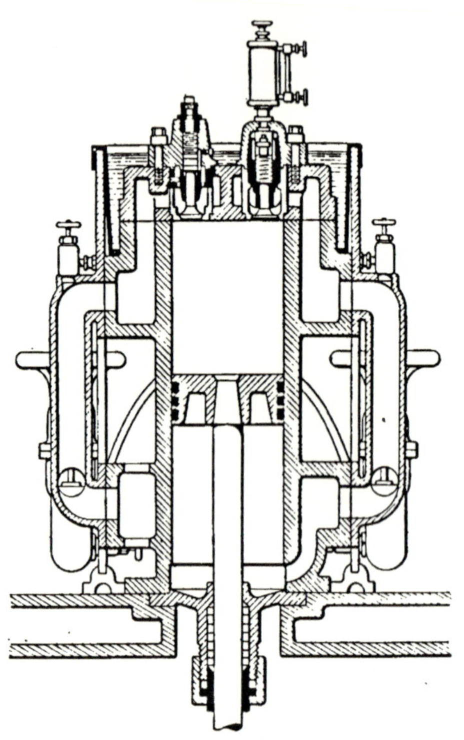

Fig. 24. Cylinder of Great Lakes Refrigerating Machine.

Great Lakes. In this machine, which is made by the Great Lakes Engineering Works, there is no valve in the piston. Separate suction and discharge valves are provided in the head of the machine, as shown in Fig. 24, there being two valves of each kind: Sight-feed lubricators are provided, as shown, and the cylinder has a specially arranged water-jacket around the upper end. One of the special features is the arrangement of the suction and discharge passages, which are connected to the piping system of the plant by a system of manifolds and by-passes that permit of handling the gas in any way desired. Like the York machine this compressor is single-acting, gas being drawn in through the right-hand valve, on the down-stroke, as seen in the illustration, and discharged through the other valve on the up-stroke. The two suction and the two discharge valves work, each pair, as one valve, they being made in pairs owing to the fact that there is not room enough within the head to give the proper diameter for the necessary valve opening for a single-suction and a single-discharge valve.

Carbon Dioxide Machines. Owing to the difficulty in getting sound castings suitable to withstand the pressure necessary to liquefy

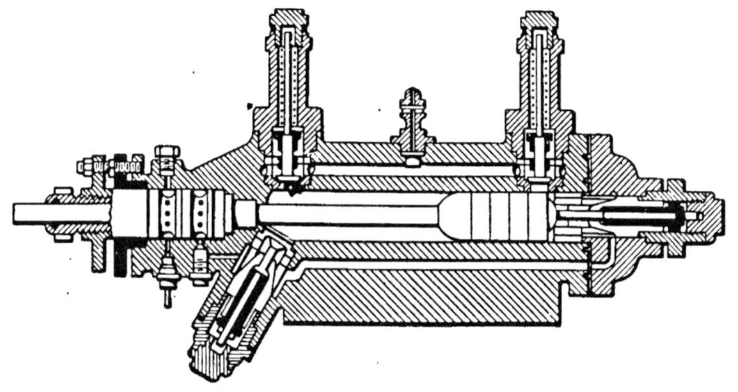

Fig. 25. Sectional View of Cylinder of Carbon-Dioxide Machine. Suction Passages are so arranged that coal gas passes around cylinder before entering suction valve.

carbon dioxide, manufacturers in the United States have largely adopted soft forged steel for the cylinders. With summer temperature of water the pressure may be as much as 1,000 pounds or more, and it is seen at once that the cylinders and piping must be very strong. The diameter of the gas cylinder must be small as compared with that of the steam cylinder. In some cases the compressors are made to compress the gas in stages. The gas leaves the first cylinder at a pressure of from 400 to 600 pounds per square inch, and is cooled before entering the second cylinder where it is compressed to the final pressure. Owing to the difficulty in keeping stuffing boxes tight with the high pressures, compressors are usually made single-acting, but some manufacturers have been successful with the double-acting machine.

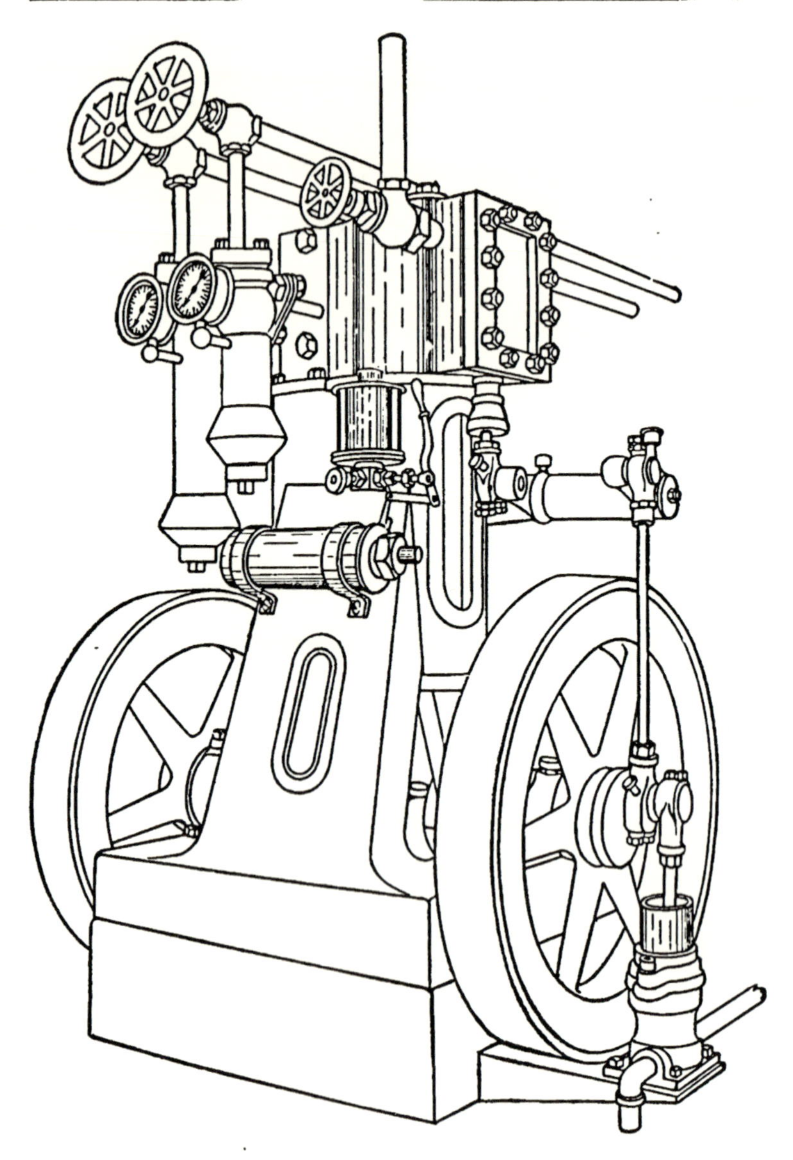

Fig. 26. Small Vertical Carbon-Dioxide Machine.

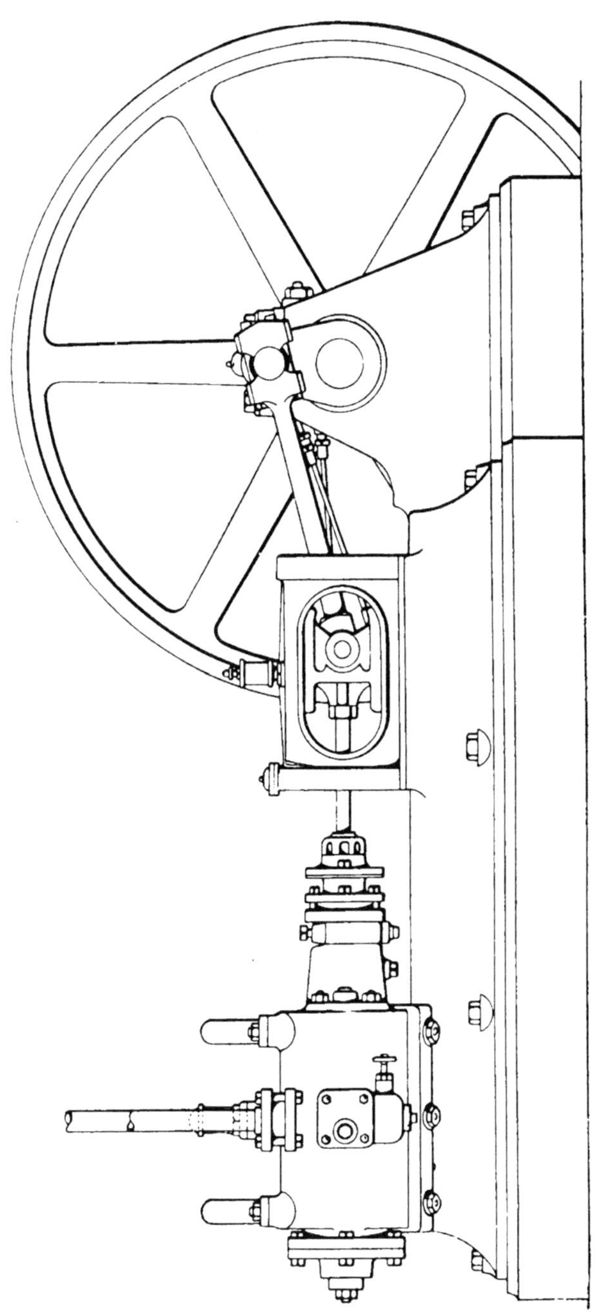

Fig. 27. Kroeschell Horizontal Carbon-Dioxide Machine.

Ordinarily the length of the stroke should be about four times the diameter of the cylinder, and, if the piston is to be kept tight, it should be at least two and one-half times the cylinder diameter. Fig. 25 is a sectional view of the typical cylinder in which it will be seen that the suction passages are arranged to pass the cool gas around the cylinder before entering it. Although the suction valves are usually placed in a horizontal position, they are easily closed by light springs as they are small and have little inertia. Guides are used to keep the valves in line with the seat; and the discharge valves, being set vertical, easily come to a true seating. Suction valves should have an area of about one-half that of the piston and the area of the discharge valves should be about one-seventh that of the piston.

Owing to the limited space on the crank end of the cylinder, it is usually necessary to have two suction valves for this end. For the discharge valves, the seats should be beveled at from 70 to 80 degrees, while for the suction valves the seats should be beveled from 60 to 75 degrees; and the seats for both valves are from 0.1 to 0.12 of the disk diameter in width. On the suction valve the lift should be about 0.33 of the disk diameter while that for the discharge valve should be about 0.28 of the diameter. Spring tension is 8 or 9 pounds on the suction and 10 or 11 pounds on the discharge valve. Stuffing boxes are made on the same general principles as those for the double-acting ammonia compressors, bearing in mind that the greater pressures call for more compartments. Cup leather packings are used except for the outer packing which is merely a wiper for the rod. Small machines are usually made vertical, Fig. 26, which shows a direct-connected vertical unit built by the Brown-Cochrane Co. Above 2 tons capacity, the machines are usually made horizontal, as shown in Fig. 27, which is an illustration of the double-acting compressor built by Kroeschell Bros., Chicago, Ill.

Small Refrigerating Plants. In recent years large consumers of ice have created a demand for small plants to be used on their premises and thus do away with the "ice man." As a result, machines are now made in capacities ranging upward from ¼-ton refrigerating duty. Such apparatus is used for a number of purposes. Where the consumer can obtain cheap electric power and is able to stand the first cost of the apparatus, there is some economy in operating a refrigerating plant by electricity. Where electric power is not avail-

12-TON LINDE REFRIGERATING MACHINE.

Operated by an Electric Motor.

Fred W. Wolf Company.

able and a special engine equipment must be used—gasoline engines as a rule—there will be no economy, and the matter of installing such a plant must be decided on other grounds. Where temperatures below 32° are required, the installing of such a plant is a necessity. Hospitals, restaurants, cafés, and saloons use the small refrigerating plant to advantage because they can keep the coolers drier, colder,

Fig. 28. Exterior of Brunswick Machine.

and more sanitary than by the use of ice. Residences and country clubs use such machines, owing to the fact that ice cannot be obtained readily. Where it is a mere question of refrigeration of cooler boxes there is an economy in using the refrigeration direct instead of melting ice. The machine thus used gives about twice as much cooling for a given amount of power expended as is secured by using the ice made by refrigeration.

As the small machine must be operated by servants and other unskilled persons, it is made as near automatic as possible, this being particularly the case with the machines designed for household use.

In the larger hotels and clubs the machine will be looked after by the engineer of the steam plant, and for machines above 1 ton refrigerating duty, it is generally sufficient to have a reliable source of water supply and a thermostat in the cooler to regulate the operation of the motor. Arrangements should be made so that the power and water are turned off or on simultaneously. Lubrication can be looked after by the attendant, but for machines smaller than 1 ton, as used in residences, it is advisable to have automatic lubricating devices and, in fact, have the whole machine practically take care of itself.

Fig. 28 is an exterior view of the complete Brunswick refrigerating machine made in New Brunswick, N. J., in sizes of 200 pounds

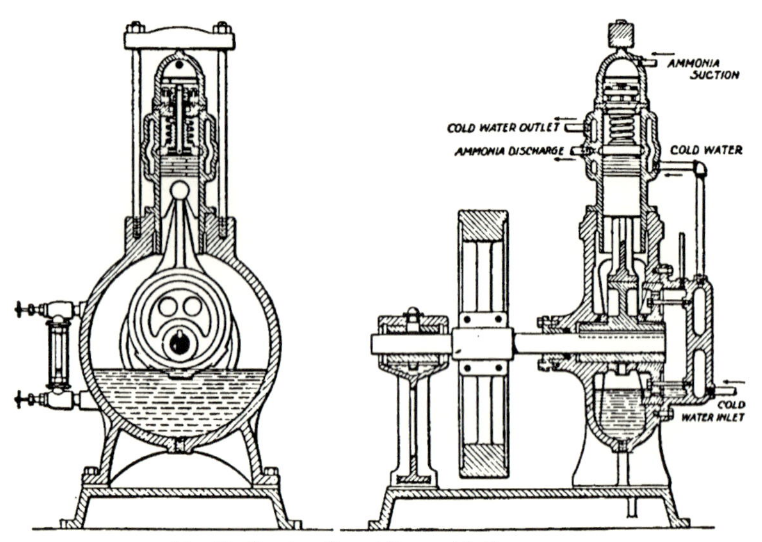

Fig. 29. Construction of Brunswick Compressor.

to 10 tons refrigerating duty or half as much ice-making capacity. Power is furnished by a motor belted to the band wheel seen in the illustration at the right-hand end of the shaft. The compressor is entirely self-contained in an enclosed crank case, which contains oil for lubricating purposes. An idea of the construction may be had from the sectional line drawings in Fig. 29. The machine is single-acting, the suction and discharge valves being of special design and made of steel, with the suction valve carried on the discharge valve and seating on its face. This construction makes it possible to have the discharge valve the full diameter of the cylinder so that it becomes a lifting head similar to the safety heads of large single-acting machines. Thus, there is no clearance and all the gas in the cylinder is forced out at each stroke, the piston, in fact, passing beyond the discharge port in the side of the cylinder and at once shutting off the port so that there can be no back slip of gas. As the piston reverses, it is followed by the discharge valve, which rests on its upper end, and as this valve comes to its seat with a slight impact there is no chance for the suction valve, which seats on its face, to get stuck and not open promptly. The lift of the suction valve is limited by a nut on the stem; but the discharge valve may lift as much as necessary to pass any obstruction that may get into the cylinder. Other details of the construction are made plain by the illustration which shows the eccentric (used instead of a crank on the shaft), and the arrangements made for lubrication. There are a number of automatic and semi-automatic machines on the market, in all of which the arrangements are more or less similar, the smallest units being self-contained with all parts mounted on a common base.

COMPRESSOR LOSSES

Having described the compressor and its parts, let us take up the losses due to the improper working or assembling of the parts of the machine, before proceeding with the description of the rest of the plant. As has been stated in a general way, the economy of the compressor lies in its filling at the nearest possible point to the evaporating pressure, and then compressing and discharging at the lowest possible pressure, as much of the entire contents of the cylinder as possible. If the compressor piston does not travel close to the upper end—of a single-acting machine—or the machine has excessive clearance, the compressed gas remaining in the cylinder re-expands on the downward stroke of the piston, and the gas from the evaporator will not be taken into the compressor until the pressure falls to, or slightly below, this point, and the loss due to this fault is equal to the quantity of gas thus prevented from entering the compressor plus the friction of the machine while compressing the portion of the gas thus expanding.

If we make a full discharge of the gas and there is a leaky outlet valve in the compressor, the escape and re-expansion into the compressor affects not only the intake of the gas at the beginning of the return stroke, but continues to affect the amount of incoming gas during the entire stroke and the capacity of the machine will be correspondingly reduced. If the inlet valve is leaky or a particle of scale or dirt becomes lodged on its seat, as the piston moves upward the portion of the gas which may escape during the period of compression is forced back to the evaporator and a corresponding loss is the result. A piston which does not fit the compressor, faulty piston rings, or a compressor which has become cut or worn to the point of allowing the escape of gas between the cylinder and piston has the same effect as the ill conditioned suction valve. The loss due to leaky or defective cylinders, joints, or stuffing boxes, are not included under this head, as these more generally effect the loss of the material than the efficiency of the compressor.

AMMONIA CONDENSERS

The ammonia condenser, or liquefier, as briefly stated in the description of the system, is that portion of the plant in which the gas from the evaporator, having been compressed to a certain point, is cooled by water and thereby deprived of the heat which it took up during evaporation; consequently it is reduced to its initial state, that is— liquid anhydrous ammonia. Condensers for other refrigerants are constructed in the same general way as those for ammonia, due regard being had to the pressures to be carried. Let us consider the general principles governing the action before describing the types.

On account of its duty having been performed, the ammonia as it leaves the evaporating coils is a gas at low temperature, usually 5° to 10° below that of the brine, or other body upon which it has been doing duty, yet it is laden with a certain amount of heat, although at a temperature not ordinarily expressed by that term. It is a well-known fact that we cannot obtain a refrigerating agent which can absorb heat from a body colder than itself, and it is therefore necessary to bring the temperature of the ammonia gas to a point at which the flow of heat from the one to the other will take place. This is done by withdrawing part of the heat in the ammonia in the following manner: The cold gas is compressed until its pressure reaches such a point that at ordinary temperatures it will condense to liquid form; as it leaves the compressor it is very hot because of the fact that it still contains nearly all of the heat it had when it left the evaporator, in only a small portion of the space occupied before. Thus when it reaches the condenser it is much warmer than the cooling water and will readily give up its heat to the cold water—so much that its latent heat is absorbed by the water and it condenses into anhydrous ammonia.

The temperature of water if pumped from surface streams will average about 60° F., and since we cannot expect to get the ammonia any colder than this, it must be compressed until the boiling point corresponding to the pressure obtained is at about 75° F. In Table XII, p. 31, we find that this temperature corresponds to a pressure of 141.22 pounds per square inch (absolute), or 126.52 pounds per square inch (gauge).

Thus if the gas is compressed until the gauge reads 126.52 and then passed into a condenser where the temperature of the water is less than 75° F., the water will absorb the latent heat and we have accomplished our object which was to remove some of the heat contained in the ammonia. In this condition it is drained from the condenser into the ammonia receiver to again repeat the cycle of operation.

The forms of condensers may be divided into three classes—the submerged, the atmospheric, and the double-pipe. Of each of these classes a number of different types and constructions are in use. To illustrate the general principles, however, it is only necessary to present one of each type.

Submerged Condenser. The submerged condenser consists of a round or rectangular tank with a series of spiral or flat coils within, joined to headers at the top and bottom with proper ammonia unions. In Fig. 30 is shown a sectional elevation of a popular type of submerged condenser. A wrought iron or steel tank *A* is formed by plates from $\frac{3}{16}$ to $\frac{5}{16}$ inch thick, of the necessary dimensions to contain the coils, and sufficiently braced around the top and sides to prevent bulging when filled with water. A series of welded zigzag pipe coils *B* are placed in the tank and joined to headers *C* with ammonia unions *D*. The ammonia gas enters the top header through the pipe *E*, and an outlet for the liquefied ammonia is provided at *F* with a proper stop valve. Water is discharged or admitted to the tank at or near the bottom and overflows at outlet *M*. It will be seen that in this type of condenser a complete reverse flow of the current is effected, the gas entering at the top and the liquid leaving at the bottom, while the water enters at the bottom and leaves at the top. This brings the cold water in contact with the cool gas, and the warm water in contact with the incoming or discharged gas from the compressor, thereby presenting the ideal condition for properly condensing ammonia.

Owing to the necessarily large spaces between the coils and the distance between the bent pipes, the portion of water coming in contact with the surface of the pipes must be small compared with the total amount passing through; it is, therefore, uneconomical as regards amount of water used. With water containing a large amount of floating impurities the deposit on the coils is considerable and not easily

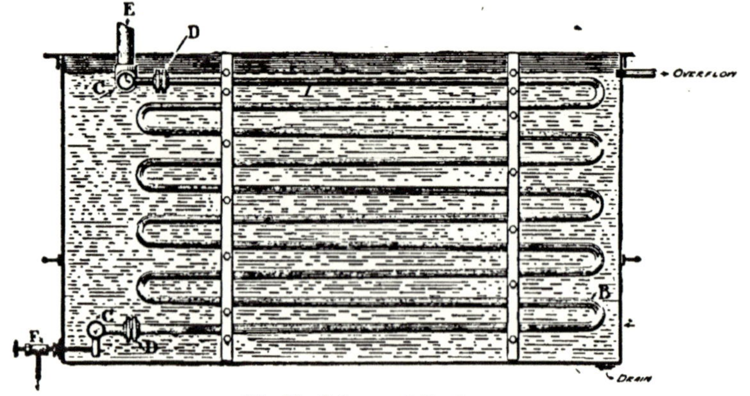

Fig. 30. Submerged Condenser.

removed owing to the limited space between the coils; and furthermore, the dimensions of the tank necessary to contain the requisite amount of pipe for a plant of considerable size is so great and its weight, when equipped with coils and filled with water, requires such a strong support, that its use is now limited to certain requirements and localities.

A better shape for a condenser of this type is one of considerable height or depth, rather than low and broad. This is owing to the fact that the greater length of travel of the water and gas in opposite directions, the greater the economy. The number of coils used should be such that the combined internal area of the pipes equals or exceeds the area of the discharge pipe from the compressor. The circular submerged condenser is similar to the above described except that the tank is circular and the coils bent spirally.

In the circular type of submerged condenser the pipes are $1\frac{1}{4}$ to 2 inches in diameter, and the separate coils are made in lengths up to 350 feet. A number of coils are used in a single condenser, the inlets and outlets being connected to manifolds with valves provided to shut off any individual coil. Where the water comes to the condenser at 70° and leaves at 80°—a range of 10 degrees—about 40 square feet of condensing surface, corresponding to 64 running feet of 2-inch pipe or 90 feet of $1\frac{1}{4}$-inch pipe are allowed per ton of refrigeration. Less surface than this means excessive condensing pressure. Siebel gives the following empirical formula for calculating the square feet of cooling surface F required in submerged condensers:

$$F = \frac{hk}{m\,(t - t')}$$

In this formula, h = the heat of vaporization of 1 pound of ammonia at the temperature of the condenser; k = the amount of ammonia passing through the condenser in one minute; $m = 0.5$ = the number of heat units transferred per minute per square foot of iron surface where the pipe contains ammonia vapor and is cooled by water; t = the temperature of the ammonia in the coils; t' = the mean temperature of the inflowing and outflowing cooling water.

The heat taken up by the ammonia, in producing refrigeration, added to that corresponding to the work done on the ammonia in the compressor, less any heat expended in superheating the gas, is equal theoretically to the heat of vaporization of ammonia at the temperature of the condenser and is the amount of heat that must be removed by the cooling water. This then gives a gauge on the amount of cooling water that should be used in the plant. For finding the number of pounds A of cooling water, Siebel gives the formula:

$$A = \frac{hk \times 60}{t - t'}$$

in which the notation is the same as in the formula above, except that t is the temperature of the outgoing cooling water, and t' that of the incoming water. The result is converted into gallons by dividing by the factor 8.33. Usually from $\frac{3}{4}$ to 3 gallons of water are required per minute per ton of refrigeration in 24 hours.

Atmospheric Condenser. This type of condenser most generally used is made of straight lengths of 2-inch extra strong, or special pipe, usually 20 feet long, screwed, or screwed and soldered into steel return bends about 3½-inch centers and usually from eighteen to

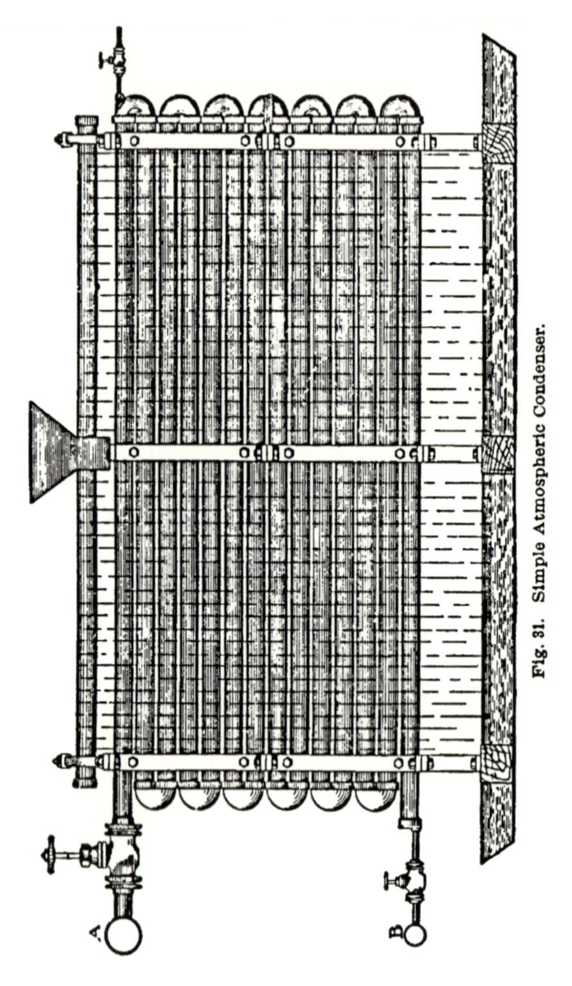

Fig. 31. Simple Atmospheric Condenser.

twenty-four pipes high. The coil is supported on cast or wrought iron stands and placed within a catch pan, or on a water-tight floor, having a proper waste water outlet, and supplied with one of the several means of supplying the cooling water over their surfaces.

Stop valves, manifolds, and unions connect with the discharge of the compressor and the liquid ammonia supply to the receiver.

In the manner of making the connections to this type of con-

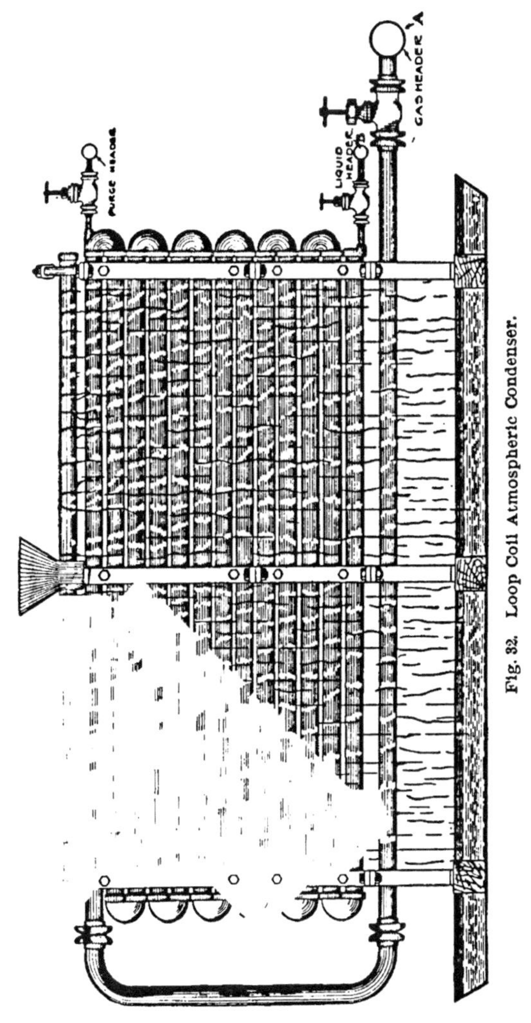

Fig. 32. Loop Coil Atmospheric Condenser.

denser and the taking away of the liquefied ammonia as well as in the devices for supplying the cooling water, a great variety exists. Fig. 31 represents a side elevation of an ammonia condenser with the discharge or inlet of the gas from the compressor entering at the top A, and the liquid ammonia taken off at the bottom B, while the water is supplied over the coils flowing down into the catch-pan or water-tight floor, where it accumulates and is taken away by any of

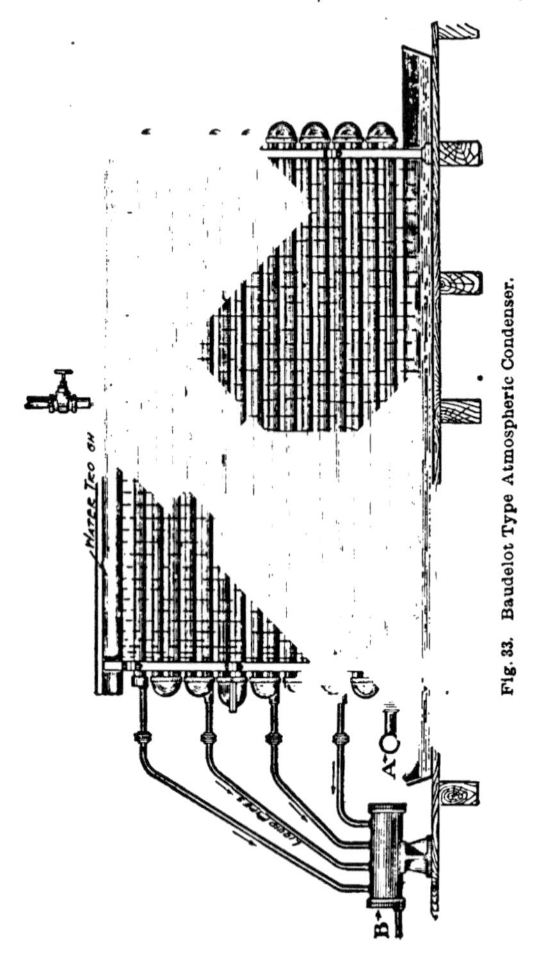

Fig. 33. Baudelot Type Atmospheric Condenser.

the usual means. It will be noticed that the flow of the water and the gas with this type of condenser is in the same direction, the coldest water coming in contact with the warmest ammonia. The temperature governing or determining the point of condensation will be that at which the ammonia leaves the condenser, or the temperature in the bottom pipe from which the liquid ammonia is withdrawn. Owing to this arrangement it is not favorable to a low condensing pressure or economy in the water used. Fig. 32 represents a type in which an attempt is made to eliminate this undesirable feature, and in which it is expected to use the waste from the condenser proper, in taking out the greater part of the sensible heat from the gas leaving the compressor.

The construction of this condenser is identical with that shown in the preceding figure except that its uppermost pipe is continued down and under the pipes forming the condenser proper; it passes backward and forward in order that a large proportion of the heat may be removed by the water from the condenser proper, before the ammonia enters the condenser. A supplemental header is sometimes introduced in connection with this pipe for removing any condensation taking place in it.

A third type of this condenser is shown in Fig. 33. In this type a reverse flow of the gas and water takes place. The gas enters the condenser through a manifold or header *A* at the bottom and continues its flow upward through the pipes to the top; at several points drain pipes are provided for taking off the condensation into the header *B*. The condensing water flows downward over the pipes. This type of condenser is the most nearly perfect of its class.

The atmospheric condenser is a favorite, and possesses many features that make it preferable to the submerged. Its weight is a minimum, being only that of pipe and supports and a small amount of water. The sections or banks may be placed at a convenient distance apart to facilitate cleaning and repairs. The atmospheric effect in evaporating a portion of the condensing water during its flow over the condenser, makes use of the latent heat of the water in addition to the natural rise in its temperature.

The various devices for distributing the water over the condenser are numerous. Fig. 34 represents the simplest and most easily obtained—a simple trough with perforations at the bottom for allowing the water to drip over to the condenser.

Fig. 35 is a modification of the one shown in Fig. 34. This is intended to prevent the clogging of the perforations, by allowing the water to flow into the space at one side of the partition, and then through a series of perforations into the second, and thence through a second set of perforations in the bottom to the pipes in the condenser.

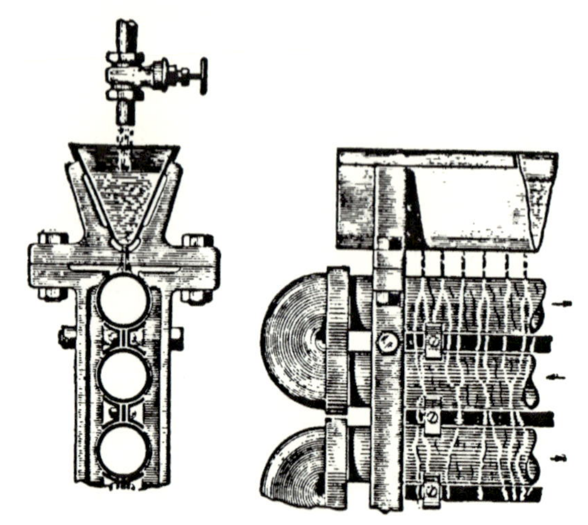

Fig. 34. Simple V-Shaped Water Trough.

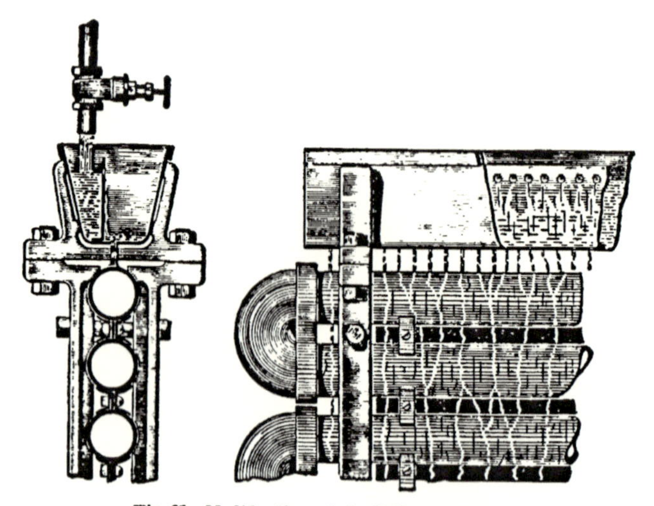

Fig. 35. Modification of the V-Shaped Trough.

Fig. 36 is a type of trough, or water distributor, designed to overcome the objections to a perforated form of trough, and the consequent difficulties due to the clogging or filling of the perforations. As will be readily understood from the illustration, this is also made of galvanized sheet metal with one side enough higher than the other to cause the water to overflow through the V-shaped notches or openings along the top of the straight or vertical side of the trough, and down and off the serrated bottom edge to the pipes.

The object of the serrated edges, as will be apparent, is the more even distribution of the water, owing to the fact that while it would be practically impossible to obtain a uniform flow of water over a

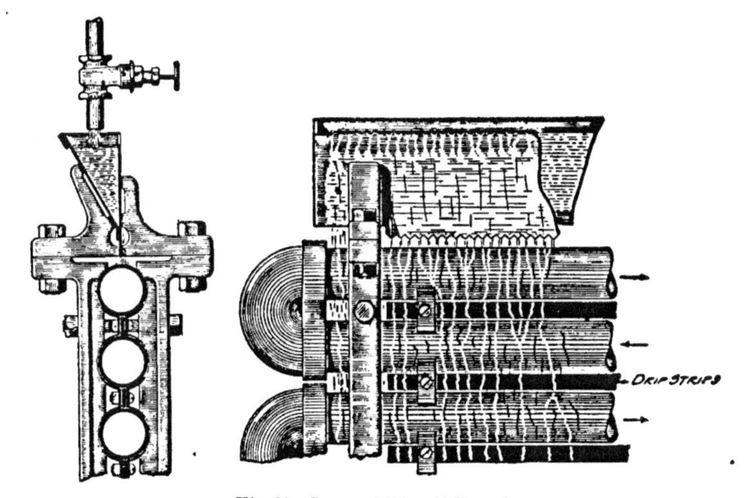

Fig. 36. Serrated-Edge V-Trough.

straight and even edge of a trough, particularly if the amount is limited, it is an easy matter to regulate the flow through the V-shaped openings.

Fig. 37 is termed the *slotted water pipe*. It is a pipe slotted between its two ends, from which the water overflows to the series of pipes below. It is good practice to lead the water supply to a cast-iron box at the center of the condenser, into the sides of which is screwed a piece of pipe—usually 2-inch—reaching the ends of the condenser, and having its outer ends capped, which may be removed while a scraper is passed through the slot from the center towards the ends while the water is still flowing, thereby carrying off any deposit within the pipe. This forms a very durable construction, and one not liable, as with the galvanized drip trough, to disarrangement or bending out of shape due to various causes. It is impossible, however, to obtain the uniform flow of water over the condenser with this, as with the serrated or perforated troughs, particularly if the supply is limited, from the fact as stated in describing the overflow trough, viz, the impossibility of obtaining a sufficiently thin stream of that length.

Double-Pipe Condenser. This is a modern adaptation of an old idea, given up owing to its complex construction and the imperfect facilities available for its manufacture. It has come into use with great rapidity, and has brought forth many novel ideas of principle and construction. It combines the good features of both the atmospheric and the submerged types, having small weight and being accessible for repairs. It has the downward flow of the ammonia and the upward flow of the water, effecting a complete counter flow of the two, minimizing the amount of water required and taking up the heat of condensation with the least possible difference between the ammonia and the water.

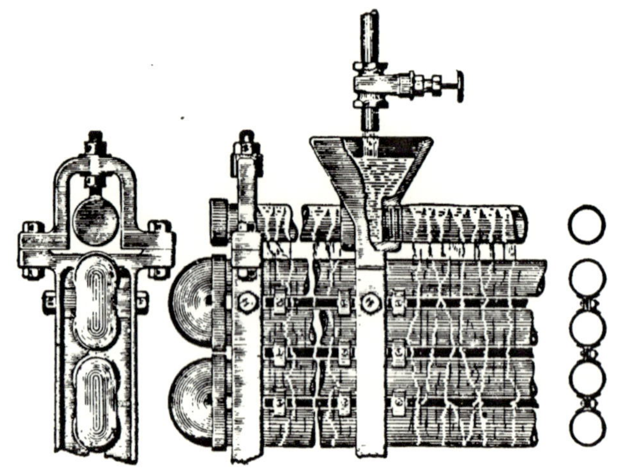

Fig. 37. Slotted Water Pipe.

The two general forms of construction are a combination of a $1\frac{1}{4}$-inch pipe within a 2-inch pipe, or a 2-inch pipe within a 3-inch. The water passes upward through the inner pipe, while the gas is discharged downward through the annular space; or, the position of the two may be reversed, the ammonia being within the inside pipe while the water travels upward through the annular space.

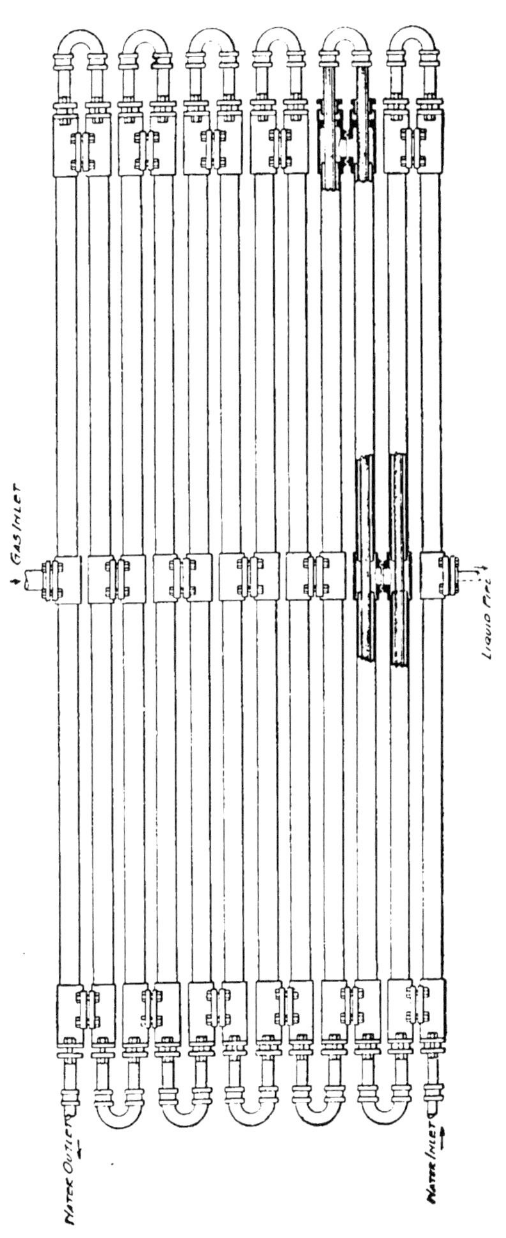

Fig. 38. Type of Double-Pipe Condenser.

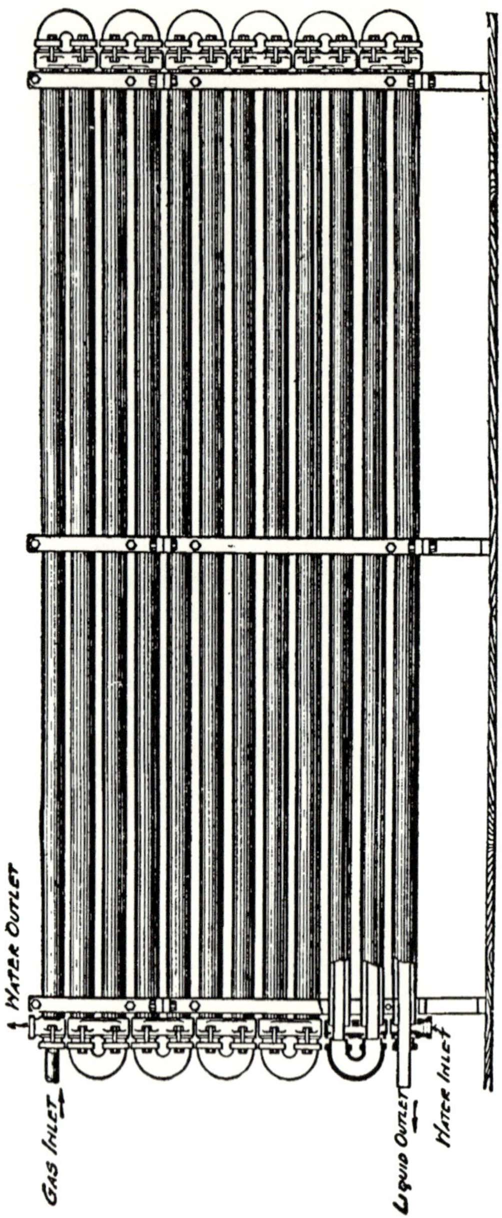

Fig. 39. Type of Double-Pipe Condenser.

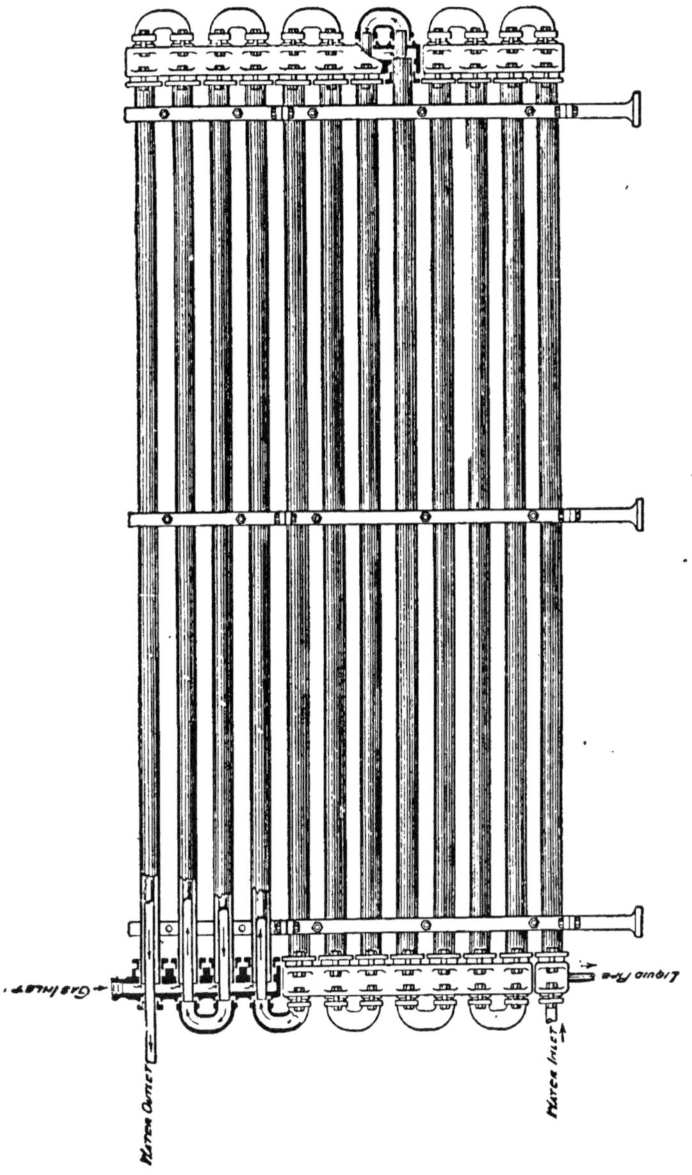

Fig. 40. Type of Double-Pipe Condenser.

They are also constructed in series, in which the gas enters a number of pipes of a section at one time, flowing through these to the opposite end to a header or manifold, at which point the number of pipes is reduced, and so on to the bottom with a constantly reduced area. The theory of this construction is, that the volume of the gas is

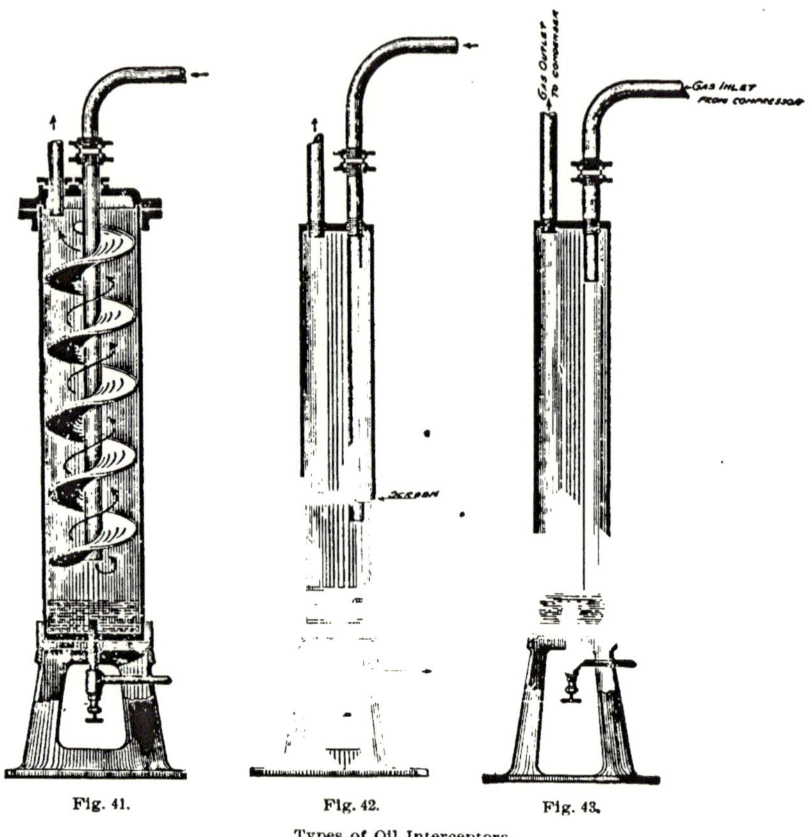

Fig. 41. Fig. 42. Fig. 43.

Types of Oil Interceptors.

constantly reduced as it is being condensed. Figs. 38, 39, and 40 illustrate a general range of the various types in use.

It is usual in the construction of this type to make each section or bank twelve pipes high by about 17½ feet long; they are rated nominally at ten tons refrigerating capacity each, although for uneven units the construction is made to vary from 10 to 14 pipes in each.

Oil Separator or Interceptor. This is a device or form of trap, placed on the line of the discharge between the compressor and the condenser to separate the oil from the ammonia gas. It is to prevent the pipe surface of the condensing and evaporating system from becoming covered with oil, which acts as an insulator and prevents rapid transmission of heat through the walls.

There are many forms of this device in common use, including the plain cylindrical shell with an inlet at one side or end and an outlet at the other, an almost endless variety of baffle plates, spiral conductors, and reverse-current devices. The object is similar to that obtained in the steam or exhaust separator, and generally speaking, that which would be effective in one service would be so in the other. Figs. 41, 42, and 43 illustrate three of the most common types in use; from these the student will understand the general principles.

COOLING TOWERS

In cities and other localities where water is scarce or of high temperature, it becomes important to conserve the supply brought to the plant. In order to do this it is necessary to cool the water after it has passed over the condensers and coolers of the refrigerating plant. This is done by evaporation of a part of the water under atmospheric pressure so that the remaining water will be cooled, owing to the abstraction of heat that becomes latent in the water evaporated. The process may be employed to advantage, even where there is plenty of water, in order to save pumping; or where the water is taken from muddy streams to a settling basin, to avoid the use of more muddy water than absolutely required. The efficiency of any apparatus for this purpose depends on the extent of the water surfaces exposed and on the amount of air brought in contact with the water. Also to some extent the pressure of the air and, to greater extent, the dryness of the air are factors having their influence, but acting alike on any apparatus however constructed.

Apparatus designed to cool water for re-use in refrigerating plants are known as *cooling towers* and of these there are many types on the market. All such towers may be divided into two classes according to whether the circulation of air is by natural or artificial currents. Also towers are classified as to material used, whether steel or wood. Both materials are used for towers operating with both kinds of air circulation. An efficient tower can, in fact, be made

Fig. 44. Atmospheric Wood Cooling Tower.

very cheaply by throwing brush into a framework arranged to prevent the brush packing, thus leaving space for air currents. Where wood is used, the tower may be anything from a cheap slatted structure made of rough lumber to an elaborate tower such as shown in

Fig. 45. Steel Forced-Draft Cooling Tower.

Fig. 44. Here No. 1 lumber is used and treated by a preservative process before being put into the tower, and the structure is painted with mineral paint when finished. This illustration represents the tower constructed for the Bauer Ice Cream and Baking Co. of Cincinnati, O., by the Triumph Ice Machine Co., for a 75-ton refrigerating plant using city water. Fig. 45 shows a circular steel tower using forced air circulation installed by the Triumph people for the Cincinnati Ice Co.

A good form of the atmospheric or *natural draft* steel tower is made by B. Franklin Hart Jr. and Co. of New York, as shown by Fig. 46, complete with spray preventors. In towers using natural

Fig. 46. Cooling Tower Installed at a Large Brewery in Monterey, Mexico.

air currents, care should be taken to set the apparatus clear of all buildings, etc., so that there is free access of the air. Such towers should be designed with about 12 square feet of cooling surface for each 5 gallons of water to be cooled per minute, with which surface the temperature of the water will be lowered from 7 to 15 degrees depending on conditions. From 1½ to 7 per cent of the water passed over the tower is evaporated, and with losses by leakage, etc., this may be increased to 10 or 12 per cent, which represents the amount of make-up water that must be supplied. With artificial draft from 5 to 25 horse-power is required to operate the fans, according to the size of the tower and the type of fan used. Either suction or pressure fans may be used in such towers, as it is not definitely settled whether best results are had by forcing air in at the bottom of the tower or drawing it down from the top. Most towers, however, force the air in at the bottom, thereby getting the flow in the opposite direction to that of the water.

EVAPORATORS

Evaporators may be divided into two classes. The first is operated in connection with the *brine* system. In this evaporator salt brine, or other solution, is reduced in temperature by the evaporation of the ammonia or other refrigerant, and the cooled brine circulated through the room or other points to be refrigerated. In the second, the *direct-expansion* system, the ammonia or refrigerant is taken directly to the point to be cooled, and there evaporated in

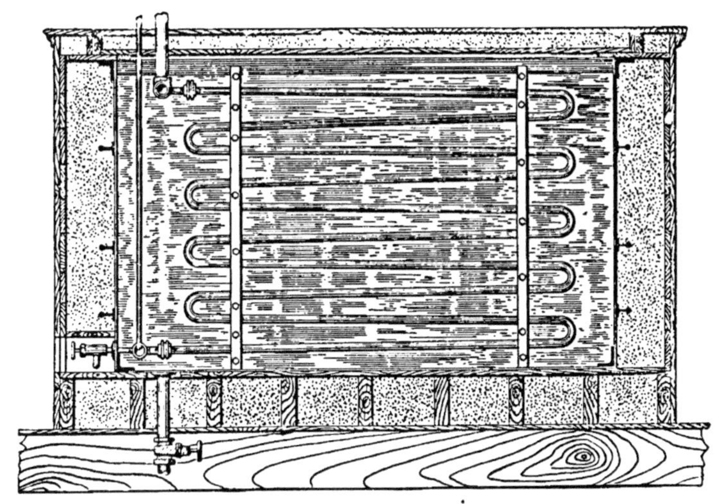

Fig. 47. Rectangular Brine-Cooling Tank.

pipes or other receptacles, in direct contact with the object to be cooled. Which of the two systems is the better, is a much disputed and debated point; we can state, in a general way, that both have their advantages, and each is adapted to certain classes of duty.

The cooling of brine in a tank by a series of evaporating coils—one of the earliest methods—is common to-day. A description of the many methods of construction and equipment would require much space. Let us, therefore, discuss the two most general types, viz, the rectangular with flat coils, and the double-pipe cooler—the spiral-coil cooler being practically obsolete.

Brine Tank. Fig. 47 shows a sectional view of a brine-cooling tank. Flat or zigzag evaporating coils are connected to manifolds or headers; the pipe connections leading to and from these manifolds for the proper supplying of the liquid ammonia and the taking away or return of the gas to the compressor are also shown. For coils of this type, 1-inch or 1¼-inch pipe is preferable, owing to the impossibility of bending larger sizes to a small enough radius to get the required amount into a tank of reasonable dimensions. It is possible to make coils of this construction of any desired length or number of pipes to the coil, "pipes high," the bends being from 3½ in. to 4 in. centers for 1-inch pipe, and 4 in. to 5 in. for 1¼-inch pipe. It is preferable to make the coils of moderate length—not less than 150 feet in each—and there is no disadvantage, other than in handling, in making them to contain up to 500 or 600 feet each. It will be observed that there is a slight downward pitch to the pipes with a purge valve at the lowest point of the bottom manifold, which is valuable and an almost necessary provision. This valve is for removing foreign matter that may enter the pipes at any time. By opening the valve and drawing a portion of their contents, the condition of cleanliness can be determined without the necessity of shutting down and removing the brine and ammonia for inspection. The coils are usually strapped or bound with flat bar iron about ¼ inch × 2 inches, or a little heavier for the longer coils, and bolted together with ½-inch square-head machine bolts. The coils are painted with some good water-proof or iron paint.

The brine tank is usually constructed of iron or steel plates, varying from $\frac{3}{16}$ inch to ⅜ inch in thickness; the average being ¼ inch for tanks of ordinary size. The workmanship and material for a tank of this kind should be of the very best; without these the result is almost certain to be disastrous to the owner or builder. The general opinion with iron workers, before they have had experience, is that it is a simple matter to make a tank which will hold water or brine, and that any kind of seam or workmanship will be good enough for the purpose. On the contrary the greatest care and attention to detail is necessary. It is customary, and good practice, to form the two side edges at the bottom by bending the sheets, thereby avoiding seams on two sides; while for the ends an angle iron may be bent to conform to this shape and the two sheets then riveted to the flanges of the angle iron. The edges of the sheets should be sheared or planed bevel, and after riveting, calked inside and out with a round-nosed calking tool. The rivets should be of full size, as specified for boiler construction, and of length sufficient to form a full conical head of height equal to the diameter of the rivet, and brought well down onto the sheet at its edges. An angle iron of about 3 inches should be placed around the top edge and riveted to the side at about 12-inch centers.

One or more braces, depending on the depth, should extend around the tank between the top and bottom, to prevent bulging; without these it would be impossible to make the tank remain tight, as a constant strain is on all its seams. A very good brace for the purpose is a deck beam. Flat bar iron placed edge-wise against the tank with an angle iron on each side and all riveted through and to the side of the tank with splice plates at the corners, or one of each pair long enough to lap over the other, makes a good brace. Heavy T-iron is also used to some extent. It is usual to rivet the bottom of the tank and a short distance up the sides, then test by filling with water; if tight, lower the tank to its foundation and complete the riveting and calking. It may then be filled with water and tested until proven absolutely tight, when it may be painted with some good iron paint; it is now ready for its equipment of coils and insulation.

A washout opening with stop valve should be placed in the bottom at one corner; for this purpose it is well to have a wrought iron flange, tapped for the size of pipe required, riveted to the outside of the bottom. If the brine pump can be located at this time, it is well to have a similar flange for the suction pipe riveted to the side or bottom of the tank, as a bolted flange with a gasket is never as durable as a flange put on in this manner.

After the tank has been made absolutely tight and painted, the insulation may be put around it, the insulated base or foundation having been put in previous to the arrival of the tank. The insulation should be constructed of joists 2 in. or 3 in. × 12 in., on edge, and the space should be filled in with any good insulating material and floored over with two thicknesses of tongued and grooved flooring with paper between the thicknesses. In putting the insulation on the sides and ends of the tank, place joists, 2 in. × 4 in., so that they rest on the projecting edges of the foundation about 2 feet apart. The upper ends should be secured to the angle iron at the top of the tank, its upper flange having been punched with ⅝-inch holes, 18 in. to 24 in. centers, and to which it is well to bolt a plank, having its edge project the required distance to receive the uprights. Between the braces around the tank, blockings should be fitted to secure the framework at the middle, as the height of some tanks is too great to depend on the support at the top and the bottom alone.

After the framework has been properly formed and secured to the base and tank, take 1-inch flooring, rough, or planed on one side, and board up on the outside of the uprights. Fill in the space, as the work progresses, with the insulating material which may be any one of the usual materials. Granulated cork is about the best, all things considered, although charcoal, dry shavings, sawdust, or other non-conductors may be used with good results. When the first course of boards is in place, it is well to tack one or two thicknesses of good insulating paper against the outer surface, care being taken that the joints lap well and that bottoms and corners are filled and turned under at the junction with the bottom insulation. It is then in shape for the final or outer course which is very often made of some of the hard woods in 2½-inch or 3-inch widths, tongued, grooved, and beaded. It is finished off with a base board at the bottom, moulding at the top, and given a hard wood finish in oil or varnish. If the tank is located in a part of the building in which appearance is of no importance, the outer course may be a repetition of the first, except that the boards are put on vertically instead of horizontally.

It is well to make the top of the tank in removable sections to facilitate examination or cleaning; for this purpose make a number of sections, about 2½ to 3 feet wide, of the length or width of the tank, using joists about 2 in. × 6 in. placed on edge, floored over top and bottom, and filled in with the selected insulating material. It is also well to have a small lid at one end of each, preferably over the headers or manifolds, which will allow of internal examination of the tank to ascertain the height or strength of brine without removing the larger sections. The tank is now fully equipped and ready for testing and filling with brine.

For a *circular* tank, the general instructions regarding construction and insulation may apply as with the rectangular tank just described; therefore only its special features will be considered. If the tank is small and there is sufficient head room above it for handling the coils there cannot be serious objection to this type, as its cost is lower than that of the rectangular tank. This is often an

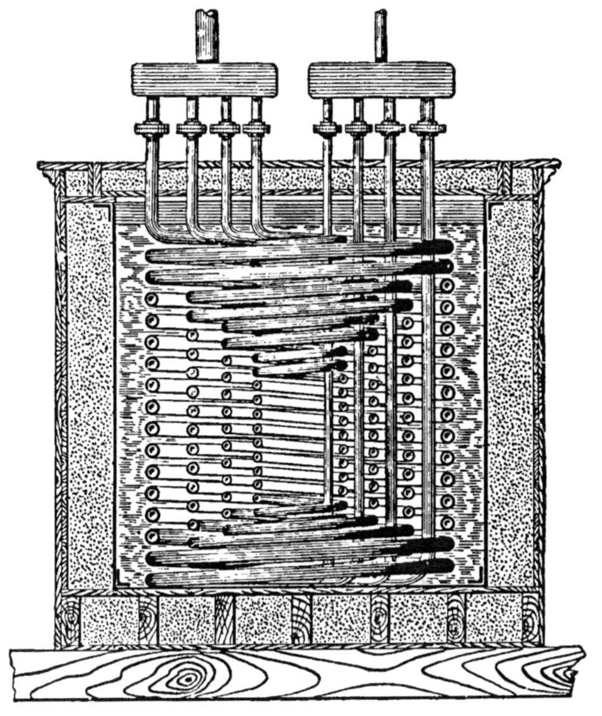

Fig. 48. Circular Brine-Cooling Tank.

important item in a small installation, but when the tank is of considerable size and the coils large it is not as readily handled and taken care of as the other type. The usual construction of a nest of coils for a round tank is to bend the inside coil to as small a circle as possible, which, if it be of 1-inch or $1\frac{1}{4}$-inch pipe, may be 6 to 8 inches. Increase each successive coil enough to pass over the next smaller until the required amount of pipe is obtained. The ends may then be bent up or out and joined to headers at the top and the bottom and the tank insulated in the manner previously described; it is then ready to test and charge with ammonia and brine. Fig. 48 represents an evaporator of this type.

Other constructions of tanks and coils are too numerous to describe in detail, and with one exception may be properly classed in one of the preceding types. The one exception referred to is illustrated in Fig. 49. It is quite common and is adopted for large pipe; it is often called *oval*, although not of that shape, but rather a combination of the flat and circular form. It has some good

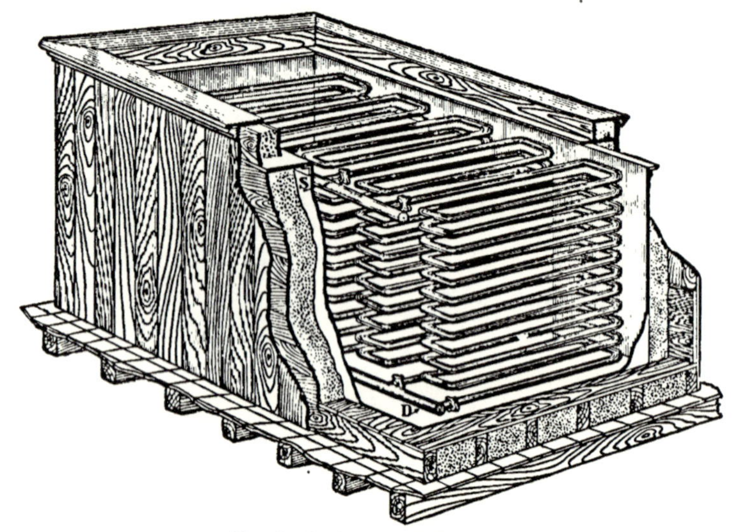

Fig. 49. Oval Brine-Cooling Tank.

features; it allows the maximum amount of pipe in the smallest space and a large amount of pipe in a single coil.

Brine Cooler. The brine cooler at present is a popular and efficient method of cooling brine for general purposes. Owing to mechanical defects and the impossibility of obtaining a brine solution which would not freeze, it was abandoned only to be taken up again, and with the aid of modern ideas and better material it has become highly successful. The great advantage of the brine cooler over the tank method of cooling brine is the fact that the brine and gas are both in circulation, passing through the double-pipe cooler in opposite directions so as to get the greatest efficiency as well as

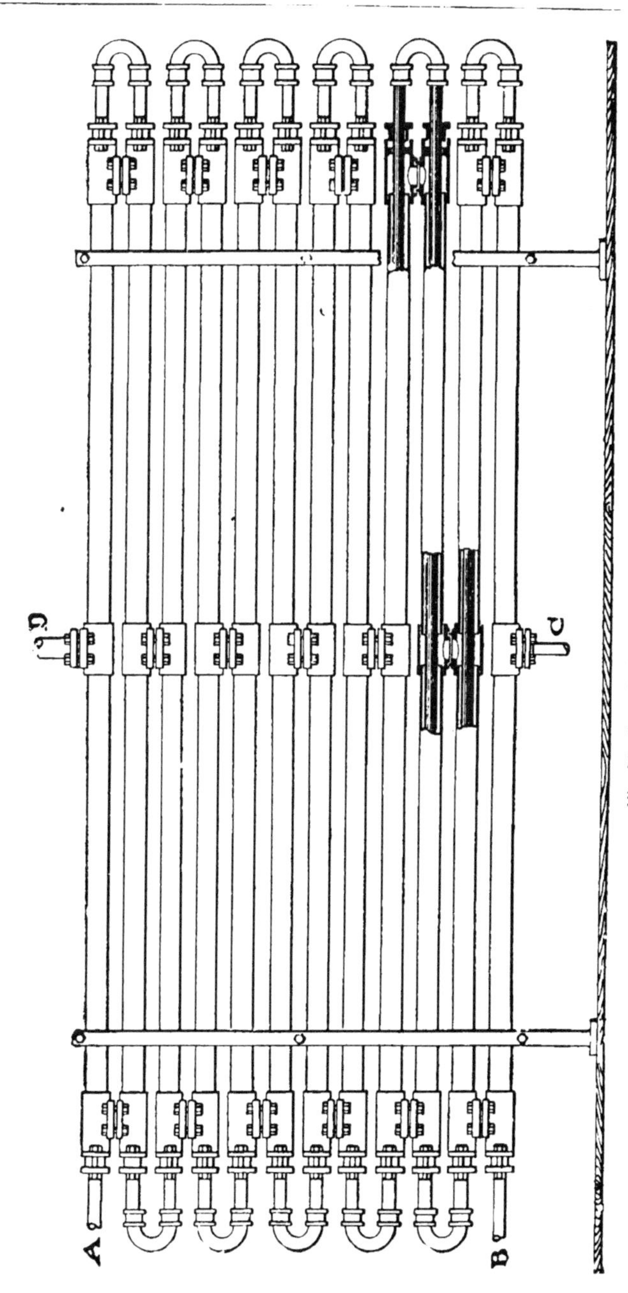

Fig. 59. Double-Pipe Brine-Cooler.

the most rapid transfer of heat and resulting rapid cooling. The *double-pipe* cooler is almost universally used at the present time in preference to the *spiral enclosed shell* cooler, notwithstanding the fact that the latter is more easily insulated and has a larger space for evaporation of the gas. Double-pipe coolers must be set up in

TABLE XVII

Properties of Calcium Brine Solution

Deg., Baume 60° F.	Deg. Salometer 60° F.	Per Cent Calcium by Weight	Lbs. per Cu. Ft. Sol.	Lbs. per Gallon	Specific Gravity	Specific Heat	Freezing Point F.	Ammonia Gauge Pressure
0	0	0	0	0	1	1	32	47.31
1	4	.943	1.25	1/6	1.007	.996	31.1	46.14
2	8	1.886	2.5	1/3	1.014	.988	30.33	45.14
3	12	2.829	3.75	1/2	1.021	.98	29.48	44.06
4	16	3.772	5	2/3	1.028	.972	28.58	43
5	20	4.715	6.25	5/6	1.036	.964	27.82	42.08
6	24	5.658	7.5	1	1.043	.955	27.05	41.17
7	28	6.601	8.75	1 1/6	1.051	.946	26.28	40.25
8	32	7.544	10	1 1/3	1.058	.936	25.52	39.35
9	36	8.487	11.25	1 1/2	1.066	.925	24.26	37.9
10	40	9.43	12.5	1 2/3	1.074	.911	22.8	36.3
11	44	10.373	13.75	1 5/6	1.082	.896	21.3	34.67
12	48	11.316	15	2	1.09	.89	19.7	32.93
13	52	12.259	16.25	2 1/6	1.098	.884	18.1	31.33
14	56	13.202	17.5	2 1/3	1.107	.878	16.61	29.63
15	60	14.145	18.75	2 1/2	1.115	.872	15.14	28.35
16	64	15.088	20	2 2/3	1.124	.866	13.67	27.04
17	68	16.031	21.25	2 5/6	1.133	.86	12.2	25.76
18	72	16.974	22.5	3	1.142	.854	10	23.85
19	76	17.917	23.75	3 1/6	1.151	.849	7.5	21.8
20	80	18.86	25	3 1/3	1.16	.844	4.6	19.43
21	84	19.803	26.25	3 1/2	1.169	.839	1.7	17.06
22	88	20.746	27.5	3 2/3	1.179	.834	— 1.4	14.7
23	92	21.689	28.75	3 5/6	1.188	.825	— 4.9	12.2
24	96	22.632	30	4	1.198	.817	— 8.6	9.96
25	100	23.575	31.25	4 1/6	1.208	.808	—11.6	8.19
26		24.518	32.5	4 1/3	1.218	.799	—17.1	5.22
27		25.461	33.75	4 1/2	1.229	.79	—21.8	2.94
28		26.404	35	4 2/3	1.239	.778	—27.	.65
29		27.347	36.25	4 5/6	1.25	.769	—32.6	1″ Vac.
30		28.29	37.5	5	1.261	.757	—39.2	8.5″ "
31		29.233	38.75	5 1/6	1.272		—46.3	12
32		30.176	40	5 1/3	1.283		—54.4	15
33		31.119	41.25	5 1/2	1.295		—52.5	10
34		32.062	42.5	5 2/3	1.306		—39.2	4
35		33	43.75	5 5/6	1.318		—25.2	1.5

insulated rooms at considerable expense, but the comparative simplicity of the construction and the fact that all parts are open and subject to inspection, has made this form of cooler the choice of practically all refrigerating engineers in recent years.

Owing to the fact that salt brine may freeze and burst the pipes of the cooler, it should not be used if avoidable. Calcium chloride brine is preferred for several reasons, but particularly on account of the fact that its freezing point for ordinary densities is 54° below zero F.; while that of salt brine is about 0° F. The construction of

TABLE XVIII

Properties of Salt Brine Solution

Degrees on Salom.	Percentage Salt by Weight	Pounds Salt per Cu. Ft.	Pounds Salt per Gallon	Specific Gravity	Specific Heat	Freezing Point F.	Ammonia Gauge Pressure
0	0	0	0	1	1	32	47.32
5	1.25	.785	.015	1.009	.99	30.3	45.1
10	2.5	1.586	.212	1.0181	.98	28.6	43.03
15	3.75	2.401	.321	1.0271	.97	26.9	41
20	5	3.239	.433	1.0362	96	25.2	38.96
25	6.25	4.099	.548	1.0455	.943	23.6	37.19
30	7.5	4.967	.664	1.0547	.926	22	35.44
35	8.75	5.834	.78	1.064	.909	20.4	33.69
40	10	6.709	.897	1.0733	.892	18.7	31.93
45	11.25	7.622	1.019	1.0828	.883	17.1	30.33
50	12.5	8.542	1.142	1.0923	.874	15.5	28.73
55	13.75	9.462	1.265	1.1018	.864	13.9	27.24
60	15	10.389	1.389	1.1114	.855	12.2	25.76
65	16.25	11.384	1.522	1.1213	.848	10.7	24.46
70	17.5	12.387	1.656	1.1312	.842	9.2	23.16
75	18.75	13.396	1.791	1.1411	.835	7.7	21.82
80	20	14.421	1.928	1.1511	.829	6.1	20.43
85	21.25	15.461	2.067	1.1614	.818	4.6	19.16
90	22.5	16.508	2.207	1.1717	.806	3.1	18.2
95	23.75	17.555	2.347	1.182	.795	1.6	16.88
100	25	18.61	2.488	1.1923	.783	0	15.67

the double-pipe brine cooler is shown in Fig. 50, in which it is seen that one pipe is within the other, the brine being discharged by the pumps into one or more pipes as at *A* and issuing at *B*. This connection leads from the main to the point to be refrigerated, and the ammonia is expanded or fed into the annular space between the two pipes and takes up the heat of the brine in evaporating, issuing as gas from the opening *D* at the top of the cooler. From thence the ammonia flows to the compressor and passes through the cycle of *compression*, *condension*, and *return to the liquid ammonia receiver* as before. The ammonia evaporating between the two pipes will naturally absorb as much heat from the outside surface as from the inner or brine, if allowed to do so, and it therefore becomes necessary to insulate the outside of the cooler or to build an insulated room in which the cooler is erected.

Although preferable in all cases, calcium brine is a necessity for very low temperatures. The proper density or strength of either salt or calcium brine is determined by the temperature to which it is necessary that the brine be reduced. In determining the proper strength for different requirements, Tables 17 and 18 are of value. It should be remembered, however, that a difference of from 5 to 10 degrees F. exists between the temperature of the brine and that of the evaporating ammonia; and that while the strength of the brine may appear ample for the temperature carried, the lower temperature of the liquid ammonia may cause it to solidify within or upon the surface of the evaporator, thus causing it to separate, or freeze, and act as an insulator, preventing the transmission of heat through the surface. It is, therefore, necessary in examining into the strength of the brine to consider it with reference to the evaporating pressure of the ammonia as well as its own temperature. In the last columns of Tables 17 and 18, the gauge pressures, corresponding to the freezing point of the brine for different strengths, are given.

The usual and proper instrument for determining the strength of brine is the Baumé scale already described in discussing aqua ammonia and its strength, but an instrument known as the *salometer* is sometimes used. This instrument is similar in appearance to the Baumé hydrometer, the difference being the way in which the scale is graduated, which in case of the salometer is from 0 to 100, the lower point being that at which the tube stands when floating in pure water, while the 100 point is that at which it stands in a saturated solution of salt brine, *i. e.*, a solution which can be made no stronger, owing to the fact that the water has dissolved all the salt it will take up. In the table giving the properties of calcium brine, the two scales are compared so that the student should have no trouble in converting readings of one scale to the corresponding readings of the other. It will be seen that the scales are to each other as 1 to 4. In testing the strength of brine a sample is drawn into a test tube and the temperature adjusted to 60 degrees, when the reading of the scale is taken at the surface of the liquid.

In *making brine* it is well to fit up a box with a perforated false bottom, or, a more readily obtained and equally effective mixer may be made by taking a tight barrel or hogshead, into which is fitted a false bottom four to six inches above the bottom head, and which is bored with one-half inch holes. Over the false bottom lay a piece of coarse canvas or sack to prevent the salt falling through. A water connection is made in the side of the barrel near the bottom, between the bottom head and the false bottom, and a controlling valve placed nearby to regulate the amount of water passing through.

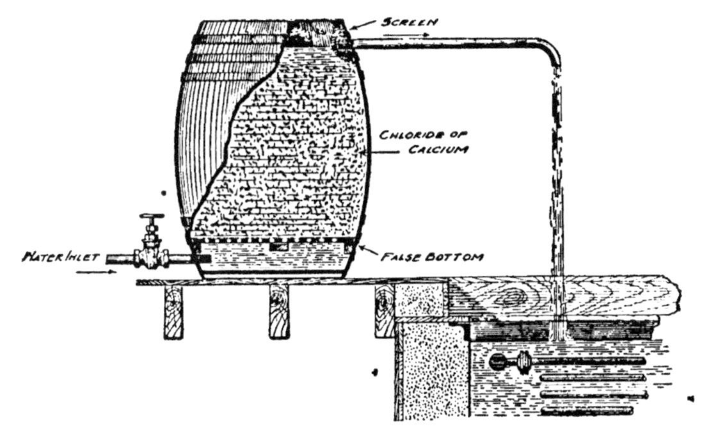

Fig. 51. Apparatus for Preparing Brine.

An overflow connection is made near the top of the cask, with its end so placed that the brine will flow into the tank, and a wire screen placed across its end inside the cask with a liberal space between it and the opening, to allow of cleaning, as shown in Fig. 51. The cask or barrel is now filled with the calcium or salt, which dissolves and overflows into the brine tank.

A test tube and Baümé scale, or salometer, should be kept at hand, and frequent tests made; the strength of the brine may be regulated by admitting the water more or less rapidly. After the first charge it is well to allow the mixer to remain in position for future requirements. A connection should be made between the return brine line from the refrigerating system and the cask with a controlling valve; by the use of this valve the strength or density of the brine may be increased without adding to the quantity. The cask should be kept full of the calcium or salt, and a portion of the return flow of brine should be allowed to pass through the cask, which will dissolve the contents and flow into the tank.

Calcium is usually obtained in sheet-iron drums, holding about 600 pounds each, and is in the shape of a solid cake within the drum. It is advisable to roll these onto the floor or top of the tank in which the brine is to be made and pound them with a sledge hammer before removing the iron casing, this process breaking it up into small pieces without its flying about the room. After breaking it up the shell may be taken off and the contents shoveled into the mixer.

It is also sold and shipped in liquid form, in tank cars, generally in a concentrated form, on account of freight charges, and diluted to the proper point upon being put into the plant. Where proper railroad facilities exist, this is probably the most desirable way of obtaining the calcium.

Salt is sold and may be obtained in a number of forms. The usual shape for brine is the bulk, or in sacks of about 200 pounds each. Where it is possible to handle salt in bulk, direct from the car to the tank, this is most generally used on account of the price, being about $1.00 per ton less than if sacked. If it is necessary for it to be carted or stored before using, the sack form is preferable. The coarser grades of salt are used for this purpose, No. 2 Mine being the grade commonly used. The finer salts are higher in price, without a corresponding increase in strength of the brine formed.

As a rule the freezing point of brine should be equal to, or slightly below, the temperature of the evaporating ammonia, rather than the temperature of the coldest brine, as is common. In referring to the table of salt brine solution, page 121, we find that if we wish to carry a temperature of 10° F. in the outgoing brine, it is necessary that the temperature of the evaporating ammonia be from 5 degrees to 10 degrees below this point, in order that the transfer of heat from the brine to the ammonia will be rapid enough to be effective, which would mean that the ammonia would be evaporating at a temperature of practically 0 degrees F. To prevent the brine freezing against the walls of the evaporator, its strength or density should be made to correspond with this, or from 95 degrees to 100 degrees on the salometer.

In examining into the causes of failure in a plant to perform its usual or rated capacity, it is advisable, unless there is every evidence that the trouble is elsewhere, to make an examination of the brine and determine whether its strength and condition is suited to the duty to be performed.

AUXILIARY APPARATUS

Ammonia Receiver. The ammonia receiver, or storage tank, is a cylindrical shell with heads bolted or screwed on, or welded in each end, and provided with the necessary openings for the inlet and outlet of the ammonia, purge-valve, and gauge fittings. They may be vertical or horizontal; the former type is generally used on account of the saving of floor space, while the horizontal is necessary when the condenser is located so low as to make the flow of the liquid ammonia into the vertical type impossible. A convenient location for the receiver in a plant in which the condenser is located above the machine room, is against the wall, or at one side of the room on a bracket or stand at one side of the oil interceptor, the sizes of the two being generally the same. They are then more readily under the control of the engineer than if at some out of the way place.

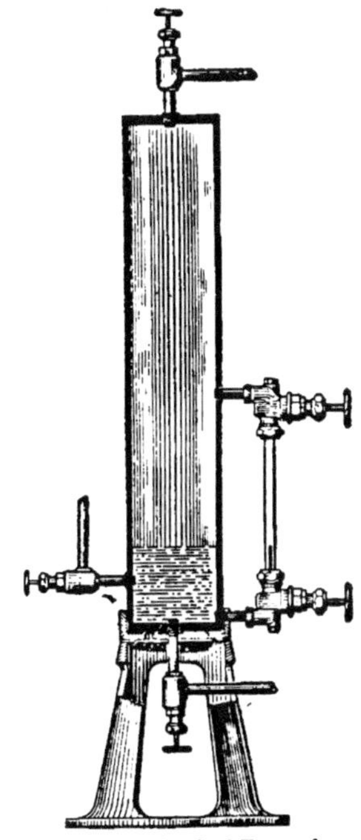

Fig. 52. Vertical Type of Ammonia Receiver.

Fig. 52 illustrates a receiver of the vertical type with the usual valves and connections for the proper equipment. The liquid ammonia enters at the top and is fed to the evaporator from the side near the bottom. The space below this opening has been provided for the accumulation of scale, dirt, or oil, and means are furnished for drawing it off through the purge valve in the bottom.

Pipes. Extra strong, or extra heavy pipe is the generally accepted pipe for connecting the various parts of the refrigerating system. Wrought-iron pipe is generally preferred to steel. Frequently, however, and particularly for the evaporating or low-pressure side of the system, a special weight or grade of pipe is used, also standard or common pipe is sometimes employed for this purpose. Without knowing the particular conditions under which this is to be used, or the relative value of the material, or manner in which the pipe is made, it is always better to use and insist on having the standard extra strong grade. The threads should be carefully cut with a good sharp die, making sure that the top and bottom of the threads are sharp and true. With this precaution, and an equally good thread in the fitting, it is not difficult to form a good and lasting joint. Particular care should also be taken that the pipe screws into the fitting the proper distance, and forms a contact the entire length, rather than to screw up against a shoulder without a perfect fit in the thread. This latter often causes leaky joints some time after the plant has been operated; the temporary joint formed either by screwing in too deep against the shoulder, or by ill-fitting threads, very often passes the test, and is used for some time until, after the alternate effects of heat and cold, and the chemical action of the ammonia cause it to break out. It is a safe rule that no amount of solder or other doctoring that is not backed up by a good fitting thread to support it can make an ammonia joint. This is particularly true of the discharge or compression side of the plant.

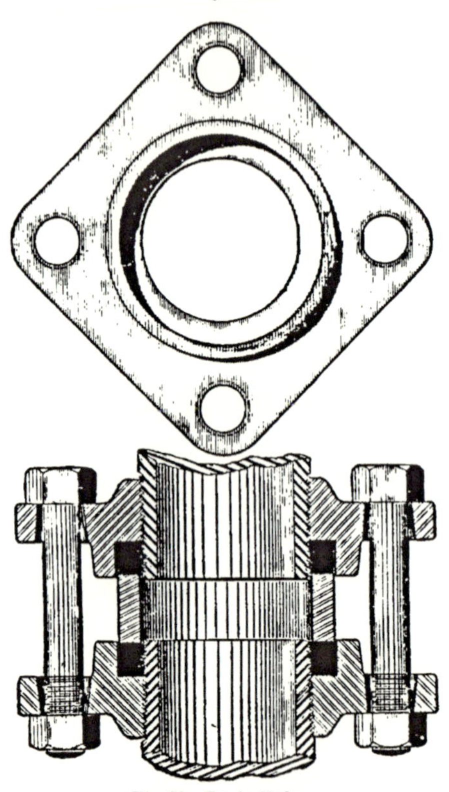

Fig. 53. Boyle Union.

The manner of making these joints may be divided into those having a compressible gasket between the thread on the pipe and the fitting into which it screws, and the screwed joint formed by a threaded pipe screwed into a tapped flange or fitting. The latter may be divided into those having a soldered joint, or one in which the union is formed by the threads only, with some of the usual cements to assist in making a tight joint.

The two most prominent types of gasket fittings are shown in Figs. 53 and 54. The former is known in this country as the *Boyle Union,* and is extensively used. As will be observed, the drawing together of the two glands by the bolts, compresses the gaskets,

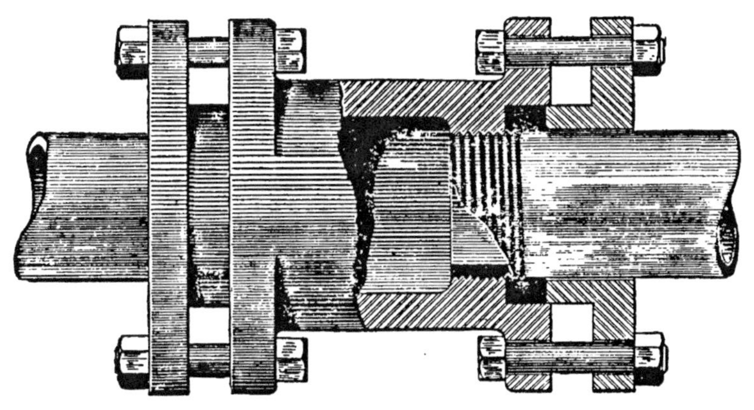

Fig. 54. Gland Type of Union.

usually rubber, against the threaded sides of the pipe, the bottom and sides of the recess in the flanges, and the edges of the ferrule between the two gaskets.

Figure 54 represents a union or joint quite frequently used, although not as commonly as the former. In this the pipe is threaded and screwed into the body of the fitting, in such a way that it does not form an ammonia-tight joint; leakage is prevented by a packing ring compressed by the gland against the pipe thread and the walls of the recess.

In Fig. 55—a type of ammonia coupling—the contact between the pipe and fitting is made to withstand the leakage of the gas without the aid of packing or other material other than solder or some of the usual cements; the two flanges are bolted together with a tongue and grooved joint having a soft metal gasket. This makes a permanent and durable fitting.

Other fittings of the class—as ells, tees, and return bends—are usually provided with one of the above methods of connecting with the system, and the different types described may be obtained of the builders of refrigerating machines.

Valves. The valves for the ammonia system of a refrigerating plant are of special make and construction, being of steel or semi-steel, with a soft metal seat which may be renewed when worn, and metal gaskets between the bonnets and flanges.

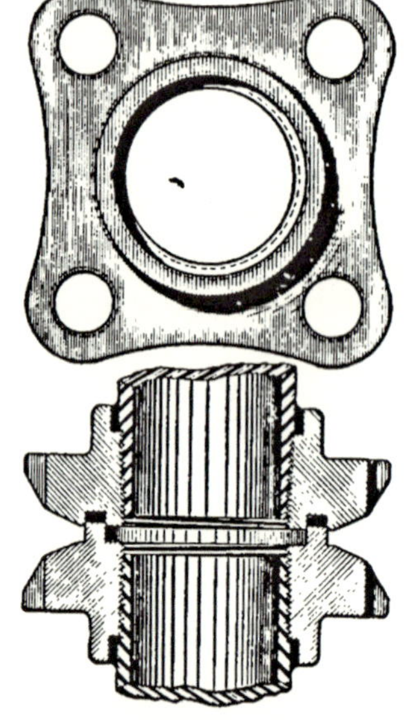

Fig. 55. Metal Gasket Coupling.

The usual types are *globe, angle,* and *gate,* subdivided into screwed and flanged. Fig. 56 is a generally adopted type of the flanged globe ammonia valve, while Fig. 57 represents the angle valve of the same construction. This seems to represent the best elements of a durable and efficient valve.

For a valve or cock requiring a fine adjustment, as is frequently the case in direct-expansion systems, particularly where the length of the evaporating coil or system is short, a V-shaped opening is desirable. Fig. 58 represents a cock for this purpose which will be found to be effective and meet the most exacting requirements.

Pressure Gauges. Two gauges are necessary for an ammonia plant of a single system; one to indicate the discharge or condensing pressure, and one for the evaporator or return gas pressure to the compressors.

Owing to the action of ammonia on brass and copper, the gauges for this purpose differ from the ordinary pressure gauge in that it is made with a tube and connections of steel instead of brass, and this construction is the general choice of gauge makers; in other respects the construction is similar. For machines of small capacity instru-

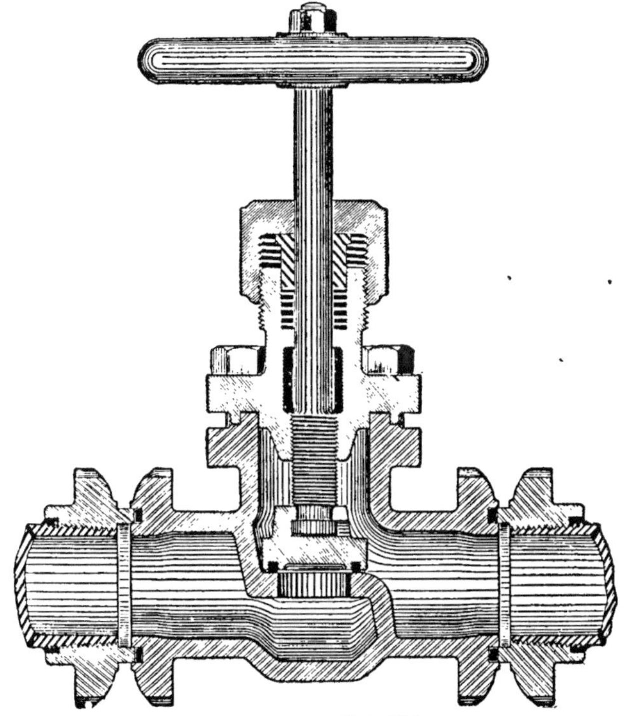

Fig. 56. Ammonia Globe Valve.

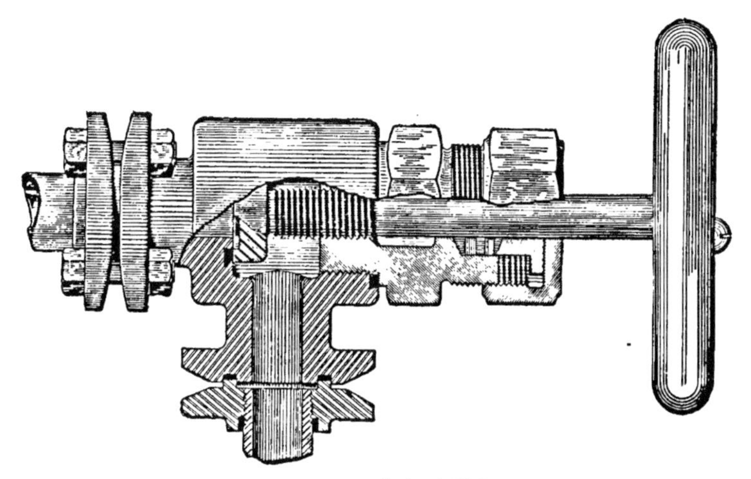

Fig. 57. Ammonia Angle Valve.

ments with 6-inch dials are common, while for larger plants 8-inch is the generally adopted size. The graduation for the high-pressure gauge is usually to 300 pounds pressure, and if a compound gauge is used, it is made to read to a vacuum also. This latter is only needed on certain occasions and frequently omitted from the high-pressure gauge. Owing to the necessity of removing the contents of the system at certain times, and usually through the evaporating side of the plant, the gauge for this portion of the system is graduated to read from a vacuum to 120 pounds pressure.

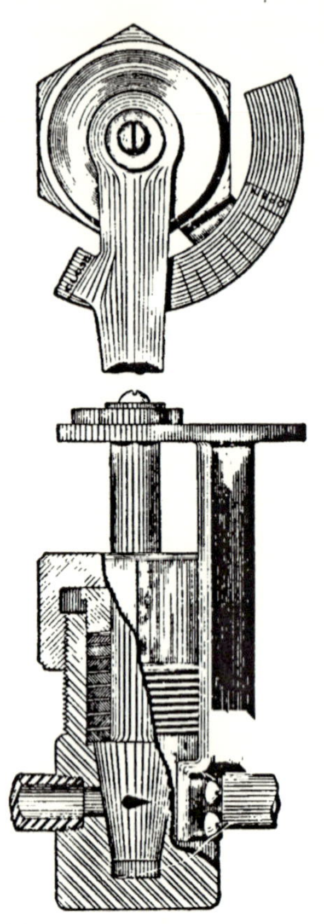

Fig. 58. V-Port Expansion Cock.

In connecting the gauges to the system, it is customary to locate the opening in the discharge and return gas lines near the machine within the engine room, placing a stop valve at some convenient point and carrying a line of $\frac{1}{4}$- or $\frac{1}{2}$-inch extra strong pipe to the gauges, making the joints with the usual ammonia unions. On account of the possibility of leakage of ammonia gas from the gauge tube, it is often considered advisable to fill the gauge pipe with oil—of the kind used for lubricating the ammonia compressor—for a short distance above the gauges, upon which the pressure of the gas will act, causing the gauge to move properly but without allowing the ammonia gas to enter the gauge. This is an application of the same principle as the steam syphon or bent-pipe arrangement in use with steam gauges, for the purpose of keeping the heat and action of steam from the gauge mechanism by the retaining of water in the gauge con nection.

Other gauges used about the refrigerating plant are of the ordinary pressure or vacuum types and do not need a special description, as their construction and manner of applying to the different parts of the system are well known to the engineer. It may be well, however, to caution the user on the importance of testing the gauges often enough to be sure they are accurate, as serious damages may result from a wrong indication of pressure.

METHODS OF REFRIGERATION

Various systems are in use for applying the cooling effect produced by a refrigerating machine in general work and in the special applications of refrigeration. All systems may be classified as either *direct*—in which the gas is expanded in pipes located so as to permit of direct abstraction of heat from the bodies to be cooled; or *indirect*—in which brine is cooled and then circulated in pipes, or by other means, so as to absorb heat from the bodies it is desired to cool. Methods of making and cooling brine have been described and it is only necessary to provide a brine pump and the necessary brine piping. The pump should be bronze fitted and the pipe should be of wrought iron. In some cases cold brine is made to cool air and the air is circulated through the cold stores. Direct refrigeration is applied by two methods, the most general being *simple evaporation of liquid in the pipes*, there being nothing but gas in the piping, as the supply of liquid is regulated by the expansion valve. In the other method—the so-called *flooded system*—the pipes or evaporators are almost full of liquid, or at least have the inner surfaces wet. Regulation is effected by the inlet valve and the gas trap of special design on the outlet, this trap being arranged to return the gas to the compressor and at the same time guard against allowing the liquid to reach the suction of the machine in any quantity that might be dangerous.

Of course in the wet system of operation devised by Prof. Linde a certain amount of liquid is passed into the suction, but this is and must be under entire control. There are reasons for and against each of the two methods of applying refrigeration, but generally the indirect or brine system, is best for small plants, while large plants should use the direct-expansion system. The chief reason for retaining the brine system in large plants thus far has been the fear of ammonia leaks that mean damage to goods in store. Such leaks are comparatively rare where reasonably good pipe work is used, and the man who uses any other kind of piping deserves to foot up the loss rather than be behind the times with his whole system. Small plants must use brine to guard against shut-downs as the refrigeration stored in a comparatively large body of brine is of considerable value in keeping temperatures down at such times. Also the small machine may be shut down at night and the brine pump kept going to circulate the brine, thus effecting considerable saving in operation costs. The same results can be had to a certain extent by having shallow pans of brine placed over the pipes in the coolers. Where very low temperatures are required, as in the case of fish and poultry freezers, the direct-expansion system is a necessity.

PROPORTION BETWEEN THE PARTS OF A REFRIGERATING PLANT

There is necessarily a certain ratio or proportion between the several parts of a refrigerating plant, as there is between the boiler engine, and parts of a steam or power plant, in order to obtain the most economical results. It is first necessary that the evaporator be provided with heat-transmitting surface sufficient to conduct 284,000 B. T. U. from the brine to the ammonia, for each ton of refrigeration to be performed. Without going into a theoretical calculation of this amount, we shall state, in both lineal feet of pipe and square feet of pipe surface, the commercial sizes and amounts ordinarily in use.

The coil surface in a brine-tank system of refrigeration, should contain approximately 50 square feet of external pipe surface, to each ton in refrigerating capacity of the plant, when it is to be operated at a temperature of 15 degrees F. This is an ample allowance and will be found under general working conditions to give readily the required capacity. While tests have been made in which 40 square feet of pipe surface has been found sufficient for one ton of refrigeration, it will be safer to use the former amount, owing to the varied conditions under which a plant may be operated. This would amount in round figures to 150 lineal feet of 1-inch pipe, 115 feet of 1¼-inch 100 feet of 1½-inch, or 80 feet of 2-inch pipe. For each ton in refrigerating capacity, the pipe surface of the brine tank should vary from 40 to 60 cubic feet, depending on the amount of storage capacity desired.

The submerged type of ammonia condenser should contain

approximately 35 square feet of external surface which nearly corresponds to 100 lineal feet of 1-inch, 80 feet of 1¼-inch, 70 feet of 1½-inch, and 56 feet of 2-inch pipe.

The atmospheric type of condenser should contain 30 square feet of external pipe surface which corresponds to 87 lineal feet of 1-inch, 69 feet of 1¼-inch, 60 feet of 1½-inch, and 48 feet of 2-inch pipe.

The double-pipe type of condenser, as usually rated, contains 7 square feet of external pipe surface for the water circulating pipe and about 10 square feet of internal pipe surface for the outer pipe, and corresponds to approximately 20 lineal feet each of 1¼-inch and 2-inch sizes for each ton of refrigerating capacity.

The above quantities are based on a water supply of average temperature—60 degrees—and quantity. In cases of a limited supply or higher temperature than ordinary, a greater amount should be used.

The ammonia compressor should be of such dimensions that it will take away the gas from the brine cooler, evaporating coils, or system, as rapidly as formed by the evaporation of the liquid ammonia; and unless the temperature at which the plant is to be operated be known, it is impossible to determine the volume of gas to be handled and the necessary size of the compressor.

As stated before, the unit of a refrigerating plant is usually expressed in tons of refrigeration equal to 284,000 B. T. U. Up to the present time, however, a standard temperature at which this duty shall be performed has never been established, and therefore the rating of a machine, evaporator, or condenser by tonnage is a merely nominal one and misleading to the purchaser, a range of as great as 50 per cent very often existing in the tenders for certain contracts. Upon the basis, however, of the average temperature required of the refrigerating apparatus that of 15 degrees F. is probably the mean; and at this temperature in the outgoing brine, it is necessary to take away from the evaporator nearly 7,000 cubic inches of gas per minute for each ton of refrigeration developed in twenty-four hours. This may be considered as a fair basis for the rating of the displacement of the compressor or compressors of the plant, unless a specific temperature is stated at which the plant is to operate. At 0 degrees F. it is necessary to calculate on approximately 9,000 cubic inches, while at 28 degrees F. about 5,000 will be the required amount.

For example, if we have two single-acting compressors 12 inches diameter by 24 nches stroke, operating at 70 revolutions per minute, we would have 113.09 (inches, area of 12-inch circle) × 24 (inches stroke) × 2 (number of compressors) × 70 (revolutions) ÷ 7,000 (cubic inches displacement required) = 54.28 (tons refrigeration per 24 hours of operation); while if the same machine is to be operated at or near a temperature of zero and we divide the product by 9,000, we have a capacity of 42.22 tons only, in the same length of time. The above quantities are given as approximate only, but they have been deduced from the average results obtained from years of practice and will be found reliable under average conditions. It is to be hoped, however, that a standard will soon be adopted which will rate machines or plants by cubic inches displacement at a certain number of revolutions or a stated piston speed, and the cooling of a certain number of gallons of brine per minute through a certain range of temperature.

TESTING AND CHARGING

Having described the different parts of the refrigerating plant and their relations to one another, let us consider the process of testing and charging, or introducing the ammonia into the system. After the connections are made between the different parts, whether the system is brine or direct expansion, it is necessary to introduce air pressure into it to determine the state of the joints. This may be done in sections or altogether. It is customary, however, to put a higher pressure on the compression side of the plant than on the evaporator owing to the difference in the pressure carried in operation. Adjacent to each compressor is placed a main stop valve, on both the inlet and outlet sides, while on either side of these it is customary to place a by-pass or purge valve.

Before starting the compressor, the main stop valve—or valves if there be two—on the inlet or evaporating side of the compressor, is closed, the small valve between the compressor and the main stop valve opened, and all of the other valves on the system opened except those to the atmosphere. The compressor may then be started slowly, air being taken in through the small by-pass valves and compressed into the entire system. It is well to raise a few pounds pressure on the entire system before admitting water into the compressor water jackets or other parts of the system, because if a joint were improperly made up, it would be possible for the water to enter the compressors, or coils of the condenser or evaporator, and serious damage or loss of efficiency in the plant occur, which it might be impossible to locate afterwards. While if pressure exists within the system when the water is admitted, its entrance into the coils or system is impossible while the pressure exists, and the leak is at once visible and may be remedied before proceeding further.

In starting the test it is also well to try the two pressure gauges and see that they agree as to graduation; as it has happened that owing to a leakage between the discharge pipe and the high pressure gauge, an enormous pressure has been pumped into the system causing it to explode, with a consequent result of loss of life and property. If, however, the pressures are found to be equal on the two gauges it is safe to assume that they are recording properly and their connections are tight. After these preliminaries it is safe to put an air pressure of 300 pounds on the compression side of the plant, care being taken to operate the compressor slowly, not raising the temperature of the compressed air too much, as, with the utmost care in making up joints and in selecting material, certain weaknesses may exist and under such high pressure it is well to proceed with caution.

After the desired pressure has been reached, the entire system should be gone over repeatedly until it is absolutely certain that it is tight. Parts which can be covered with water, such as a submerged form of condenser or brine tank with evaporating coils, should be so covered that the entire surface may be gone over at once and with almost absolute certainty. The slightest leakage will cause air bubbles to ascend to the surface. This leakage may be traced by allowing the water to flow from the tank while the air pressure is still on the coils or system, marking the points where the bubbles occur. The coils may then be taken up or repaired when empty. For parts which cannot be covered with water, it is customary to apply with a brush a lather of such consistency that it will not run off too readily; upon coming in contact with a leak, soap bubbles are formed, and by tracing to the starting point the leak may be located. After the compression system has been subjected to a pressure of 300 pounds and found to be tight, the air may be admitted through the liquid ammonia pipe to the evaporating side of the plant, care being taken that the pressure does not rise above the limit of the gauge—which, as previously stated, is usually 120 pounds—and the same process of testing as applied to the opposite side of the plant gone over.

Many engineers require the vacuum test as well as the foregoing, and although if the former is gone over thoroughly, there can be little chance of leakage afterwards, it is better to be over-exacting than otherwise in the matter of testing and preparation of the plant, thus preventing the possibility of leaks that may prove disastrous. Open the main stop valves on the inlet line and close the main valves above the compressor on the discharge line, closing the by-pass valves in the suction line, and opening those in the discharge line between the main stop valve and the compressor. Have all the other valves on the system open as before for testing. Starting the compressor draws in the air, filling the system through the compressor and discharging it at the small valve left open. Assuming the system to be tight, continuing the operation will finally exhaust the air, or nearly so, when the small valves should be closed and the pressure gauges watched to determine whether or not leakage exists.

Assuming that the system and apparatus is tight in every particular and that it is otherwise ready to be placed in operation, we are now ready to charge the ammonia into the plant.

If the air has been exhausted from the system in testing, this usual step need not be taken before charging, and it is only necessary to put the machine in proper condition to resume the pumping of the gas, and to attach a cylinder of ammonia to the charging valve to enable the refrigeration to be commenced. The main stop valves above the compressor, which were closed in expelling the air, should now be opened, and by-pass and other valves to the atmosphere closed. Close the outlet valve from the ammonia receiver and start the machine slowly, at the same time opening the feed valve between the drum of ammonia and the evaporator. The anhydrous liquid ammonia will flow into the evaporator through the regular supply pipe, the gas resulting from evaporation being taken up by the compressors and discharged into the condenser and finally settling down into the receiver, when a sufficient quantity has been introduced to form a supply there. Upon closing the valve between the drum from which the supply is being drawn, and opening the outlet valve from the receiver, the process of refrigeration by the compression system is regularly in operation.

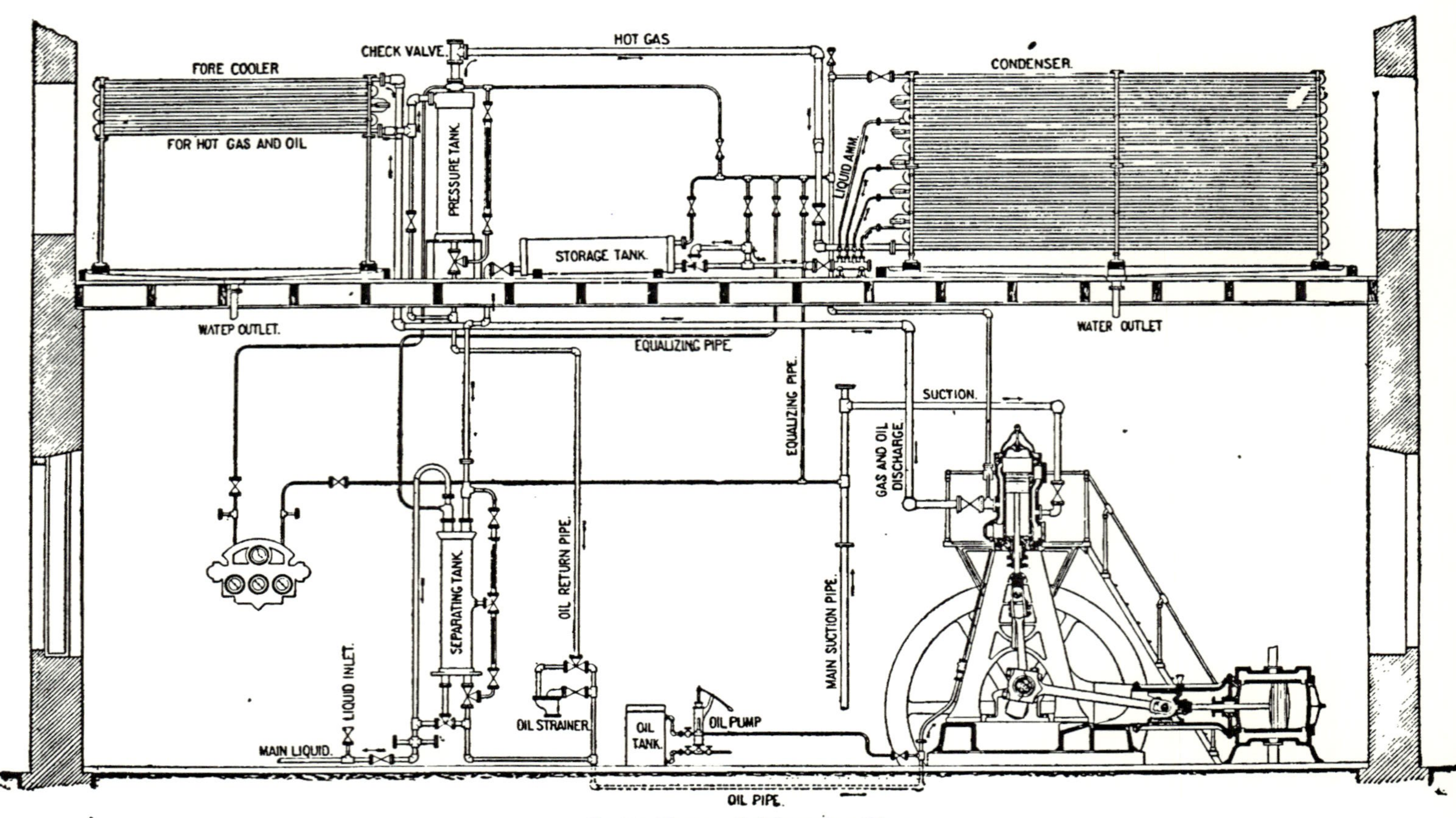

De La Vergne Refrigerating Plant.

REFRIGERATION

PART III

OPERATION AND MANAGEMENT OF THE PLANT

Assuming that the plant has been properly erected, tested, and charged with ammonia of a good quality—and if a brine system, with brine of proper strength or density, as already explained—it only remains to keep the system or plant in that condition. As all forms of mechanism are liable to disarrangement and deterioration from various causes, repairs and corrections from time to time must be made to keep them in good condition. Let us now consider the most important points requiring attention.

It is absolutely necessary for the good working of any type of plant or apparatus that it be kept clean. As a steam boiler must be clean to obtain the full benefit of the fuel consumed, so must the surfaces of the condenser and evaporator be clean to obtain the proper results from the condensing water and evaporation of the ammonia or other refrigerant.

For satisfactory work, the system should be purged of any foreign element present in the pipes, such as air, water, oil, or brine. Foreign matter is the most common among internal causes for loss of efficiency, and the valve openings which have been shown and described should be used for cleaning the system.

Oil is used as a lubricant in nearly if not quite all compressors, and the quantity should be the least amount that will lubricate the surfaces and prevent undue wear. This is considerably less than the average engineer is inclined to think necessary, and consequently a coating forms on the walls of the pipes or other surfaces of the condensing or evaporating systems, and a proportionate decrease in the duty is obtained. It is also necessary that the oil be of such a nature that it is not saponified by contact with the ammonia. Such a change would choke or clog the pipes, coating their surfaces with a thick paste which causes a corresponding loss as the amount increases.

The purge valve in the bottom of the oil interceptor may be opened slightly about once each week, and the oil discharged from the compressors drawn off into a pail or can, unless a blow-off reservoir is provided. After the gas with which it is charged has escaped, the oil should be practically the same as when fed into the compressors. If, however, the oil is not of the proper quality it will remain thick and pasty, or gummy, showing it to have been affected by the ammonia. Its use should not be continued.

By opening the purge valves, which are usually provided at the bottom manifold or header of the brine tank and the bottom head of the brine cooler, oil or water, if there is any in that part of the system, may be drawn off. These valves, however, should not be opened unless there is some pressure in that part of the system, as air would be admitted if the pressure within the apparatus is below that of the atmosphere. Air may enter the system through a variety of causes and its presence is attended with higher condensing pressure and a falling off in the amount of work performed. For the removal of air from the apparatus, a purge valve is placed at the highest point in the condenser or discharge pipe from the compressors near the condenser, which may be tried when the presence of air or foreign gases is suspected. This should be done after the compressor has been stopped. When the condenser has fully cooled and the gases separated, a small rubber hose or pipe may be carried into a pail of water and the purge valve or valves slightly opened. If air or other gases exist in the system, bubbles will rise to the surface of the water so long at it is escaping; while, if ammonia is being blown off, it will be absorbed in the water and not rise to the surface.

To prevent the possibility of air getting into the system the evaporating pressure should never be brought below that of the atmospheric, or 0 degree on the gauge, as at such times, with the least leakage at any point, it is sure to enter. Should it become necessary to reduce the pressure below that point, it is well first to tighten the compressor stuffing boxes and allow the pressure to remain below 0 degree only the shortest possible time, as not only air may enter, but if it be the brine system and a leak exists, brine also will be drawn in.

From the foregoing it is evident that in order to obtain satisfactory results, the interior of the system must be kept clean by purging at the different points provided for this purpose; and it need only be added in this connection, that when the presence of oil or moisture becomes apparent in any quantity, the coils or other parts should be disconnected and blown out with steam until thoroughly clean, and afterwards with air to make certain that condensation from the steam does not remain. After this the parts may again be connected and tested ready for operation.

If the plant be a brine system, it is necessary that the brine be maintained at a proper strength or density to obtain satisfactory results; for if it becomes weakened, it freezes on the surfaces of the pipes or evaporator, thereby acting as an insulator and preventing the rapid transmission of heat through the walls.

It is of great importance to know at all times whether or not the gas taken into the compressors is fully discharged into the condenser, as the slightest loss at this point is certain to make itself felt in the operation of the plant. The compressor and valves seldom need be taken apart to determine their operation. The engineer should be able to discern when the compressors are working at their best, by placing the hand on the inlet and outlet pipes or on the lower part of the compressors so as to detect slight change from normal temperature. Should the inlet pipe to one compressor be warmer than that to the other (of a pair), or the frost on the pipe from the evaporator reach nearer one compressor than the other, it is then certain that the one with the higher temperature, or, from which the frost is farthest, is not working properly or doing as much duty as the other; and it is equally certain that some condition exists which prevents the complete filling and discharge of its contents; possibly it has more clearance or leaky valves.

The most common difficulties experienced with ammonia condensers are those of keeping the external surfaces clean and free from deposits, and preventing the accumulation of air or foreign gases within. Deposits on the surface are usually of two kinds—one, a soft deposit which may be washed off with a brush or wire scraper such as is used for cleaning castings in a foundry; the other, a hard deposit which must be loosened with a hammer or scraper. It is hardly necessary to explain in detail the methods employed in cleaning the condenser as this is a matter that each engineer will be able to accomplish in his own way. It should not, however, be overlooked, and with a condensing pressure higher than ordinary, this should be the first point to be examined after the water supply.

Air and foreign gases due to decomposition of the ammonia or other causes, find their way into the condenser and make themselves manifest generally in a higher condensing pressure, or a falling off in the duty to be obtained from the plant. They should be blown off through the purge valve at the top of the condenser in the manner already described.

It is possible, through leakage of the coils or other parts of the apparatus, that the ammonia may become mixed with brine or water, thereby retarding its evaporation and interfering with the proper or usual operation of the plant. If this is suspected, a sample may be drawn off into a test glass through the charging valve or purge valve of the brine tank or ammonia receiver and allowed to evaporate, in which case the water or brine will remain in the glass and the relative amount be determined. Through careful evaporation and continued purging of the evaporator at intervals, this may in time be eliminated, and care should be taken to prevent future recurrence.

Loss of Ammonia. This should be constantly guarded against. It is watchfulness which determines between a wasteful and an economical plant in this particular, and the engineer who allows the slightest smell of ammonia to exist about the plant is certain to be confronted with excessive ammonia bills; while he who is constantly on the alert and never rests until his plant is as free from the smell of ammonia as an ordinary engine room, will be referred to as the one who ran such and such a plant without addition of more ammonia for so many years.

The escape of ammonia into the atmosphere is readily detected; but where a leakage occurs in a submerged condenser, brine tank, or brine cooler it is necessary to examine the surrounding liquid to determine whether or not it exists. For this purpose various agents are employed, and may be obtained of druggists or from the manufacturers of ammonia. *Red litmus paper* when dipped into water or brine contaminated with ammonia will turn blue. *Nessler's solution* causes the affected water to turn yellow and brown, while *phenolphthalen* causes a bright pink color with the slightest amount of ammonia present.

The stopping of a leakage of ammonia in the brine tank or

cooler may be possible while the plant is in operation, by shutting off the coil in which it occurs, or, if the point is accessible a clamp and gasket may be put in place temporarily.

Purging and Pumping out Connection. A common cause of failure to operate properly and effectively is the introduction of some foreign substance into the system. This will be readily understood and appreciated by engineers and those familiar with the requirements of a steam boiler. Clean surfaces on the shell or tubes are necessary for the maximum evaporation of water, or for the transfer of heat through the walls of pipe or other forms of heat-transmitting surface. The most common difficulty encountered in a refrigerating plant is oil, either in its natural condition, or saponified by contact with the ammonia, water, or brine. It enters the system in many ways; through leakage, condensation in blowing out the coils or system, foreign gas arising from decomposition of the ammonia through excessive heat and pressure, or the mingling of air which may enter the system through pumping out below atmospheric pressure, or the air may have remained in the system from the time of charging, never having been fully removed. It is also probable, though hard to determine with certainty owing to the various conditions surrounding the operation of plants, that impurities are introduced with the ammonia, either in the form of liquid, gas, or air, which afterward become impossible to condense.

The oil in a system forms a covering or coating on the evaporating surface which acts as an insulation and prevents the ready transfer of heat through the walls of the evaporator. The presence of water or brine causes an absorption of a portion of the ammonia into the water or brine, forming aqua ammonia which raises the boiling point of the ammonia and causes material loss in the duty. Air or other non-condensable gas in the system, excludes an equal volume of the ammonia gas, thereby reducing the available condensing surface in that proportion.

For the purpose of cleaning the system and removing the different impurities which may appear, purge and blow-off valves and connections are provided. One of these is placed at or near the bottom of the oil interceptor, which is located between the compressors and the condenser; it is used to draw off the oil used as a lubricant in the compressor and which is precipitated to the bottom. This oil should not be allowed to accumulate to any great extent as it may be carried forward to the condenser by the current of gas.

If the liquid ammonia receiver be placed in a vertical position it is customary to place a purge valve in the bottom for drawing off oil or other impurities. The supply of liquid to the evaporator is taken off at a short distance above the bottom say, 4 to 6 inches.

The next point for the removal of impurities is at the bottom of the brine cooler, or the lower manifold of the coil system in a brine-tank refrigerator. Tests at these points may be made as often as necessary to determine the state of cleanliness of the system. If the system is charged with any of the common impurities, they should be blown out and the system cleansed at the earliest possible moment, as they cause a decided loss.

Air and foreign gases accumulate in the condenser because the constant pumping out of the evaporating system tends to remove them from that part of the system to the condenser. This point, therefore, is the most natural place for their removal. For this purpose it is customary, on the best condensers, to place a header or manifold at the top at one end, and connect each of the sections or banks with a valve opening. A valve is also placed at each end of the header, and a connection made from one end of this header to the return gas line between the evaporator and the compressors. By closing the stop valves on the gas inlet and liquid outlet of any one of the sections and opening the purge or pumping-out line into the gas line to the compressors, the section or tank may be emptied of its contents for repairs or examination and then connected up and put into service without either shutting down the plant, or losing a material quantity of ammonia. For purging of air or gas, the valve between this header and the machine should be closed, and the valve on the opposite end opened to the atmosphere, the valves on each section in turn opened slightly while the foreign gases are expelled. This process should not be used while the compressor is in operation, as the discharge of the ammonia into the condenser would keep the gas churned to the extent that it would become impossible to remove the foul gases without removing a considerable portion of the ammonia also.

For this reason it is customary before blowing off the condenser to stop the compressor and allow the water to flow over the condenser until it is thoroughly cooled. Sufficient time should elapse for the ammonia to liquefy and settle towards the bottom, while the air and lighter gases rise to the top, at which point they may be blown out through the purge valve to the atmosphere. If doubt exists as to whether ammonia or impurities are being blown out, attach a piece of hose to the end of the purge valve and immerse its other end in a pail of water. If it is air, bubbles will rise to the surface, while if it is ammonia, it will be absorbed into the water; the mingling of the ammonia with the water will cause a crackling sound, and the temperature of the water will increase owing to the chemical action.

ICE-MAKING PLANTS

One of the most important applications of refrigeration is in the production of artificial ice. Thus refrigeration, which in former times was produced only by the melting of ice, is now produced artificially and used in making ice. In order to freeze water, it is only necessary that its temperature be lowered to the freezing point and the latent heat of liquefaction abstracted. In practice, to get rapid freezing, the temperature of the ice formed is carried below the freezing point, so that calculations of the heat to be abstracted must cover this. Assuming that the water supply has a temperature of 60° F., 28 B. T. U. will have to be removed from 1 pound of water to reduce it to freezing temperature. Then, since the latent heat of ice is 142.65, this number of heat units must be abstracted to freeze the pound of water, and since the temperature of ice is usually about 20° F.—or 12 degrees below freezing—and its specific heat 0.5, we have 6 B. T. U. to be removed on this account. Thus altogether 176.65 B. T. U. must be removed from one pound of water to freeze it.

Of the many more or less impracticable schemes that have been devised to freeze ice, only three are to any extent in use at the present time. These are known as the *can*, the *plate*, and the *cell* systems. The latter is used in England but not to any extent in the United States, where the great majority of the plants are on the can system with a few working on the plate plan. Indeed, the can system is most in use the world over, in spite of the fact that there are a number of disadvantages connected with its use. As good ice can be made by one system as by the other when both are operated properly, but it costs more to make ice with the can system owing to the purifying apparatus that must be employed if the ice is to be made clear and firm. The plate plant costs from 30 to 75 per cent more to construct and requires considerably larger buildings and more ground space. This means greater fixed charges, but the disadvantage is offset in part, at least, by the fact that the plate plant will make from 10 to 14 tons of ice per ton of coal burned, while the average for good can plants is from 6 to 8 tons. The practical skill required to operate the two plants is about the same, but somewhat more technical knowledge is required in the case of direct-expansion plate plants. As a rule it never pays to build small plate plants and in the larger plants a high grade of equipment, consisting of compound condensing engines and power-driven handling devices, must be employed to get the economy that will justify the building of such a plant. Such machinery requires a high degree of skill for proper operation. Fixed charges and depreciation are greater with the plate system. Where a large plant is to be operated by hydroelectric or other cheap power, it will pay to build a plate plant, the system requiring no steam to freeze the water, as in the case of the can plant.

Can System. As the name implies, this system uses cans in which the ice is frozen, the cans being filled with water and partially immersed in a mechanically-cooled non-freezing brine bath. Freezing proceeds from the four sides and the bottom, and the impurities in the water, which have not been removed during the first stages of the freezing process, are finally frozen into the center of the block. The central opaque core formed in this way is undesirable, and it is to the necessity for eliminating it that all the complications of the can system are due. Distilling, reboiling, and filtering apparatus must be employed except where porous opaque ice is not particularly objectionable, as in packing fish and icing cars. By freezing ice at a comparatively high temperature, say 25°, clear ice can be made by the can system with natural water, but it is not practicable to use so high a temperature, owing to the length of time required to freeze the ice and the comparatively large tanks that would have to be employed. On this account ice is frozen at from 12° to 20°, the usual working temperature being from 14° to 16°.

At very low temperatures the ice crystals are formed so rapidly that they do not have time to solidify so that the block is rather porous, being made up of the separate crystals. With a freezing temperature of 15°, the tank for holding the cans should be large enough to contain from 2 to 2½ times the number of cans necessary to make up the daily output. Thus what is called a *15-ton tank* will really contain cans sufficient to hold about 38 tons of ice. This factor by which the number of the cans in the tank is increased is known as the *tank surface.* The time of freezing depends, of course, on the thickness of the ice, as the first inch of thickness is formed much quicker than the rest of the block. Thus, with a temperature of 20°, a 1-inch thickness can be frozen in an hour or less; while a 4-inch thickness will require about 10 hours, the cooling being from one side only. For this temperature, the time in hours required to freeze can be found approximately by adding 1 to the *inches of thickness,* multiplying this sum by the thickness in inches, and dividing the result by 2. Thus to freeze ice 8 inches thick from one side, we have $8 (8 + 1) \div 2 = 36$ hours. For a temperature of 15° the results obtained by the rule should be decreased by 20 per cent for all thicknesses under 8 inches, and by 25 per cent for thicker ice.

Can Plant Equipment. The complete can ice plant is made up of a steam boiler plant, a refrigerating or ice-making machine, distilled water system, and freezing tank with accessories. In addition to this equipment, it is customary to provide ice storage rooms and the necessary brine cooling and circulating apparatus. Pumping apparatus is also required to supply water to the plant, and in cases where water is scarce or obtained at great expense, cooling towers are employed. The steam boiler plant will not be described in detail as it should differ in no way from a first class steam power-plant equipment. It consists of a good boiler with fixtures and stack, a boiler feed pump, an injector, and a feed-water heater together with water softening or purifying apparatus in case the water is bad. For small plants, ordinary return tubular boilers are about the most practical type and should be installed so as to have a reserve unit if possible, particularly if the water is bad. Larger plants may use the more expensive water-tube boilers to advantage but there is little need of these except where condensing or compound engines are employed as in plate plants.

The engines are of standard manufacture and in no-wise specially constructed for the ice plant. The refrigerating machine has already been discussed in detail so that it only remains to describe the distilling and freezing apparatus and show how all the apparatus is assembled to form the complete ice plant.

Distilling Apparatus. This consists of an oil separator, a back pressure or relief valve, an exhaust steam condenser, a reboiler and skimmer, a hot filter, a cooling coil, a gas forecooler, and a cold filter. Fig. 59 shows the course of the steam from the engine through the various parts of the apparatus. From the engine cylinder *A*, the steam passes directly to the oil separator, or, if a receiver is used, to the receiver, and then to the separator. The separator should be of

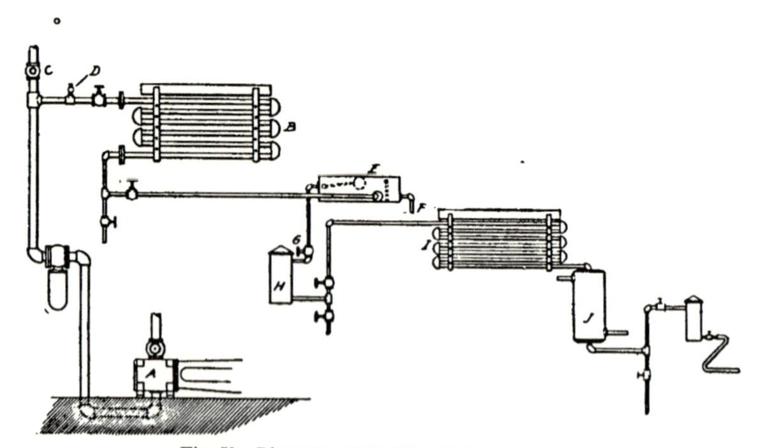

Fig. 59. Diagram of Distilled-Water Apparatus.

ample size and never less than the size of the pipe to which it is connected. In the separator the oil and priming water that may have come over are separated out and the purified steam carried to the exhaust steam condenser *B*. Any steam that is not condensed immediately escapes through the relief valve *C*; and the vent cock *D* serves to rid the steam of any air and other gases that may be present. Water from the steam condenser passes to the reboiler *E*, which is provided with a skimming diaphragm near one end over which scum and light impurities can pass off to the waste pipe *F*. A float valve regulates the water level in the reboiler and a live steam coil furnishes the heat for boiling. After the water is boiled, it passes through the pipe G to the hot filter H, and thence to the cooling coil I over which cooling water runs. From the cooling coil the water passes to the tank J which contains a coil of pipe connected at its two ends into the suction line of the compressor. Thus the return gas from the expansion coils passes through the coil, and the water in the tank, which is known as the forecooler, is cooled down to a temperature of 45° to 50°. From the forecooler, the water goes through the cold filter and thence passes through a hose to the can filter and thence to the cans.

Steam Condenser. This usually consists of pipe coils over which water is run as in the atmospheric type of ammonia condenser. Either 1¼-inch or 2-inch pipe may be used, care being taken to have sufficient coils to give a condensing area equivalent to about twice the area of the exhaust pipe. This will be found satisfactory where the oil separator is of such design as to act as a receiver, or where a receiver is connected in the exhaust pipe. The idea is to avoid throwing back pressure on the engine, and if the exhaust pipe is long or has a number of unavoidable bends, the aggregate area of the condenser pipe coils must be made larger in proportion. About 80 square feet of pipe surface should be allowed for each ton of ice making capacity in 24 hours. This means 128 running feet of 2-inch or 180 running feet of 1¼-inch pipe per ton of capacity. Many plants do not use this amount of pipe, and in cases where the cooling water has a low temperature the use of less pipe surface may be justifiable.

In addition to the pipe coil condenser, there are a number of special designs on the market, most of them designed to economize space. These generally consist of some form of receiving tank made of galvanized iron, water being run over the outer surfaces of the tank. It is claimed that the thin metal gives rapid and economical transfer of heat and that the cooling water is used to the best advantage so that less of it is required. A steam condenser of this type made by the Triumph Ice Machine Co. is shown in Fig. 60. In some cases local conditions make it desirable to use the regular standard surface condensers, this being particularly the case where the ice plant is operated in combination with a power plant.

Hot Skimmer and Reboiler. Cleansing of the water formed by condensing the steam is done by driving off all volatile matter, during the process of boiling, and skimming such of the impurities as

Fig. 60. Triumph Steam Condenser.

cannot be volatilized. These processes may be carried on in two separate pieces of apparatus or the hot skimmer and reboiler may be combined as shown in Figs. 59 and 61. The combined apparatus requires fewer pipe connections and is somewhat more simple and inexpensive. As seen in Fig. 61, the skimming is accomplished by a heavy galvanized-iron diaphragm near one end of the rectangular tank containing the water to be boiled. The impurities and refuse water flow over this diaphragm and out through the pipe connection made to the end of the tank. Distilled water is brought to the tank by the connections *A*, the pipe being extended into the tank and perforated so that the water escaping through the holes is evenly dif-

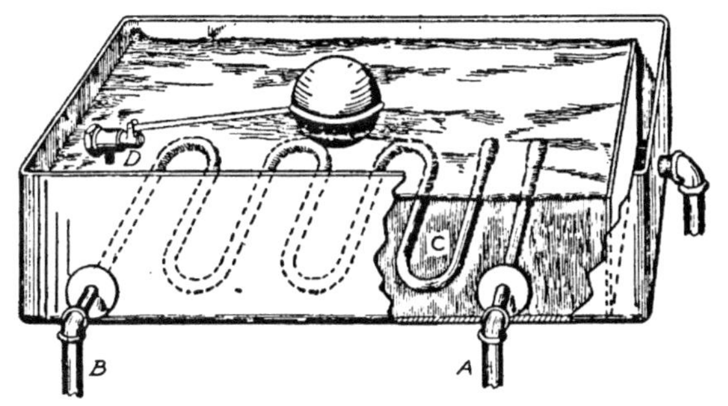

Fig. 61. Reboiler and Skimmer.

fused and, rising to the surface level of the water in the tank, gets rid of any entrained air.

Live steam is supplied to the zigzag coil in the bottom of the tank by means of the connection *B*, and being condensed in its passage through the coil is allowed to escape through the perforations shown at *C* in the last turn of the coil. It will be seen from this that all water enters the tank at the end nearest the skimming diaphragm and must pass to the other end of the tank before going through the outlet valve *D* to the hot filter. During this passage the water is thoroughly boiled by the heat from the live steam coil and is freed from any impurities and air it may have contained. The outlet valve is controlled by a float and is so constructed that it is wide open while the tank is full and closes gradually as the water level falls until, when the level is about 3 inches above the outlet, the valve is entirely closed. Thus there is no chance for air to be drawn off with the water as would be the case if the valve remained open until the water level should fall as low as the outlet opening.

Filters. Filters usually consist of vertical cylindrical tanks made of heavy sheet iron or cast iron with the interior surfaces well galvanized. A perforated false bottom supports the filtering material in place, crushed quartz, sand, or good charcoal being used for this purpose. Quartz is preferable for the hot filter but charcoal is frequently used for this as well as for the cold filter. All pipe connections are made to the side of the tank so that the covers can be removed for cleaning and recharging and, in addition to stop cocks and valves on the pipe connections, a by-pass with proper valves should be provided so that cleaning can be done without interfering with the operation of the plant. The frequency with which the filtering material must be renewed depends altogether on local conditions, and varies from every week or ten days to once a season. Under average conditions renewal once a season or, at most, twice will be found satisfactory.

The water may run through the filter from the top down, as is usually done, or the direction of flow may be reversed according to the preferences of the engineer in charge. In the average filter, the depth of the filtering material is about 5 feet. The surface area required depends on local conditions. Filters are ordinarily from 30 to 40 inches in diameter and one hot filter of this size, 7 feet high over all, is about right for a 15-ton plant. For the cold filter, 1 square foot of surface will suffice for a plant of the same size. In small plants the cold filter is often a very small affair known as a *sponge filter* and may consist of nothing more than a sheet-metal cylinder about 20 inches long and 8 or 10 inches in diameter with proper connections at its two ends for the water pipes. Charcoal, grass sponges, or other filtering material is placed in one end and the other end is filled with alternate layers of cotton and cloth of fine weave. This arrangement is considered very effective in catching rust and other material that gives *red core* ice.

Cooling Coils and Gas Cooler. After leaving the hot filter, the distilled water goes to the cooling coils constituting the *flat cooler,* in the manner already described. These coils are built like the steam condenser but are usually of a smaller sized pipe. In fact the coils are nothing more than an atmospheric condenser used to cool water instead of to condense steam. About 4 square feet of pipe surface should be allowed for each ton of ice-making capacity. As the cooling coils may be compared to the atmospheric condenser, so also the gas forecooler may be likened to the submerged condenser. In the case of the forecooler, however, the cooling medium—expanded gas—flows through the coils and the water to be cooled fills the tank, whereas with the condenser the water filling the tank does the cooling and the gas inside of the pipe coil is condensed.

For small plants the tank for the forecooler should preferably be cylindrical and the pipe coil be made in the form of a spiral without joints. Such a cooler is illustrated in Fig. 62, showing the Triumph construction. Larger plants have the cylindrical or rectangular tank as best suits local conditions. The combined area of the coils in every case should not be less than the area of the suction pipe of the compressor and should preferably be from $1\frac{1}{2}$ to 2 times greater. This proportion will ordinarily give from 4 to 5 square feet of cooling surface per ton of ice, which is about right for average conditions. Care should be taken to see that all the connections of the distilled water apparatus are of block tin or galvanized iron so that no trouble will be had with rust. The reboiler and other vessels should be made of galvanized iron or have the surfaces in contact with the water thoroughly galvanized. Valves should be of composition. Connections should be made to blow out all parts of the system with live steam as occasion may require, andblow-off cocks should be provided at convenient points.

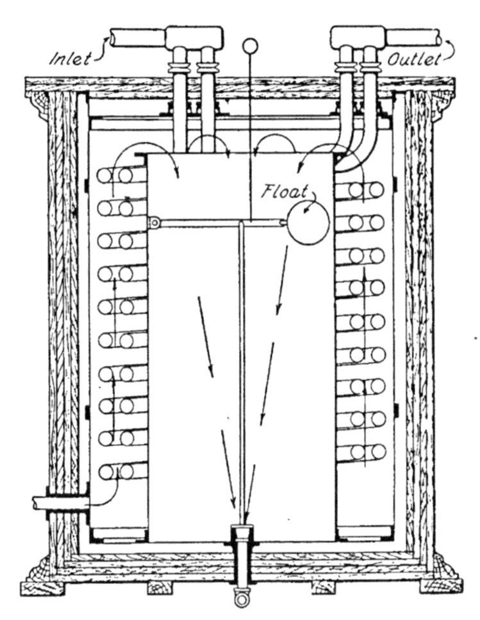

Fig. 62. Distilled Water Storage Tank.

Freezing Tank. The freezing tank with its accessory apparatus is the center of operations in the ice plant. The complete equipment includes *ice cans*, with covers to be placed over them in the spaces of the floor grating that holds them in position; a *brine agitator*; a *crane* with geared hoist and can lift; an *ice can dump* with thawing apparatus; a *can filler* with hose to reach any part of the tank room floor; *expansion coils* with headers and valves; a *brine hydrometer* and a *thermometer*. The tank itself is made of metal or wood and should be well insulated, one method of doing which has been de-

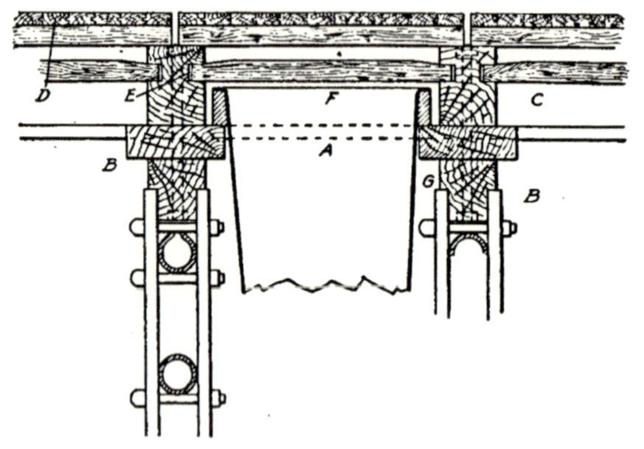

Fig. 63. Construction of Tank Grating and Covers.

scribed in discussing brine cooling tanks. For tanks up to 30 inches deep, $\frac{3}{16}$-inch steel is thick enough, but for deeper tanks up to about 4 feet the thickness should be $\frac{1}{4}$-inch. The sides and ends should be well braced with angle irons and an angle-iron rim should be put around the top. Sufficient holes should be punched in this rim to make sure that the grating can be bolted securely in position.

Expansion Coils. The expansion coils should be of extra heavy welded pipe running the full length of the tank if possible and held in position, a coil between each two rows of cans, by iron straps. These straps also support the grating, as seen in Fig. 63. The inlet of each coil is fitted with an expansion valve and each of the outlets is provided with a stop valve. Thus any coil may be cut out of operation, if it is found to be leaking, without interfering with the operation of the plant. All of the coils of the tank are connected to a manifold at the inlet end, the connection being made so that the expansion valve is between the coil and the manifold. A similar manifold is used to connect the outlets of all the coils with the suction of the compressor. Ordinarily the *bottom feed* is used, the liquid ammonia entering the bottom pipe of each coil and passing off to the suction manifold from the upper pipe of the coil; but this method of feeding is reversed where the *wet* system of operation is used. Thus there are two methods of feeding the coils, each of which has its advantages and disadvantages.

There is also a third system of operation known as the *top feed* and *bottom expansion* which is a combination of the two methods just described. At the feeding end of the coils a manifold is connected to each alternate coil and the ammonia is fed downward in these coils as in the wet system. The bottom ends of all the coils are connected to a common manifold so that the liquid after flowing down through half of the coils, rises and evaporates through the other coils and finally passes to a third or suction manifold which is connected to the upper ends of the coils not connected to the feeding manifold. The gas passes from this third manifold to the suction of the compressor. There should be 220 lineal feet of 2-inch pipe or 350 feet of $1\frac{1}{4}$-inch pipe for each ton of ice to be made in 24 hours, due regard being had for the temperature of the brine and the most economical capacity of the machine. It is true that many tanks are installed with much less pipe surface than this, but the plants so installed are necessarily operated extravagantly, as the back pressure must be carried very low to get capacity. This low pressure calls for more coal, and is the cause of increased depreciation of the apparatus.

Ice Cans. Ice cans are made in 50-, 100-, 200-, 300-, and 400-pound standard sizes, the top and bottom dimensions for each of the sizes respectively being 8 x 8 and $7\frac{1}{2}$ x $7\frac{1}{2}$ inches, 8 x 16 and $7\frac{1}{4}$ x $15\frac{1}{4}$ inches, $11\frac{1}{2}$ x $22\frac{1}{2}$ and $10\frac{1}{2}$ x $20\frac{1}{2}$ inches, $11\frac{1}{2}$ x $22\frac{1}{2}$ and $10\frac{1}{2}$ x $21\frac{1}{2}$ inches, and $11\frac{1}{2}$ x $22\frac{1}{2}$ and $10\frac{1}{2}$ x $21\frac{1}{2}$ inches. For the 50- and 100-pound can, the inside and outside depths are respectively 31 and 32 inches, while for the 200- and 300-pound cans the depths are 44 and 45 inches, and the 400-pound can has an inside depth of 57 inches with an outside depth of 58 inches. Reinforcing rings are used around the tops of the cans which are made of iron bands $\frac{1}{4}$-inch thick by $1\frac{1}{2}$ inches wide. All except the 400-pound cans are made of No. 16 steel, U. S. gauge, and these cans are made of No. 14 material.

The metal should be of good quality and of uniform thickness and all except the largest cans should be made with but one side joint. All joints are riveted on 1-inch centers, the rivets being driven close and the seams soaked with solder and floated flush. The bottoms are flanged and inverted 1 inch into the body of the can. All bands are welded and galvanized and should be punched in the middle of the long sides with $\frac{5}{8}$-inch holes placed $1\frac{5}{16}$ inches from the top of the band. Cans made of No. 16 steel should have the sides turned over at the top and bottom.

Grating and Covers. Gratings and covers for holding the ice cans in position are constructed as shown in Fig. 63. The rim of the can rests on a galvanized-iron cross-strap *A* which is mortised into the oak strip *B*. Above and below the strip *B* are strips *C* and *G*, and all three of the strips are held together by through bolts, as shown in the illustration. The whole structure is supported on the iron straps that hold the expansion coils in position, these traps being mortised into the strip *G* as shown. Grooves *E* are cut into the sides of the strip *C* and a stick *F*, having its ends set in these grooves, serves to hold the cans down so that they cannot float. The covers *D* rest on the strips *C* and in common with the other parts of the grating are made of oak, two thicknesses being used with good insulating paper between them. These boards must be thoroughly nailed together, as they are subjected to rather severe usage and have a tendency to warp out of shape. Some means should be provided for lifting them and this may be done by hollowing out hand holes at the ends or by providing regular plates and handles.

Brine Agitators. Brine agitators are of three classes, using centrifugal pumps, displacement pumps, and propellers respectively. As the object in all cases is to get a steady, uniform circulation of the brine in all parts of the tank, it is plain that the propeller is well adapted to the work and for this reason it is used in the great majority of cases. Where brine coils are placed in the ice storage house or where coolers are operated in connection with the ice plant, there is an advantage in using a displacement pump, as the brine when drawn from the tank may be pumped through the cooling coils before being returned. When this method of circulating the brine is adopted, discharge pipes must be put in the tank so that the returning brine will be distributed throughout the entire tank. One of these pipes is placed under each of the expansion coils and small holes in the pipes distribute the brine all along the coils. In this way a current is set up at each of the coils and the comparatively warm brine returned by the pump from the cooler coils is brought in direct contact with the expansion coils, with a resulting high efficiency of heat absorption by the expanding gas.

In the use of the centrifugal pump, the chief point of advantage lies in the fact that a large quantity of brine can be circulated. The brine is taken from one corner of the tank and discharged into a header on the side of the tank opposite the suction connection of the pump. The rapid circulation set up in this way causes rapid freezing which is a great advantage when ice is needed in increased quantities to meet the demands of the market in hot weather.

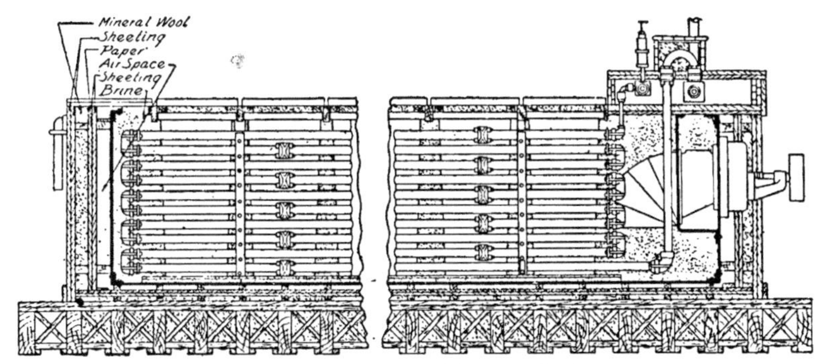

Fig. 64. Construction of Freezing Tank.

Where a propeller is used for circulating the brine, as shown in Fig. 64, which is a longitudinal section through a freezing tank, one or more wooden partitions are constructed in the brine tank between the cans and along the expansion coils for almost their entire length. The propeller is driven by a direct-connected engine or by a motor, and forces the brine to circulate by moving from the discharge side through the length of one compartment, around the end of the wooden partition and back through the other compartment to the suction side of the propeller. Thus it is seen that there are two passages essential to the operation of the system, one of the passages being open to the suction and the other to the discharge side of the propeller. Where only one propeller is used on a tank, it is advisable to use two partitions so that the suction passage of the propeller is divided into two parts. The propeller is placed near the center of the tank at one end and discharges through the middle compartment between the two partitions, and the brine, arriving at the far end of the tank, is divided into two streams that flow back by the side passages outside of the partitions to the suction of the propeller. For a 10-ton tank, a 12-inch propeller of ample size, and for a 15-ton tank an 18-inch propeller should be used. Larger tanks require more than one propeller. In planning a method for circulating the brine, one should remember that a comparatively large amount of power is required to operate the propeller system.

Crane and Hoist. The great majority of small and medium sized plants use hand cranes, consisting of a light channel iron carried on four wheels, which run on suitable rails placed at a convenient height on the side walls of the tank room. The channel iron carries a four-wheeled trolley provided with a geared hoist on the drum of which is wound the hoisting chain or rope. A *can latch*, one form of which is shown in Fig. 65, is attached to the end of the chain. The apparatus consists of a board mortised out at the ends so as to drop into the top of the can. In the middle is an eyebolt to which the hoisting chain is connected and hook latches at the two ends are adapted to catch in the holes in the sides of the can. When the can has been lifted and is to be set down on the dumping table the latches are pulled outward so as to release the hooks. With the hand crane and an apparatus of this kind, a man can handle about 15 tons of ice in a day of 12 hours. For plants of larger capacity it is advisable to use a pneumatic hoist and when this is done a special latch may be used so that two or more cans may be lifted at the same time. With this kind of hoist, one man can handle from 40 to 50 tons of ice in 12 hours. In still larger plants, special means are used to hoist the cans and dump the ice.

Dumping and Filling. The dumping and filling of the cans should be done according to some regular, well-ordered system of rotation. Numbers should be plainly stenciled or cut on the covers of the cans so that the tankmen need make no mistake in pulling the proper cans. All the cans should not be pulled from any one part of the tank at the same time, and except in large tanks it is not well to take all the cans of any one row at a single pull. As an example of what may be done, suppose that the cans are numbered in consecutive order over the entire tank. The tankman could then pull every fourth can, taking the numbers 1, 5, 9, etc. At the next pull he would take the numbers 2, 6, 10, etc.

Various styles of dumping tables are in use, varying from a simple home-made apparatus designed to handle one can at a time, to the elaborate apparatus of a large plant dumping a number of cans at one operation. The can may be dipped into a hot-water tank to thaw the ice block loose, or tepid water may be sprinkled over it to accomplish the same purpose, as is done with the style of apparatus illustrated in Fig. 66.

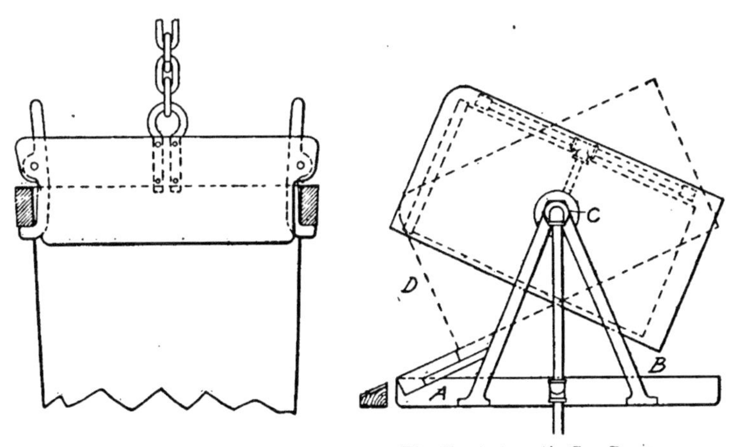

Fig. 65. Can Latch. Fig. 66. Automatic Can Dump.

The table here shown is constructed of metal and consists of a drip pan *A* on the bottom of which are riveted two pairs of supporting brackets *B*, made of pipe. Hollow trunnions *C* are connected to the water supply and sprinkler pipes and support the box *D*, in which the can is placed. The ports in the hollow trunnion are so arranged that when the box is in the position shown by the full lines, the water connection is shut off. However, as soon as the box is tilted to the dumping position shown by the dotted lines, the connection to the water supply is made and water flows over the can from the sprinkler pipes until the block of ice is thawed out. When this occurs, the weight of the can, which is not evenly balanced on the trunnions, acts to return the box to the first position, in doing which the water is shut off. Thus the operation is automatic and the tankman, having placed a can in the tilting position, gives it no further attention until he pulls and brings up another can to be put in the place of that from which the ice has been dumped. The empty can is then returned to its place in the tank and filled with distilled water from the supply hose by means of an automatic can filler which is placed in the can and the trigger pulled, after which the attendant leaves it and goes about his business. When the water rises to the desired level, it raises a float that automatically moves the trigger to shut off the water supply. By using this apparatus all cans are filled to the same level without special attention.

Layout. The layout of a plant should be given the most careful consideration as success or failure depends to a large extent on

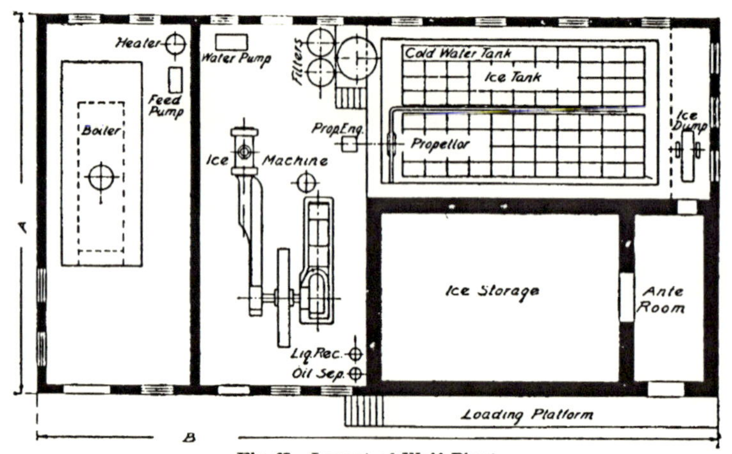

Fig. 67. Layout of Wolf Plant.

the arrangement of the different parts with reference to convenience and economy in operation. No set designs can be given which will meet the local conditions of every case, but plans of a few typical plants are given to show what should be sought for in constructing a plan suitable for any particular case. Where local conditions do not require specially constructed buildings, the whole plant may be housed in a single building of rectangular form such as that shown in Fig. 67. This design is by the Fred W. Wolf Co. and has been used as a basis of work in designing a large number of successful plants. The boilers are in the end of the building remote from the freezing tank, and the ice machine is so set that as little as possible of the heat radiated from steam pipes, etc., will get to the tank room. The overall dimensions for a plant of this kind are given in Table 19 for capacities ranging from 5 to 100 tons. Fig. 68 shows a diagram plan and elevation of a design used by the Arctic Ice Machine Co. The dimension letters refer to Table 20 which gives a complete schedule of dimensions for plants ranging in capacity from 2 to 200 tons daily output. All dimensions given in the table are in feet.

TABLE XIX

Dimensions for Fig. 67

Daily Capacity	Dimensions A	Dimensions B
5 tons	30 feet	56 feet
10 "	35 "	73 "
15 "	37 "	78 "
20 "	40 "	85 "
25 "	42 "	95 "
30 "	42 "	107 "
35 "	42 "	117 "
40 "	49 "	120 "
50 "	49 "	135 "
60 "	54 "	150 "
80 "	59 "	154 "
100 "	73 "	160 "

Fig. 69 shows another arrangement for a small factory. This is a design of the Frick Co. and is suitable for a plant having a daily output of from 6 to 10 tons. For a plant of about 35 tons capacity, the Frick Co. uses the design shown in Fig. 70, which gives sectional side and end elevations and a plan view. Another design by the same company for a 100-ton plant is shown in Fig. 71. These three illustrations give an idea of the necessary changes in arrangement for plants of different sizes. Fig. 72 shows in plan and elevation a modern ice factory as constructed by the Triumph Ice Machine Co.

Where absorption machinery is used, the arrangement of machinery may be modified if desired so as to make the plant somewhat more compact, for the same capacity, than a plant operating with compression machinery. This ability to compact the arrangement is due principally to the fact that in the absorption machine there are no moving parts, the only moving machinery being the pumps. A good plan for an absorption plant is that used by the Henry Vogt Machine Co., an isometric view of which is shown in Fig. 73.

Plate System. Although *plate ice* may be produced by freezing from two sides of the compartment containing the water and allowing the ice cakes to meet, thereby reducing the time of freezing, this has not been done to any extent. Practically all plate ice is frozen from one side only and on this account a great deal of time is

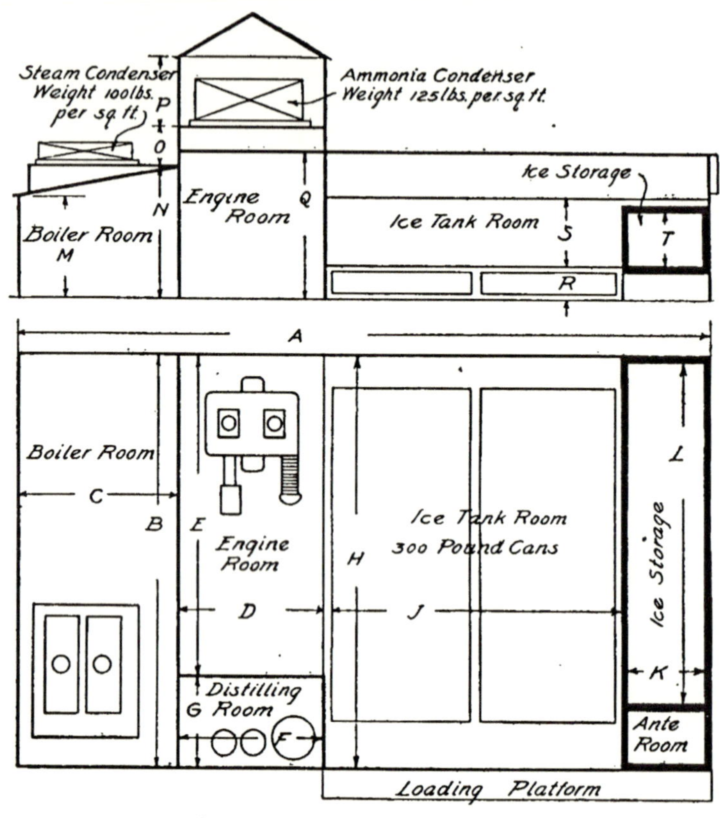

Fig. 68. Arrangement of Arctic Plant.

required, about eight or ten days being necessary to freeze 11-inch ice. The time of freezing for a 20-degree temperature is determined by a rule similar to that given for the can system, viz: *Multiply the thickness to be frozen by twice itself plus one. For a temperature of 15°, deduct one-fourth from the result thus found.* Thus for 11-inch ice we have 11 (2 x 11 + 1) = 253 hours and deducting one-fourth the freezing time at 15°, is about 190 hours. The freezing

TABLE XX
Dimensions for Fig. 68

Capacity Tons	A	B	C	D	E	F	G	H	J	K	L	M	N	O	P	Q	R	S	T
2	17	28	17	10	18	10	10	28	12	8	18	16	20	6	9	16	5	12	10
5	48	42	17	10	32	10	10	42	13	8	32	16	20	6	9	16	5	12	10
8	61	34	19	12	24	12	10	34	22	8	24	16	20	6	10	16	5	12	10
10	72	34	21	15	24	15	10	34	26	10	24	16	20	6	10	16	5	12	10
12	72	38	21	15	28	15	10	38	26	10	28	16	20	6	10	16	5	12	10
15	78	44	24	18	34	18	10	44	26	10	34	16	20	6	12	20	5	12	10
18	80	50	24	18	40	18	10	50	26	12	40	16	20	6	12	20	5	12	10
20	87	46	25	18	36	18	10	46	32	12	36	16	22	6	12	20	5	12	10
25	89	54	25	20	44	20	10	54	32	12	44	16	22	6	12	20	5	12	10
30	92	62	25	20	50	20	12	62	32	15	52	16	22	6	12	20	5	12	10
35	93	70	26	20	58	20	12	70	32	15	60	16	22	6	12	22	5	12	10
40	95	79	26	20	67	20	12	79	32	15	69	16	22	6	12	22	5	12	10
50	144	54	35	25	39	25	15	54	64	20	44	20	22	6	12	21	5	12	10
60	145	62	36	25	47	25	15	62	64	20	52	20	25	6	12	21	5	12	10
75	160	79	46	30	64	30	15	79	64	20	69	20	25	6	12	27	5	12	10
100	173	59	54	30	80	30	15	95	64	25	85	20	25	6	12	30	5	12	10
150	247	95	76	40	80	30	15	95	96	35	85	20	25	6	12	30	5	12	10
200	333	95	100	40	80	30	15	95	128	45	85	20	25	6	12	30	5	12	10

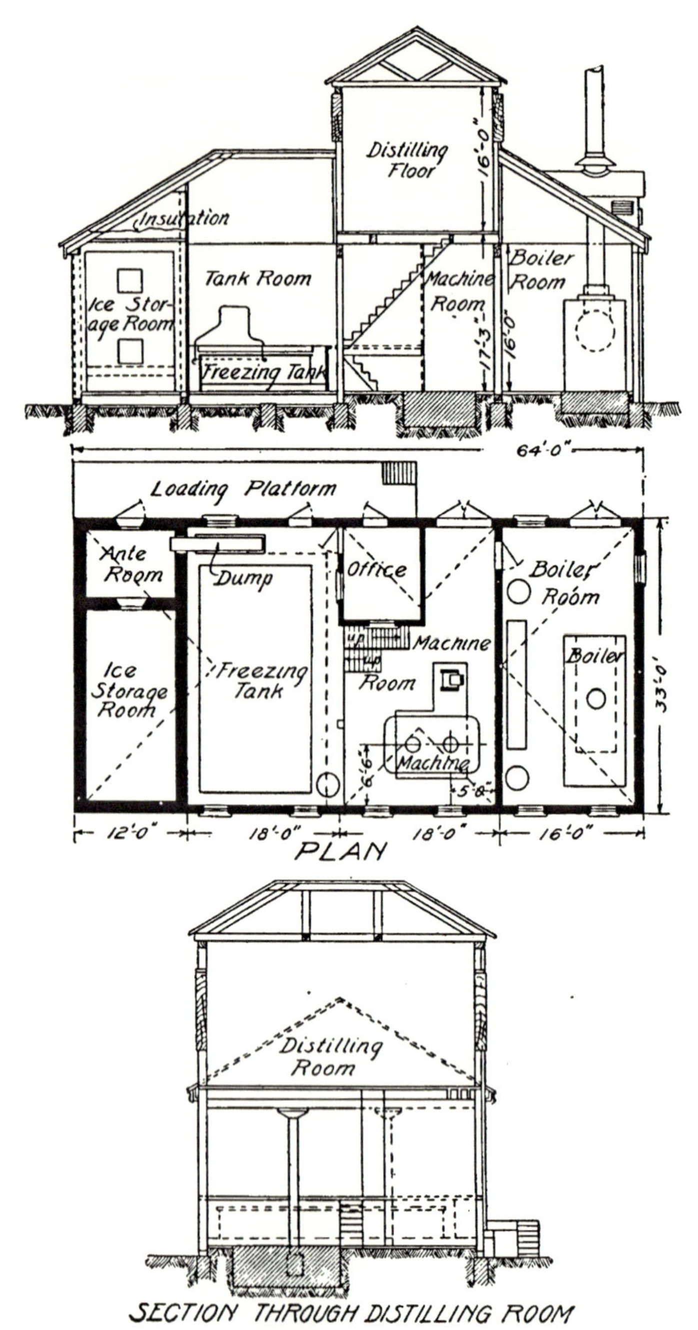

Fig. 69. Frick 10-Ton Plant.

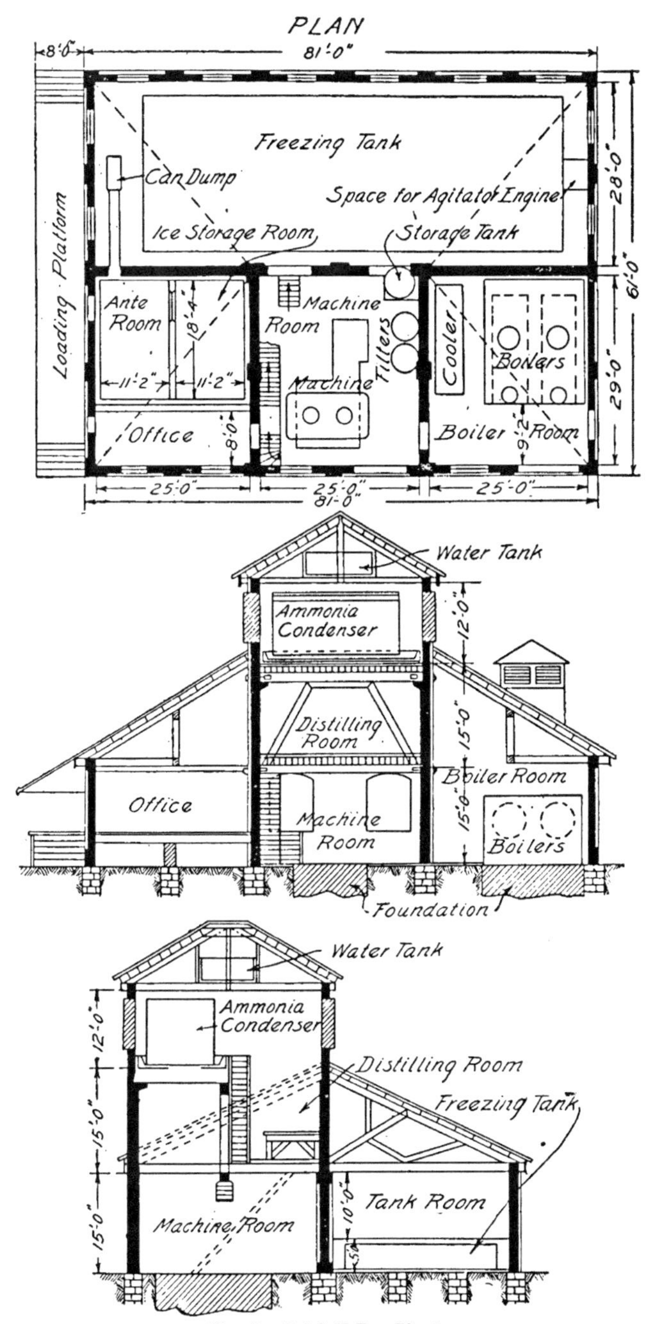

Fig. 70. Frick 35-Ton Plant.

apparatus of the plate plant consists of a tank divided into compartments and fitted with freezing plates; a forecooling tank with coils; a crane and hoists; a tilting table; cutting-up saws; water filters and pipe connections for supplying water.

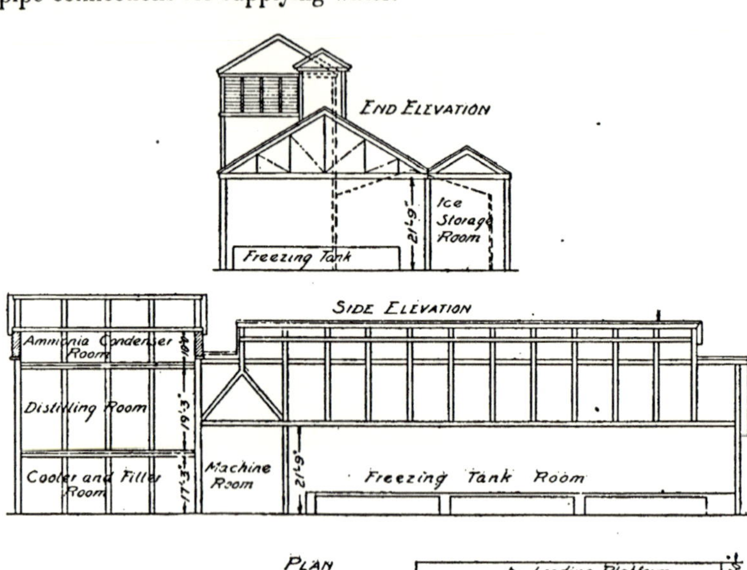

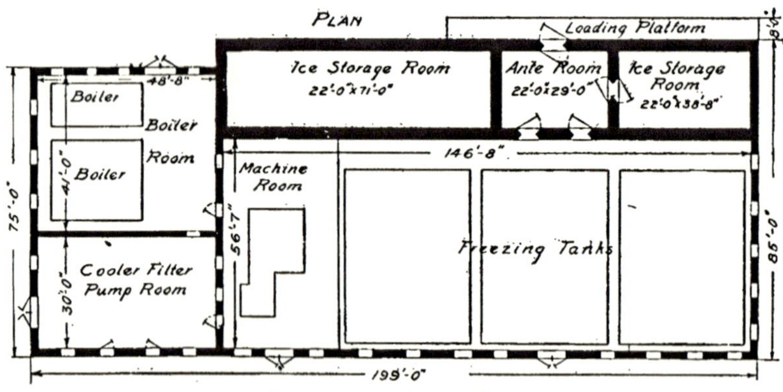

Fig. 71. Frick 100-Ton Plant.

There are two methods of operation. The first method, which is known as the *dry-plate* system, is that in which ammonia gas is expanded directly in pipe coils that make up the freezing plate, the spaces between the pipes being filled in with wood or other material to form a smooth freezing surface on the two sides of the coil. In the *wet-*

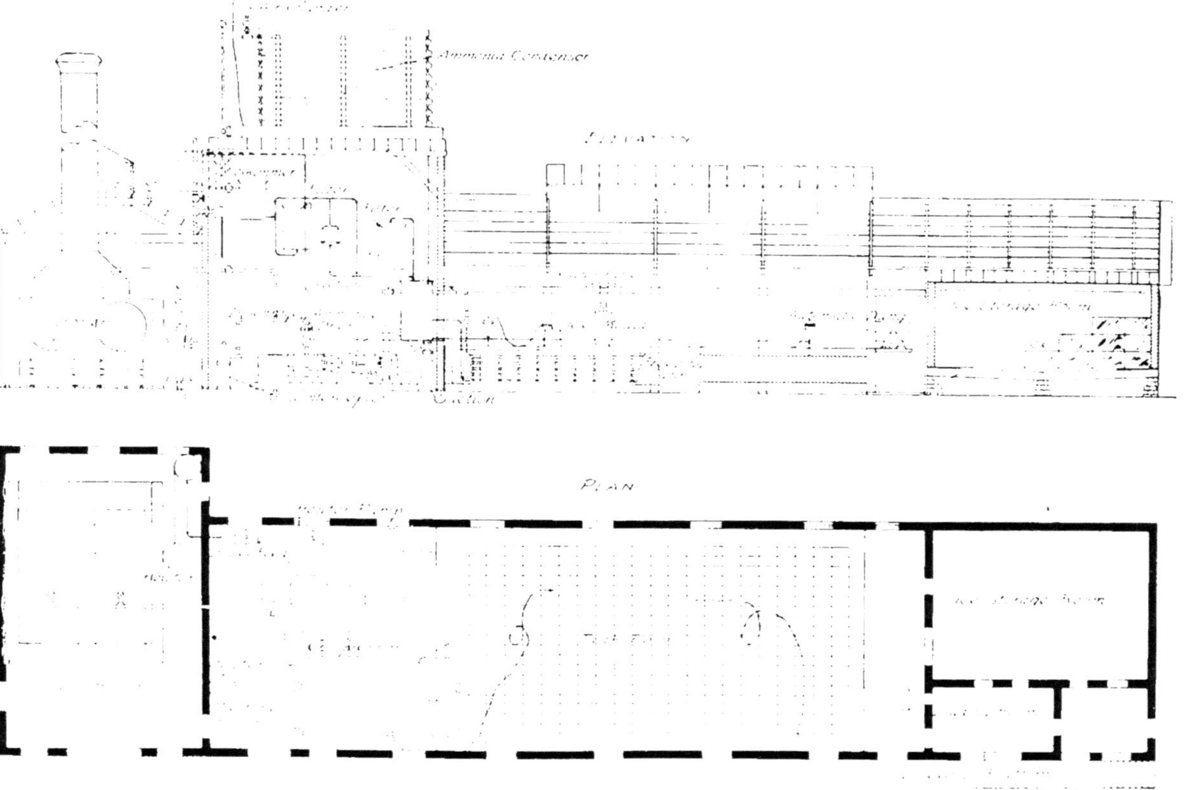

Fig. 72 Model Triumph Factory.

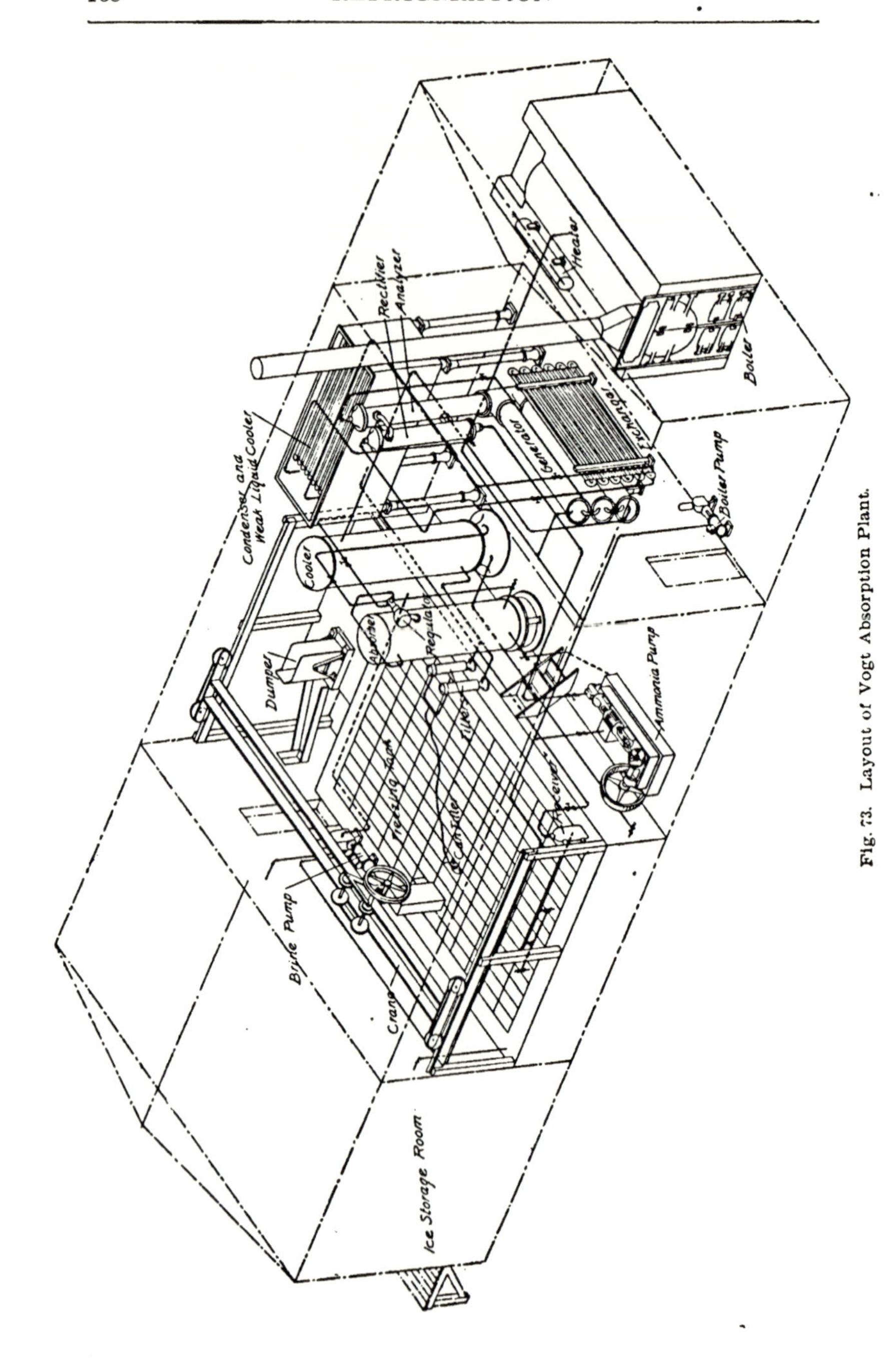

Fig. 73. Layout of Vogt Absorption Plant.

plate system, brine is used and is closed up in a metal cell or tank from 4 to 6 inches thick and of the size necessary to form the freezing plate. This imprisoned brine is kept cold by ammonia expanding in a pipe coil placed in the tank. In some cases the plate and the attached blocks of ice are removed from the tank bodily, by disconnecting the pipe connections to the expansion coil after drawing off the ammonia in the coil. Where this is done, provision is made to drain the cold brine into another of the hollow plates which is immediately placed in the tank so that the freezing process goes on while the ice is being detached from its plate and disposed of.

In another form of the wet-plate system the cells forming the freezing plates are designed to have cold brine pumped in at the top

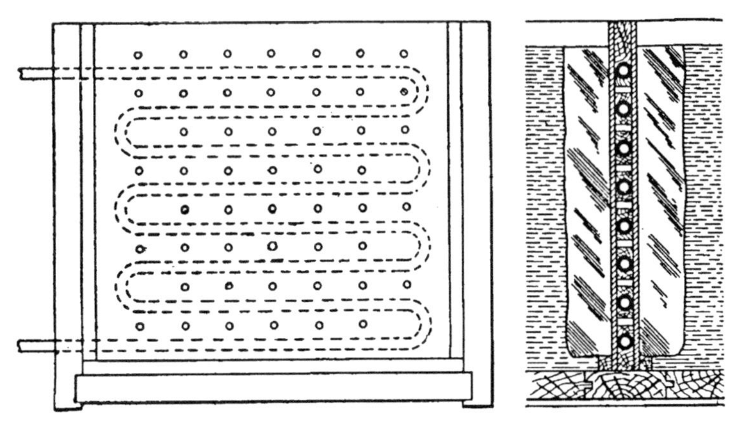

Fig. 74. Coil Plate with Wood Filling.

and run down in a thin sheet over the inner surfaces of the plate and collect at the bottom of the cell, from which it is drawn off by the brine circulating pump. The great difficulty with all applications of the wet-plate system is that of making the cells tight. It is almost impossible to roll large plates which will not show up small leaks and a few such leaks turn enough brine into the water to ruin the ice. This difficulty of obtaining tight plates stands in the way of freezing by expansion of ammonia direct into a cell on the sides of which the ice is frozen. The cells can easily be made tight against the escape of brine under small difference of pressure but to make them tight enough to retain expanding ammonia gas is impossible except at prohibitive expense. In the dry-plate system, pipe coils can be made to hold the gas where care is taken in welding the pipe and in making the joints; the chief difficulty has been that of getting suitable surfaces against which to freeze the ice, while using the coils.

In small plants the ice is allowed to freeze directly to the coils and is then cut loose, but this is wasteful of ice and of labor. Wood filling between the pipes of the coils is not stable enough to withstand the rough usage to which the plates are subjected, and on this account smooth metal plates are bolted on each side of the coil and its wood filling, as shown in Fig. 74. Plates of this kind are placed in the compartments of the freezing tank about 30 inches apart on centers, and the tank is filled with water to within about 9 inches of the top of the plate. As the ammonia is expanded directly into the coil, the cooling is very rapid and on this account great care must be taken in feeding or the ice will be frozen before the air and impurities have had time to be separated out. Another difficulty with the dry-plate system of operation is the fact that the ice forms thicker where the ammonia is fed into the coil than over the rest of the plate so that the block is not of uniform thickness. These difficulties are avoided to some extent by expanding the gas into a forecooler before turning it into the plate coils, but after all the dry plate is difficult to operate successfully.

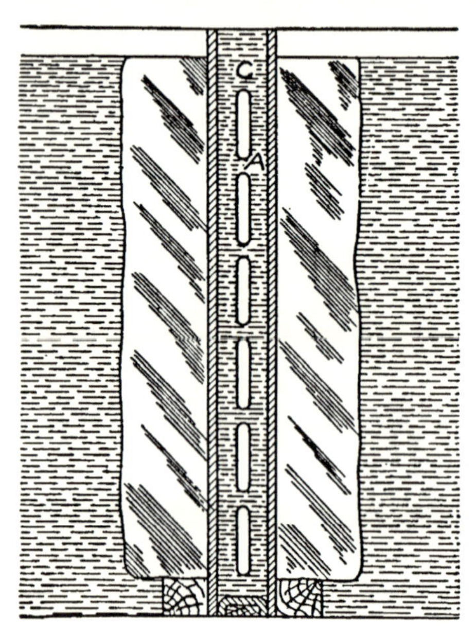

Fig. 75. Cell Plate.

This difficulty is offset, at least partially, by the fact that the dry-plate apparatus is comparatively inexpensive to install while the wet-plate system with its brine storage tank, brine pump, and extra piping for the brine is expensive to install and somewhat extravagant in operation on account of the extra transfer of heat to the brine. Fig. 75 shows a cross section of a cell such as is used with the wet-plate system. The quantity of brine in the cell acts as a kind of fly-wheel or balancer for the system and aids materially in regulating the temperature to the uniform standard required for making good ice with the water available. Different waters require different brine temperatures, more time being allowed where the water is impure. Practice with the given plant is the only way to determine the best temperature for getting good ice in a given case.

Slabs of ice frozen on the plate system may weigh from 1 to 10 tons, the maximum dimensions being about 16 x 9 feet by 12 inches

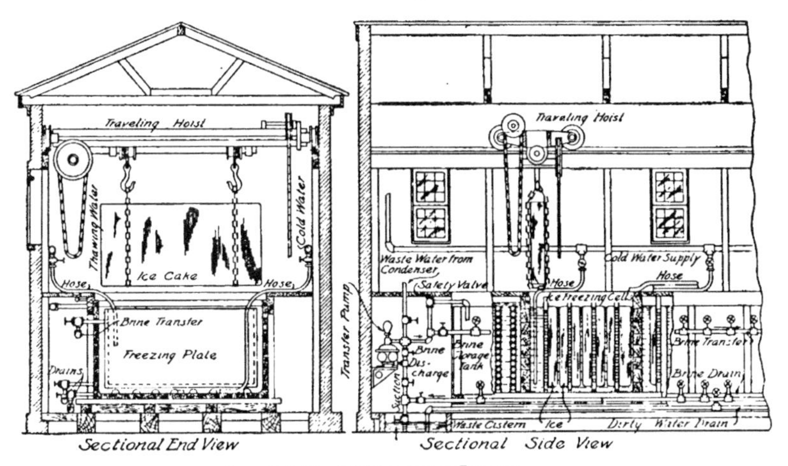

Fig. 76. Eclipse Plate System.

thick. In the United States the cakes are usually about 14 x 8 feet by 11 inches. During the process of freezing the impurities eliminated and thrown out settle to the bottom and may be washed out before refilling the compartment with fresh water, if considered necessary. Heavy traveling cranes are used to lift the ice when frozen and transport it to the cutting floor to which it is lowered from the vertical position by the tilting table. Power-driven gang saws are now made to cut the cakes up automatically into any size blocks desired, and chain conveyors take the blocks from the table to the storage room or loading platform. The whole process of lifting is done as readily as one lifts a single can from a tank, everything being done by power. Fig. 76 shows a sectional elevation of a plate plant

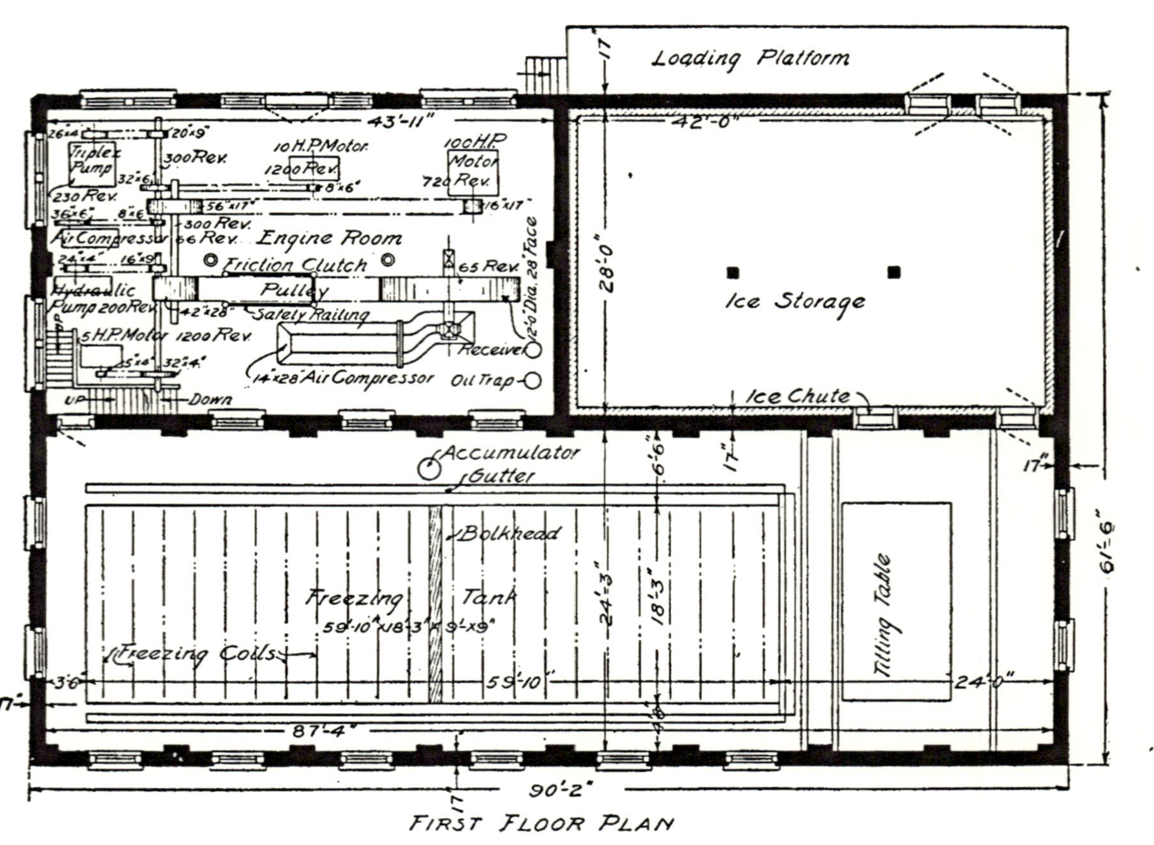

Fig. 77. Floor Plan of Vilter Plate Plant.

built on the *Eclipse* plan used by the Frick Co., while Figs. 77 and 78 show plan and sections in outline for a 25-ton plate plant as designed by the Vilter Mfg. Co., for electric-motor drive. The

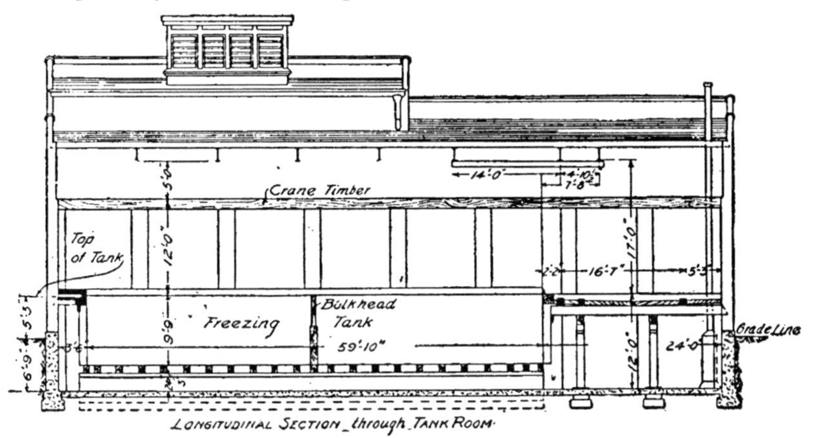

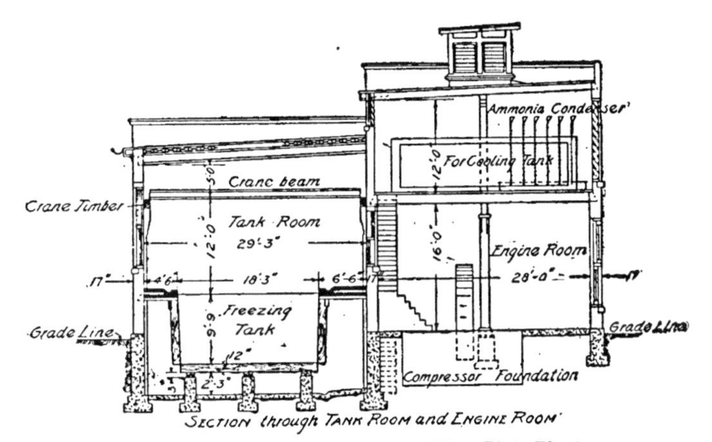

Fig. 78. Front and Side Views of Vilter Plate Plant.

drawings, it will be noted, show complete details and dimensions, while the notes are self-explanatory. There is only one bulkhead in the freezing tank, so that only two compartments are formed. It would be better to increase the length of the tank so that additional bulkheads could be inserted to divide the tank into eight compartments, each containing four double-face cells as required to freeze a day's pull of ice.

STORING AND SELLING ICE

Distribution is the one great problem of the ice manufacturer. The product of his factory is perishable and cannot be held except at considerable expense. It must be disposed of as made if the ledger is to show a profit. Some manufacturers, it is true, find it advisable to have a smaller plant than required to meet the demands of the summer trade and arrange to run all the year round, storing up the ice made during the winter so as to have it available when needed. The cost of such storage is about as much as the interest on the larger plant, and there is considerable loss of the ice put in store unless it is refrigerated. This necessitates using part of the capacity of the machine on the cooling coils of the storage rooms. Then again there is no opportunity to overhaul the plant when run continuously and the item of depreciation is larger than it should otherwise be. On the other hand the best machine and the wisest manager cannot regulate production to exactly meet demand, for two successive summer days may bring very different demands for ice. Some days the manager has ice melting on his hands or his machine *killing time,* while at other times he sits up nights wondering where he is going to get the ice.

All things considered it is best to store a moderate amount of ice and have a medium-sized plant, arranging things so that there will be opportunity to overhaul, repair, and repaint the system during the winter. Ice plants, then, are usually provided with a temporary storage room and a larger room for permanent storage. Any overplus of production from day to day goes into the small storage room and in seasons of very light demand practically all the output will go into the large store-room. In most cases, and wherever possible, it is best to have the storage rooms on a floor level lower than the top of the tank so that the ice can be passed by chute to the stores. Otherwise conveyors are used, these being of several forms, but all built on the endless chain plan. A chain having catch lugs passes along a slot in the floor of the slide or chute and carries the ice up to the storage room. In storing large quantities of ice, as in plants where fruit and produce cars are iced, it is necessary to pack the ice in tiers and for more than two tiers some form of hoisting apparatus should be used in the storage room. About 50 cubic feet of space are allowed per ton of ice to be stored, and 1 running foot of $1\frac{1}{4}$-inch brine piping should be allowed for every 6 to 10 cubic feet of space, depending on the latitude and the size of the room. Owing to the low temperature of the ice, it is safe to allow at least one-third less pipe-cooling surface than is provided for ordinary stores. A temperature of 28 degrees is ample to prevent melting but some authorities prefer to carry as low as 22 degrees, for they consider the low temperature keeps the ice firm.

Methods of packing ice vary, some preferring to allow space between the blocks for ventilation and others packing the blocks close together and exactly over each other in successive tiers—*i.e.*, no breaking of joints—with packing material provided where the store-room is not to be cooled artificially. Where spaces are left, wood spacing strips are placed between tiers and blocks in all directions. If hard dry ice is put in store at or below freezing and the temperature is kept down—or, even packed simply, without cooling coils in the rooms—it will come out in good condition without the strips, which are little used in the United States. The rooms should have ventilators so that all foul, damp air may be removed, and good drainage should be provided. In permanent store-rooms good insulation should be provided and this is done almost universally in ice plants, there being no need of packing materials except in the case of natural ice storage. Materials used for such purpose are hay, rice straw or chaff, soft wood shavings, sawdust, and similar substances. About six inches of this material is packed around the walls in storing ice where cooling coils are not used. Also a good thick layer is placed on top of the upper tier. When coils are used they should be placed in racks hung from the ceiling and provided with drip pans having proper drainage connections to the sewer. In some cases it is necessary to place the coils on the walls of the room.

Where ice taken from the stores is to be used in icing cars, it is put through a crusher and sent by barrow or chute to the car box. In delivering ice to the retail trade, care should be taken to be regular and systematic in all arrangements and dealings with customers. Go to the same house at the same time on the day when it is known that ice is wanted and don't go any other time. Endeavor to have the men satisfied so that they will stay on the job long enough to learn their business and the little traits of the customers. Have everything about the wagons and the men clean, and use tested scales with the coupon system of selling. Let each team have its regular route and let the drivers understand that they are responsible for what happens on their respective routes. In shipping small lots of ice, blocks may be packed in sawdust in bags. Where car loads are to be shipped get refrigerator cars if possible. If not, pack the ice in an ordinary car with sawdust using heavy paper around the blocks after they are in position so as not to foul the ice with the dust. The paper also helps to keep out heat. Suitable hand tools should be provided in plenty as they are worth their cost several times over in saving the time of drivers and help, not to speak of insuring accurate cut of ice which means less loss and better satisfaction to customers.

ICE-PLANT INSULATION

The most important piece of insulating work in an ice plant is on the freezing tank. Aside from this the ammonia line from the tank to the compressor as well as the suction headers, where exposed, and all brine circulation lines should be insulated. The steam lines should be covered to prevent undue radiation. For this purpose a good magnesia covering should be employed. For the ammonia and brine lines, cork pipe covering should be used. Cork is now prepared for such work with parts made to fit all pipe fittings, as well as any bends and turns. Next to the tank in importance the ice storage rooms should be protected with insulating material, more or less care being used in this work depending on the use to which the room is to be put. For a temporary small storage it is not worth while to spend a large amount of money, but for permanent storage rooms the insulation should be done well. This may be done in any of the ways usual for cold storage rooms, care being taken to provide drainage for any water that may result from melting.

Owing to the dampness, cork insulation finished with cement plaster is about the most satisfactory material for an ice storage room. On a concrete foundation, from 4 to 6 inches of Acme corkboard is laid in cement, coated with asphalt, and finished over by laying 3 inches of concrete having cement finish on top. The ceiling and walls are insulated with from 3 to 6 inches of corkboard laid in Portland cement, with asphalt between layers, and finished with Portland cement plaster. This method of insulation gives a permanent water-proof finish and is effective and durable.

Tank Insulation. There are a number of ways to insulate a tank and various insulating materials that may be used, but planer shavings, when perfectly dry, give about as good insulation as can be had at moderate cost. Cork is better but is more expensive, though by using it in granulated form for the sides of the tank the expense is reduced considerably. Other manufactured products are too expensive, as a rule, in proportion to the benefit to be derived from their use. Sawdust lies so close that there is not enough dead air space, and takes up water readily so that its insulating properties become greatly impaired. Ground tan bark is subject to the same objection. Since good insulation cuts down operating expenses, it is economy to use as thorough insulation as circumstances permit. This does not necessarily mean the most expensive job that can be had. Where shavings are used, the thickness beneath the tank should not be less than 12 inches and on the sides it should be at least 10 inches, greater thickness being desirable if space can be had.

Matched flooring is used for partitions, with tarred paper laid on the flooring when in place to make it air tight. A simple construction is obtained by coating the outside surface of the tank with asphalt or pitch, and pack shavings between it and the air-tight wall built up of matched boards on studs set on 18-inch centers. In the case of metal tanks, which are used almost universally in preference to wood, a layer of pitch and 1 or 2 inches of shavings should be laid on the floor so as to form a cushion on which the tank sets, thus preventing possible strain due to irregularities of surface. Although wood tanks have been in use for some time and cement has more recently come into use as a material for tank construction, neither of these materials may be considered as satisfactory in all respects as the steel tank. More framing must be used with the wooden tank to withstand the pressure of the water and great care must be taken to make the cement or concrete tank waterproof. For plate plants, the freezing tanks are commonly constructed of wooden compartments, the framing being made ample to withstand the pressure. Where the water is bad and it is necessary to wash out the freezing compartments often, it is well to have each compartment insulated from the others.

Cork insulation is largely used on tanks, owing to its waterproof qualities and the fact that it requires much less space than other insulation. Where this material is used, the foundation on which the tank is to be made is constructed of concrete, preferably laid on cinders, on top of which Acme corkboard is laid in Portland cement. This insulation is from 3 to 6 inches thick and is coated on top with asphalt, the tank being set directly in the asphalt. The sides are then insulated with Acme or Nonpareil board laid on in cement, with asphalt between layers and finished outside with Portland cement plaster. This insulation is usually about 6 inches thick and is quite as effective as the thicker insulating walls made with other materials. Where it is desired to reduce expense, and space is of no particular account, regranulated cork may be used around the sides packing it between the tank and a wooden wall, as in the case of shavings.

GENERAL COLD STORAGE

After ice-making, cold storage is the most important and widely used application of mechanical refrigeration. By means of low temperatures, fruits and perishable products may be preserved during periods of plenty until such time as the supply falls off and the demand increases, when the products, coming from store in good condition, may be sold at a profitable figure. Thus decay is prevented, the product is preserved in its natural form—quite another matter from goods preserved by drying or salting—and the dealer is able to make a profit. The producer at the same time is able to market his crops to better advantage than when all the goods have to be sold on a glutted market. Owing to the fact that the period of consumption for any given product is greatly increased, production can be increased accordingly with a corresponding increase in profits to the farmer and fruit grower. The owner of perishable goods is not compelled to market them for fear that they may spoil but may choose his market and hold his goods irrespective of time and distance.

For these reasons, cold storage establishments are coming into use the country over. Refrigerator cars make possible the shipping of goods to distant markets where good prices prevail, and in case markets are not satisfactory in the country of production, goods may be exported in carefully constructed storage rooms on board the large steamers. Thus it is possible to place perishable products in the hands of consumers in such quantity and at such time as the owner desires so that the market conditions are steady and the consumer is able to obtain articles of food that could not be had without the aid of refrigeration. Meats produced in Australia are shipped to England and fruits grown in California are sold in New York and Europe. On a bleak winter day one may have the freshest vegetables and fruits on his table, in any part of the world. Peaches grown in Australia have been served at millionaire feasts in New York during the months of January and February. Apples, butter, eggs, and other such products are sent abroad by the ship-load. In a word there is no perishable product that cannot be handled to advantage by means of refrigeration.

But food products are not the only things kept in cold storage. Many articles not classed as perishable are kept at low temperatures. Furs, for example, are placed in storage to prevent damage from moths and to preserve the skins. On coming out of store the luster of the furs is greatly improved, the articles becoming more valuable in some cases than before storage. Clothing and woolens are also stored. Dried fruits are kept during the warm months of summer. Seedsmen find it profitable to store their stock so as to prevent deterioration by the seed drying out, owing to evaporation of the oils. Thousands of dollars are saved in this way, as the germinating value of the seed may be kept unimpaired and the seedsman has good insurance against failure of any year's crop. Peanuts, walnuts, and other like goods are carried through the summer and, during the winter, potatoes and cabbage are put in store. Each year some new product is added to the list of articles carried in cold stores and there seems to be no limit to the growth of the ever-widening field for the cold storage industry.

Conditions for Preservation. In any cold storage room, three things are essential if the goods are to be cared for properly. The air in the rooms should be renewed frequently by a good system of ventilation; all air entering the rooms should contain an amount of moisture suitable to the temperature and the goods carried; and the temperature should be suited to the given products, not varying outside of certain limits. A *hygrometer* is used for measuring the amount of moisture in the air and this should be done accurately as too little moisture will cause the goods to be damaged by evaporation, while too much moisture will cause mold to be formed. One of these evils is about as bad as the other. Products that are dried to the mere fibre are of no value for food and those that are musty are not palatable. The lower the temperature, the less moisture the air can carry and if the air is brought into the rooms with more moisture than corresponds to the amount it can carry at the temperature of the room, the excess will be condensed and form ice on the pipes and other places. This frost should be kept off the pipes—where pipe coolers are used in the rooms—as far as possible, as ice is a good insulator and prevents the heat being absorbed by the gas or brine in the pipes. The temperature is entirely under the control of the operator and can be raised or lowered at will by circulating more brine or expanding more gas, according to the system used.

There is a wide difference of opinion among authorities as to the temperatures required for the best results with different products. A number of conditions complicate the question so that, except for a few of the articles most handled, no definite rule can be laid down. Where the goods have been shipped, consideration must be taken of the temperature before and during shipment and particularly whether or not the temperature has been excessively high or low at any time during shipment. Furthermore the condition of the goods must be taken into account, and the time they are to be kept in store, as well as the purpose for which they will be used when taken out of store. Thus for example, if decay has set in with a shipment of peaches which is being unloaded into the storage rooms and they are to be kept only a short time and then disposed of at whatever the market will stand, it would be well to get the temperature down as quickly as possible to that required for keeping this fruit—the lower working limit being preferable. On the other hand if it is a shipment of sound fruit that is to be kept some time, one could do better to work at a higher temperature and cool the fruit more gradually, it being recognized, of course, that under such conditions the refrigeration can be done more economically. In both these cases, the moisture in the air in a given cold room would have a considerable bearing on the course pursued.

Where certain rooms are used for commission trade with goods being taken in and out at frequent intervals, the refrigeration must be more effective than for the same class of goods stored over a long period of time. Altitude and local conditions affect the case to a

TABLE XXI

Temperatures for Cold Storage of Products

Products	Deg. F.	Products	Deg. F.
Apple butter	42	Huckleberries (frozen, long carry)	20
Apples	30	Japanese fern balls	31
Asparagus	33	Lard	40
Bananas	45	Lemons (long carry)	38
Beans (dried)	45	Lemons (short carry)	50
Beer (bottled)	45	Lily of the valley pips	29
Berries, fresh (few days only)	40	Livers	20
Buckwheat flour	42	Maple sugar	45
Butter	14	Maple syrup	45
Butterine	20	Meat, fresh (ten to thirty days)	30
Cabbage	33	Meats, fresh (few days only)	35
Canned fruits	40	Meats, salt (after curing)	43
Canned meats	40	Mild cured pickled salmon	33
Cantaloupes (one to two months)	33	Nursery stock	30
Cantaloupes (short carry)	40	Nuts in shell	40
Carrots	33	Oatmeal	42
Caviar	36	Oils	45
Celery	32	Oleomargarine	20
Cheese (long carry)	35	Onions	32
Chestnuts	34	Oranges (long carry)	34
Chocolate dipping room	65	Oranges (short carry)	50
Cider	32	Oxtails	30
Cigars	42	Oysters, iced (in tubs)	35
Corn (dried)	45	Oysters (in shell)	43
Corn meal	42	Palm seeds	38
Cranberries	33	Parsnips	32
Cucumbers	38	Peach butter	42
Currants (few days only)	32	Peaches (short carry)	50
Cut roses	36	Pears	33
Dates	55	Peas (dried)	45
Dried beef	40	Plums (one to two months)	32
Dried fish	40	Potatoes	34
Dried fruits	40	Poultry, (after frozen)	10
Eggs	30	Poultry, dressed (iced)	30
Ferns	28	Poultry (short carry)	28
Field grown roses	32	Poultry (to freeze)	0
Figs	55	Raisins	55
Fish, fresh water (after frozen)	18	Ribs (not brined)	20
Fish, nt frozen (short carry)	28	Salt meat curing room	33
Fish, salt water (after frozen)	15	Sardines (canned)	40
Fish (to freeze)	5	Sauerkraut	38
Frogs legs (after frozen)	18	Sausage casings	20
Fruit trees	30	Scallops (after frozen)	16
Fur and fabric room	28	Shoulders (nt brined)	20
Furs (undressed)	35	Strained honey	45
Game (after frozen)	10	Sugar	45
Game (short carry)	28	Syrup	45
Game (to freeze)	0	Tenderloin, etc	33
Ginger ale	36	Tobacco	42
Grapes	36	Tomatoes (ripe)	42
Hams (not brined)	20	Watermelons (short carry)	40
Hogs	30	Wheat flur	42
Hops	32	Wines	50

certain extent and the best rule for the practical man is to study his own plant and the conditions prevailing; being careful particularly to see that his thermometers are correct so that, when he finds some unusual temperature the best for a certain article in his plant, it will not be due to error in his instrument. As a rough guide to be used where no other information for the particular plant is available, Cooper gives the temperatures shown in Table 21, which, however, should be used with caution. Other authorities give figures varying anywhere up to 10 degrees from the data in this table, but on the whole the table may be considered about the least arbitrary of all and is based in the main on figures taken from practice.

INSULATION

In considering this topic, the student should recall the discussion of heat in the first part of this paper, where it was shown that heat is transmitted in three ways—by convection, by radiation, and by conduction—the transmission taking place in all three of these ways in cold storage plants. Owing to the low temperatures, however, there is comparatviely little transmission in storage rooms by radiation, and practically all heat passes into the rooms by the process of conduction. Where air space insulation is used there is considerable transmission by convection currents in the manner explained already and this is the principal reason why air space insulation is inefficient as compared with the other methods of insulating. When it is considered that from one-half to seven-eighths of the refrigerating work in a storage plant is required to remove the heat that leaks through the walls, the importance of good insulation is seen. The increase in cost for insulating work well done is insignificant as compared with the resulting decrease in operating cost. On the other hand, the cheaper insulation is less permanent and demands a larger amount of coal to drive the refrigerating machinery to keep out the heat from the rooms, which could have been excluded with effective insulation.

If a perfect insulation could be had there would be no need of refrigerating machinery except to cool the rooms down in the first place and to remove any heat admitted by opening doors for taking goods in and out of store. Such a condition is impossible, however, as, owing to the nature of heat, it is not possible to wholly prevent an increase in the rate of vibration in a body that has contact with a hotter body in which the rate of vibration is more rapid. Different rates of vibration tend to become equalized in adjacent bodies just as naturally as water tends to run down hill. Different materials have varying capacities for hindering or retarding such vibration but no matter what the material or how thick the walls, some transmission will take place until finally the temperatures on the two sides of the wall will be equal. During the past century scientists have made many laboratory tests as to the heat-transmitting qualities of various insulating materials, but in most cases these results are of little value to refrigerating men on account of the fact that the experiments were mostly performed with high differences in temperature and dry-air conditions, as for the insulation of steam pipes where the temperature difference is at least 275 degrees. These conditions do not apply for the low temperatures and moist air met in refrigerating establishments and it is usually the case that a good insulator, for the conditions named, is a decidedly poor one under the conditions prevailing in cold storage establishments.

Non-Conductors. To be a good insulator a material should have poor conducting power at low temperatures or with small temperature differences, and should have a high specific heat and high specific gravity. The higher the product of the two last items, the more valuable, other things being equal, is the insulation. As an example take a cubic foot of air and a cubic foot of flake charcoal used as insulation. The air weighs 0.0807 pound and since its specific heat is 0.237 it will require 0.0191 B. T. U. to increase its temperature 1° F. For charcoal, the figures for weight, specific heat, and heat required for 1 degree rise of temperature are 11.4, 0.242 and 2.758 respectively. In case, then, the machinery should be stopped for any reason, the insulating material would have to be heated up before the temperature in the storage rooms could rise, the amount of heat absorbed by the insulation itself being in proportion to the product just mentioned. In other words for each degree rise in temperature in the rooms the cubic foot of air must absorb only 0.0191 B. T. U. while the charcoal must absorb 2.758 units. This is the *reserve power* of the insulation and is of considerable importance, though it may be overbalanced by the conductivity of a materal if this be high.

TABLE XXII

Non-Conducting Power of Substances

Non-Conductors One Inch Thick	Net Cubic In. of Solid Matter, in 100	Heat Units Transmitted per Sq. Ft. per Hour
Still air	....	43
Confined air	....	108
Confined air = 310°	....	203
Wool = 310°	4.3	36
Absorbent cotton	2.8	36
Raw cotton	2	44
Raw cotton	1	48
Live-geese feathers = 310°	5	41
Live-geese feathers = 310°	2	50
Cat-tail seeds and hairs	2.1	50
Scoured hair, not felted	9.6	52
Hair felt	8.5	56
Lampblack = 310°	5.6	41
Cork, ground	....	45
Cork, solid	....	49
Cork charcoal = 310°	5.3	50
White-pine charcoal = 310°	11.9	58
Rice-chaff	14.6	78
Cypress (*Taxodium*) shavings	7	60
Cypress (*Taxodium*) sawdust	20.1	84
Cypress (*Taxodium*) board	31.3	83
Cypress (*Taxodium*) cross-section	31.8	145
Yellow poplar (*Liriodendron*) sawdust	16.2	75
Yellow poplar (*Liriodendron*) board	36.4	76
Yellow poplar (*Liriodendron*) cross-sec.	30.4	141
"Tunera" wood, board	79.4	156
Slag wool (Mineral wool)	5.7	50
Carbonate of magnesium	6	50
Calcined magnesia = 310°	2.3	52
"Magnesia covering," light	8.5	58
"Magnesia covering," heavy	13.6	78
Fossil meal = 310°	6	60
Zinc white = 310°	8.8	72
Ground chalk = 310°	25.3	80
Asbestos in still air	3	56
Asbestos in movable air	3.6	99
Asbestos in movable air = 310°	8.1	210
Dry plaster of paris = 310°	36.8	131
Plumbago in still air	30.6	134
Plumbago in movable air = 310°	26.1	296
Coarse sand = 310°	52.0	264
Water, still	....	335
Starch jelly, very firm, "	....	345
Gum-Arabic, mucilage, "	....	290
Solution sugar, 70 per cent, "	....	251
Glycerin, "	....	197
Castor oil, "	....	136
Cotton-seed oil, "	....	129
Lard oil, "	....	125
Aniline, "	....	122
Mineral sperm oil, "	....	115
Oil of Turpentine, "	....	95

Ordway gives the data in Table 22 for the non-conducting power of various substances, the figures being determined from insulation tests on steam pipes where the difference in temperature on the two sides of a 1-inch thickness of the substances was 100 degrees. Where not stated in the table, the source of heat was water at about 176° F., but in some cases steam at a temperature of 310° F. was used. It will be noted in the table that still air is one of the poorest conductors of heat, but if a body of confined air is arranged to allow the setting up of convection currents as before mentioned, the air is rendered a much better conductor. It is to prevent these currents, that the air is confined in spaces small enough to keep it still, and the value of most insulators is determined by the effectiveness with which they prevent the air from moving by separating it into such small particles that its viscosity is too great to allow of movement.

Experiments have been made to determine the amount of heat conduction through a vacuum, with results that tend to indicate a much smaller rate of conduction than through still air. In laboratory experiments liquid air when placed in a glass vessel surrounded by a vacuum retained its liquid form for several days, but no successful scheme for applying a vacuum in practical insulation work has yet been evolved. It has generally been assumed that the rate of heat transmission is in proportion to the difference of temperature on the two sides of a wall and inversely proportional to the thickness of the material; but recent experiments go to show that the rate of transmission increases with increase of temperature difference and that it does not decrease exactly in inverse proportion with increase of thickness. As most coefficients of heat transmission are determined with a temperature difference of 1 degree, it is obvious that they cannot be correct for much larger differences and should in fact be increased by at least 50 per cent in practical calculations. With the constant tendency toward the use of lower temperatures in cold storage and general refrigeration work, this matter becomes of first importance. In some cases the temperature difference may be as much as 90° F.

Aside from the loss, incident to using poor insulating material by reason of the heat that must be removed, there are other disadvantages that must be taken into consideration. The heat admitted through poor insulation raises the temperature of the outer parts of a room near the walls so that it is impossible to keep all parts of the store at the same temperature. This is undesirable in many respects and for fear of freezing the goods near the cooling pipes or air ducts, it often happens that the temperature is carried too high for best results. On the score of conductivity, the choice of insulators is limited to vegetable and animal substances. In selecting one of these a number of practical considerations must be taken into account. The material should be odorless so as not to taint the goods and, as far as possible, should be proof against moisture. In case it gets wet, it should not be of such nature as to rot easily. It should have no tendency to spontaneous conbustion and should be elastic so that it can be packed enough to avoid settling. Aside from these things, the material should be reasonably cheap and easy to apply in general work, and should be vermin proof. As far as possible the insulation should be waterproof and fireproof, but as none of the vegetable and animal materials have these qualities, special measures must be taken to make the finished work proof against fire and water. This is done by masonry work and cement plaster.

For practical purposes, the choice of an insulating material will be restricted to cork, dry shavings, mineral wool, or hair felt. Tarred papers are used on the board work used to retain the insulating material in place and each of the materials named may be used in a number of combinations and forms. In brick buildings it is customary to use a hollow tile course in the walls, or rather to lay such tile in the wall so as to form a moisture barrier. Two such arrangements are shown at the top of Fig. 79, which shows a number of composite insulation structures, tested by the Fred W. Wolf Co. Several other combinations arranged and tested by Madison Cooper are shown in Fig. 80. In both these illustrations the figures at the right represent the number of B. T. U. transmitted per square foot per day per degree difference of temperature. It will be noted that in all cases the air-space insulation makes a poor showing as compared with other forms, especially in view of the care and attention necessary to construct it properly. It is almost impossible to get workmen that will give proper attention in putting together the timber work for such insulation so that it will be tight, and much of the complaint with cold stores in the past has been due to air leaks resulting from such neglect.

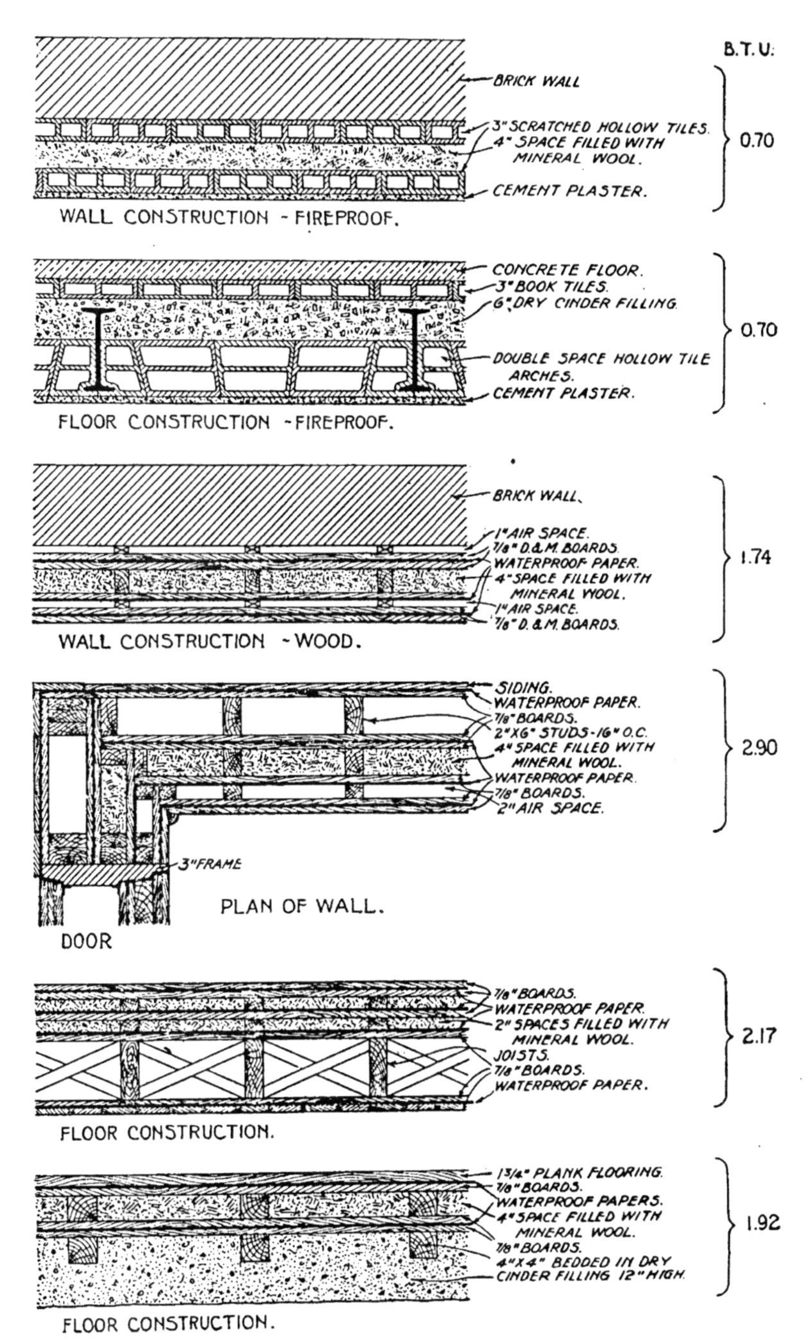

Fig. 79. Composite Insulation Tested by Fred W. Wolf Co.

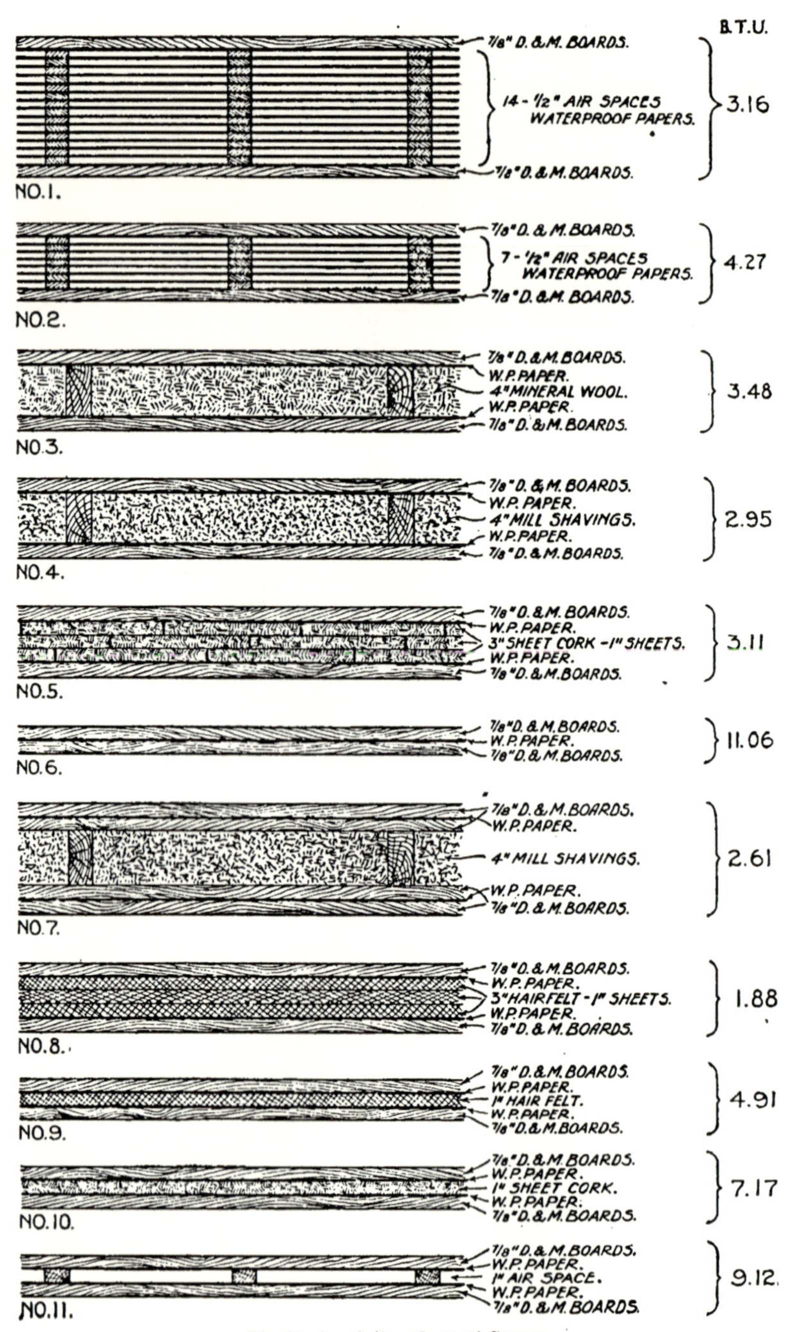

Fig. 80. Insulation Tests of Cooper.

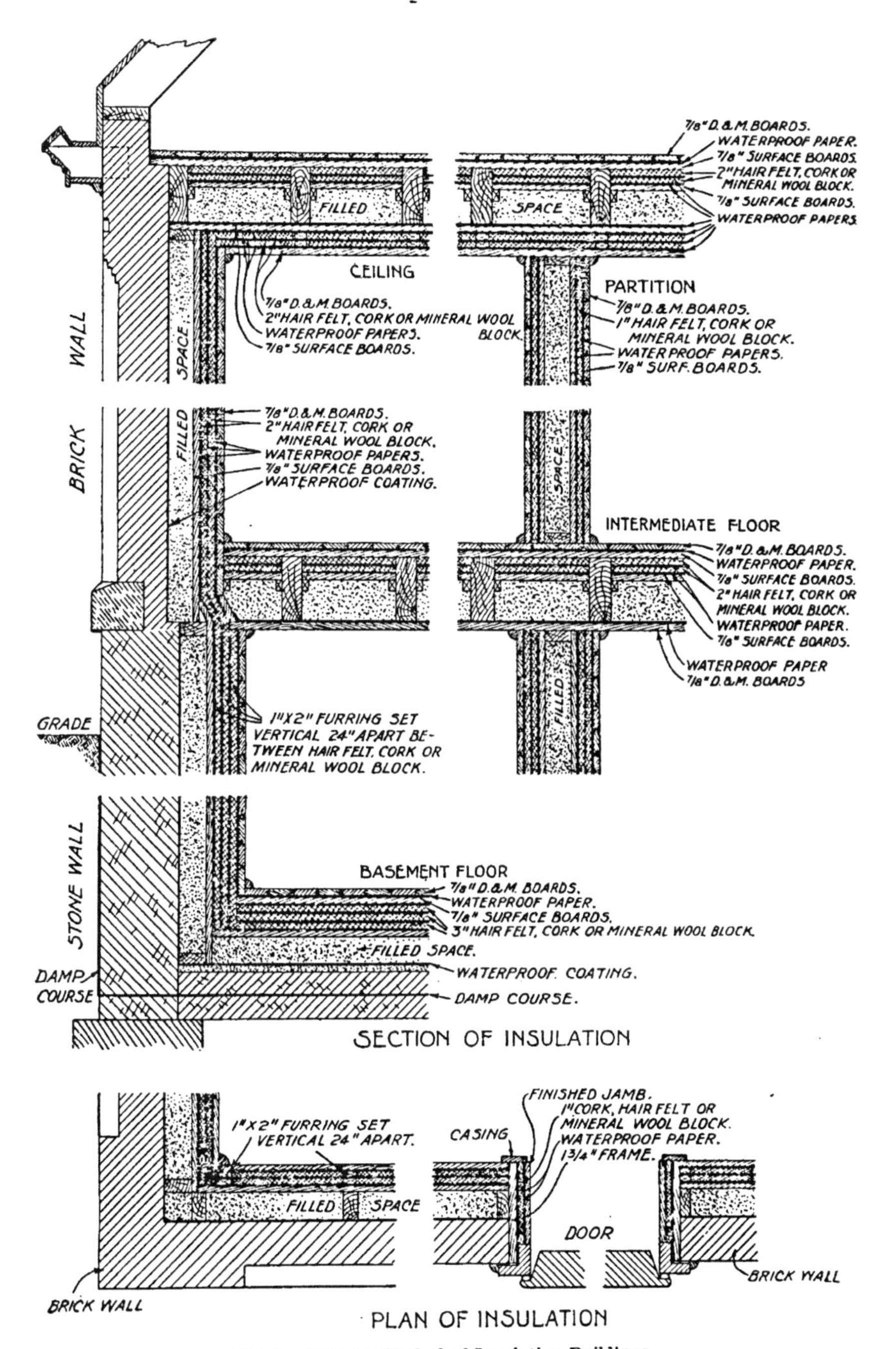

Fig. 81. Cooper's Method of Insulating Buildings.

Various methods of applying insulation to buildings are in use. Fig. 81, illustrating the construction designed by Madison Cooper, may be considered as representative of the best modern practice for ordinary cold storage buildings. The walls are made waterproof with asphalt or similar material; the filling material, either shavings, granulated cork, or mineral wool, is placed against the wall. The sheathing and paper are arranged in courses inside as shown in the figure. An 8-inch filled space with 4 inches of sheathing, etc., gives the insulation for 30 degrees. A total thickness of 13 inches is about right for 20 to 25 degrees, and for sharp freezers the filled space should be 10 inches with an additional thickness of sheet material. This form of insulation is compact and durable, the indestructible materials such as waterproof paper, sheet cork, mineral wool block, etc., being placed inside where the conditions are most severe. In many storage houses constructed within the last five years, sheet cork is used exclusively, being laid in cement directly on the brick walls in one or more layers as desired and finished over inside with waterproof cement after the manner already mentioned for ice storage rooms.

In packing any filling material, due regard should be had to getting it compact enough to preclude the possibility of settling, but at the same time not too dense, especially in the case of those substances which are fairly good conductors in the natural state. Where the building regulations of cities require fireproof construction, the problem of insulation is much complicated and it is difficult to construct any effective insulation at moderate cost. About the most practical construction is the use of corkboard cemented direct on the brick or tile walls and ceilings as already mentioned. Plaster-of-Paris blocks have been tried as a fireproofing over the filling material but with poor results as the blocks give way in fire and let the filling fall out. Especial care must be taken to insulate all steel I-beams and other structural metal parts in fireproof structures, as the high conducting power of the metal will work havoc with insulation otherwise well constructed when these beams and columns are neglected. This and the other difficulties met in fireproof structures make the insulation of such buildings much more expensive than for ordinary structures and it is not generally believed that this expense is justified, except in special cases, as where furs and valuable garments are carried in store. Other goods generally carried are not inflammable and most fires originate outside of the storage rooms or on account of improper electric wiring.

METHODS OF COOLING

In olden days, natural ice was the sole reliance of cold storage people and knowledge of the laws governing air circulation and ventilation was so meager that little success was had in using it. Mechanical refrigeration has now become so well understood in its production and application that there is little use for ice, even in northern climes where it is produced naturally, as the expense of harvesting taken with that of delivery is greater in many cases than the cost of making artificial ice and delivering it to the customer. This, taken with the fact that refrigeration by ice is very inefficient and expensive as compared with the direct application of cold produced mechanically, places ice out of consideration as a cooling medium, except for household refrigerators. Even in this application ice is being abandoned in some quarters, where owners are able to install one of the small automatic refrigerating outfits already mentioned.

For applying mechanically produced refrigeration, there are a number of methods in use, some of which are now considered obsolete, while each of the others has its advantages and disadvantages. In primitive cold stores, gravity air circulation was relied upon for mixing and cooling the air to uniform temperature throughout the rooms; but the system proved inefficient, as most of the cooling was done by direct radiation and little or no circulation of air was induced. The unsatisfactory results had thus led to a study of direct radiation. In the first cold stores, pipes containing expanding ammonia gas or cold brine were placed directly in the rooms and the cooling produced was by direct radiation of heat from the air in the rooms to these pipes. As the results were poor, various systems were devised to improve the circulation; thus leading to the *indirect-radiation* system, where the coils are located in a loft separate and apart from the room and above it to one side, so that the greater height gives a greater difference in density between the cold and hot air. The circulation is directed as desired by false ceiling and wall shields.

Modern practice considers both these radiation systems out of date, the principal objection to the indirect system aside from the fact that circulation is imperfect, being the impracticability of introducing fresh air into the rooms. There are in use at the present time five systems of cooling rooms, some of which are combinations of the elementary processes. Thus we have *direct expansion* where the refrigerant is expanded direct in the coils placed in the rooms to be cooled, the best arrangement being that with deflecting shields, etc., for directing circulation of the air in the manner just mentioned. This method can be applied successfully in large rooms where the temperature and duty to be performed is constant, such as in breweries, packing houses, and large cold storage rooms; or where very

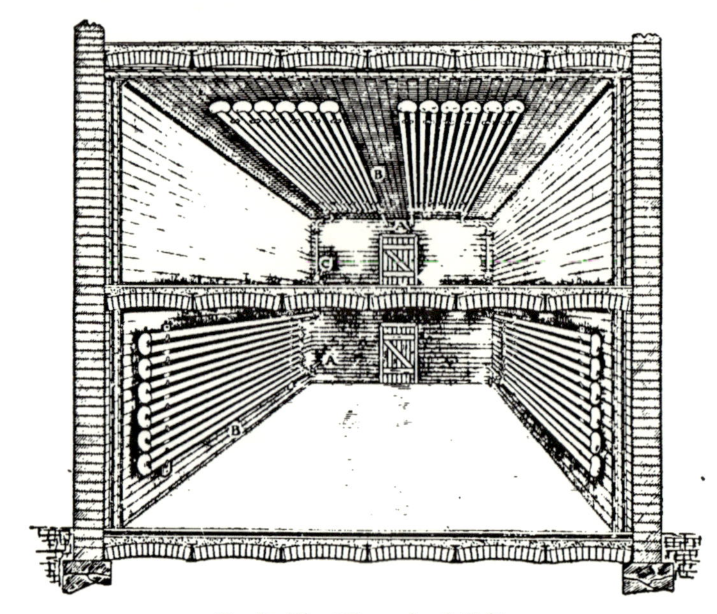

Fig. 82. Direct-Expansion Cold Store.

low temperatures are required as in sharp freezers for fish, poultry, etc., in which work direct expansion is desirable, the efficiency being much greater with this system than with any other. Fig. 82 illustrates a large room arranged for direct expansion, where the ammonia is expanded by the valve *A* into the piping *B*, the gas being returned to the compressor through the pipe *C*. Fig. 83 shows a fish freezing room on each side of which is arranged a series of pipe-coil shelves through which the ammonia is evaporated. Fish are laid in tin trays and placed on the pipe shelves until the room is filled, when the room is closed and the ammonia turned on the coils as long as necessary to freeze up the fish. As the coils are both above and below the trays and close together, the application of cold is effective and only a few hours are required for freezing.

In some cases, as has already been mentioned, it is desirable to use *brine circulation* for cooling the rooms, the arrangement for which is shown in diagrammatic form by Fig. 84, which illustrates a system using a brine-cooling tank with the brine coils set directly in the room to be cooled. This, it will be seen, is an application of *direct radiation.* The system may be used to better advantage with

Fig. 83. Fish Freezing Room.

a brine cooler connected instead of the brine-cooling tank, as the cooling with double-pipe brine coolers is more rapid and efficient than with the tank, for reasons already stated. Fig. 85 illustrates a forced-air circulation system, using direct-expansion coils in the bunker room, which arrangement is the most modern practice. Some engineers, usually of the old-time stamp, prefer to use brine coils in the bunker room; but there is no advantage in this, as a pump and brine cooler must be employed for handling the brine. Cooling is more effective with the direct-expansion coils in the bunker room. Some years ago when manufacturers were not able to put up pipe work for direct expansion, there may have been excuse for the brine-circulation system with direct radiation, but it is now altogether out of date except for those who may be wedded to "their system" to such an extent that they cannot see the light of advancement and progress.

If for any reason it is desired to use brine, much more effective cooling can be had by passing the brine down through the open space of the bunker room, after coming from the cooler, in thin sheets flowing over vertical walls or with other suitable arrangement. The air is forced through the brine and cooled after which it may be passed to the storage rooms. By this process, the brine purifies the air so hat the stores are kept sweet at all times, the gases and impurities taken up from the products in the rooms being absorbed by the brine. This system has been applied with considerable success and about the only objection is the possibility of getting too much moisture in the air going to the rooms; which cannot occur, however, if due regard is given to the temperature used in the bunker room, so that the air leaves at about the temperature of the store rooms.

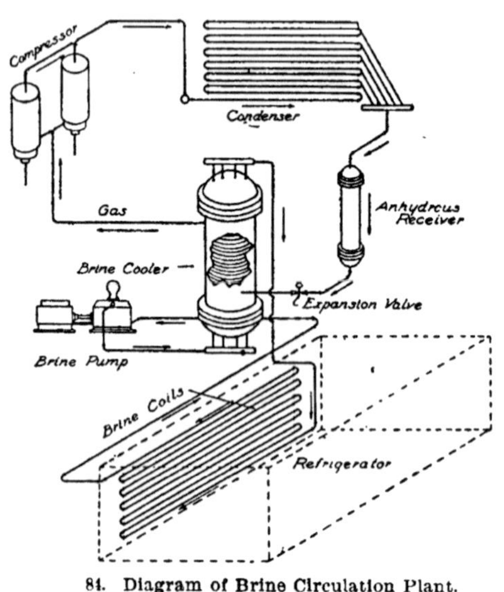

84. Diagram of Brine Circulation Plant.

Where pipe coils are used in the bunker room, calcium chloride can be used to advantage. It is placed in trays over the coils in such a manner that moisture in the air will be taken up by the calcium, forming brine that drips down over the pipes and absorbs impurities. In any system of cooling, it is important that the air be circulated so as to maintain uniform temperatures in all parts of the various rooms. Also the air must be purified or renewed at suitable intervals if the goods are to be kept in first-class condition. There is much difference of opinion as to how often the air should be changed and the cold-storage man must use considerable judgment in this matter, as much depends on the character of the goods and the length of time they have been in store. Fresh meats and vegetable products when first put in store, give off a large amount of gases and impurities which must be removed, but after having been in store for some time there is less of this action tending to contaminate the air. Ordinarily, if the entire volume of air in a room is renewed with pure fresh air once or twice a week, good results will be had. To effect this in a proper manner it is essential that some system of

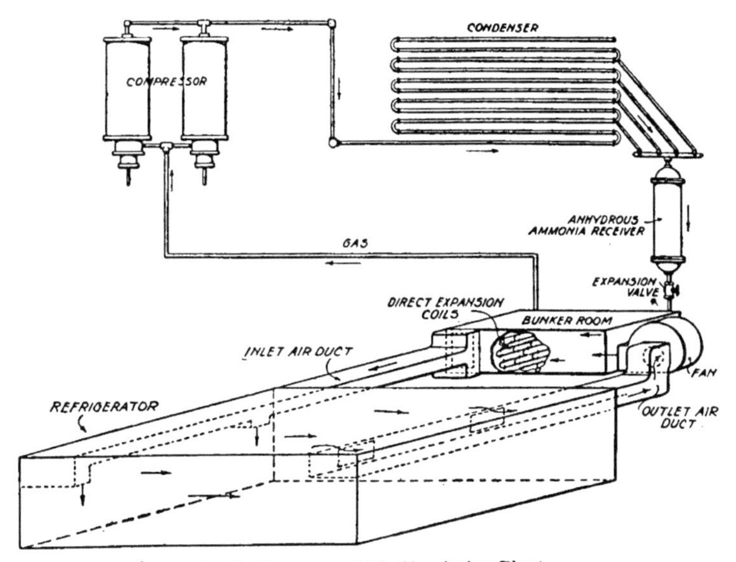

Fig. 85. Diagram of Air-Circulation Plant.

forced-air circulation be employed; and in fact such a system is necessary, aside from the question of ventilation, if proper circulation is to be had.

Although many cold-storage men are bitterly opposed to forced-air circulation, it is generally recognized by eminent authorities that such circulation is necessary if good results are to be obtained. There is a choice between the exhaust and pressure systems but the latter is so far superior that little consideration need be paid to the exhaust system of ventilating by drawing air out of the rooms. Comparatively little power is required for the air circulation and the results obtained where the forced-air system is properly installed are much superior to the results with systems now becoming obsolete. This does not mean that small fans of the type used in offices and hotel parlors should be employed, as little advantage can attach to the use of such apparatus. A properly designed fan made of light weight, corresponding to the low speed and duty required for the slow circulation of air used in refrigerating rooms, should be used and the air should be forced through bunker cooling rooms in the manner just described, fresh, purified air being drawn in as necessary.

In northern climates, it has sometimes been the practice to ventilate the rooms in fall and winter by opening windows and doors, but there is some question as to the advisability of this practice and it should be applied with the greatest judgment and caution. Where the outside temperature is about the same as that carried in the rooms on a bright clear day, or, better still, on a clear cold night, there can be no harm in filling the rooms with Nature's pure air. There should, however, be no guess work about the matter, and measurement of the moisture in the air and its temperature should be made carefully before "opening up." Where the forced-air circulation system is used and arrangements made to draw in fresh air as needed, there is little need for opening doors and windows and in fact, such openings should not be used in storage plants where possible to avoid them, owing to the difficulty of insulating them and the fact that all necessary light can be had more cheaply by using incandescent electric lamps. There is no objection to these lamps other than the heat cast off into the rooms and this is small indeed compared to the heat coming through the best insulated windows, even where four or five thicknesses of glass and air spaces are employed.

Some of the most successful cold-storage houses abroad have no openings or doors in the walls, entrance being had only through the top of the building. In such cases, hoists are used to elevate the goods to the top and lower them inside the building. In the large storage houses of receiving ports in England, where cargoes of fresh meats are received from steamers, houses built on this closed plan have been most successful. Where a general line of goods is carried, as for example, in the wholesale commission trade, it is impracticable to operate such a house, as a certain number of doors must be had. Such doors should be of the best possible construction and it is usually better to buy one of the patent doors on the market.

The manufacturers, being specialists in this line of work, may be considered more competent to turn out a good door than any ordinary carpenter who may be on a construction job. In nine cases out of ten the so-called "home-made" doors constructed on the premises will give poor results and be a constant source of annoyance, owing to sticking and failure to open.

REFRIGERATION REQUIRED

The refrigeration required depends altogether on the effectiveness of the insulation, on the character of the goods carried, on the size and shape of the rooms, the quantity of goods, and the frequency with which goods are taken in and out of store. The amount required also depends to some extent on the temperature of the goods received and to a much greater extent on the temperature at which the rooms must be kept. When all these conditions are known, Levey gives the following instructions for finding the amount of refrigeration necessary:

1. *For the room.*

Calculate the exact area of the exposed surface in the walls, floor, and ceiling of the rooms in square feet, and multiply the total number of square feet by the numbers given opposite the required temperature and divide by 281,000.

For rooms containing less than 1,000 cubic feet:

If held at zero F. multiply the exposed surface by 1,775
" " 5 deg. " " " " " " 710
" " 10 " " " " " " " 535
" " 20 " " " " " " " 355
" " 32 " " " " " " " 265
" " 36 " " " " " " " 180

For rooms containing 1,000 to 10,000 cubic feet:

If held at zero F. multiply the exposed surface by 1,250
" " 5 deg. " " " " " " 600
" " 10 " " " " " " " 300
" " 20 " " " " " " " 190
" " 32 " " " " " " " 160
" " 36 " " " " " " " 125

For rooms containing over 10,000 cubic feet:

If held at zero F. multiply the exposed surface by 1,100
" " 5 deg. " " " " " " 550
" " 10 " " " " " " " 275
" " 20 " " " " " " " 180
" " 32 " " " " " " " 140
" " 36 " " " " " " " 110

2. *For the stores.*

Multiply the amount of goods (in pounds) to be stored per day by the number of degrees the temperature is to be lowered and by the specific heat of the goods, and divide by 284,000. This will give the amount of refrigeration in tons per day necessary to hold the goods at the required temperature.

Add together the results of 1 and 2 and the total will be the amount of refrigeration in tons per day which will be required to hold the goods and the room. If the goods are to be frozen, the latent heat of freezing should be added to the heat to be removed in lowering the temperature.

COLD STORAGE

Handling Goods. The proper handling of products stored is a matter of great importance, both from the standpoint of preservation and profit. If goods are carelessly piled in a room so that there can be little circulation of air among the packages, deterioration is the result, particularly with those packages in the middle of the pile. Also there is much danger of crushing goods in the lower tiers, where a number of cases or barrels are packed one on top of the other. Thus, for example, in storing apples in barrels several tiers high, 2 x 4-inch scantling should be placed on the floor under the ends of the first tier and similar scantlings should be placed on the barrels before laying on each successive tier. In this way, the weight is taken on the heads of the barrels and not on the bilge, and danger of crushing the fruit in the lower tiers is thereby eliminated.

It is important, of course, to give attention to handling the goods in and out of store with as much facility as possible, thereby saving time and trouble, but as these matters are usually looked after by warehouse foremen who do not wish to do any work that they can avoid, it is not generally up to the manager to be on the lookout in this particular. Some classes of goods can be stored more compactly than others. It matters little how close cases of butter are piled; but articles that give off moisture, such as fruits, should not be piled too close and with too many packages in a pile. With other classes of goods, as frozen fish and poultry, for example, the more compact the goods can be stored the better, as close packing tends to check the drying and evaporating action of the air in removing the coating of ice used to prevent drying out.

In taking goods from store, there is likely to be trouble from *sweating*, as the low temperature of the goods condenses moisture present in the atmosphere. This may have a decidedly detrimental effect with some classes of goods, as eggs and fruits, so that precautions should be taken to prevent the action by cooling the goods gradually. This may be done by piling them in the receiving room with a heavy wagon tarpaulin or other similar covering placed over the pile. In fall weather, if the goods are removed in this way at night they will be all right by the next morning, but in the summer season where goods are taken out of store at low temperatures, as much as 36 to 48 hours may be necessary to get them warmed up properly. On the other hand goods taken into storage can be handled much better both from the standpoint of economy and preservation of quality by lowering the temperature gradually.

In all mechanical and engineering lines, it is a well-recognized fact that sudden changes of conditions are effected at comparatively high cost. This is particularly true of refrigerating work, where sudden temperature reductions cost much more than the same reduction effected gradually. In storing butter, which is usually kept at about zero F., it is highly advisable to place the trucks in a comparatively warm room before taking them into the sharp freezers. When this is done, much less cooling surface need be used for the freezers and the results are more economical and satisfactory in every way. There are many other points in handling goods in cold stores that cannot be taken up in brief space, but it should be pointed out that the large number of failures in small storage plants, usually operated in connection with ice plants, is due mostly to the fact that several products are stored in the same room at a common temperature which makes impossible good results, so necessary to permanent success in storage work.

Where goods are to be stored for any length of time, there are few different products that can be stored in the same room to advantage. Thus butter requires a lower temperature than chesee, while fruits are kept at a higher temperature than the cheese. If eggs and butter are kept in the same storage room with other materials, they will soon absorb the flavor of the fruits and other things in the room. Oranges, pineapples, etc., may have the most delicious flavors, but these may not be very agreeable to the palate in butter or cheese. Such products may be stored together for a short time with some success if great care is taken with the ventilation, but it is the practice of all

TABLE XXIII

Rates for Cold Storage

Goods and Quantity	First Month	Each Succeeding Month	In Large Quantities, per Month	Season Rate per Bbl. or 100 Lbs.	Season Ends
Apples, per bbl.	$0.15	$0.12½	$0.12½	$0.45	May 1
Bananas, per bunch	.15	.10	.10	......	......
Beef, mutton, pork, and fresh meats, per lb	.00¾	.00½	.00⅜	......	......
Beer and ale, per bbl	.25	.2	......	......	......
Beer and ale, per ½ bbl	.15	.1	......	......	......
Beer and ale, per ¼ or ⅛ bbl.	.10	.1	......	......	......
Beer, bottled, per case	.10	.1.	......	......	......
Beer, bottled, per bbl	.20	.2.	......	......	......
Berries, fresh, of all kinds, per qt.	.00½	.00⅓	.00⅓	......	......
Berries, fresh, of all kinds, per stand.	.10	......	......	......	......
Butter and butterine, per lb	.00⅓	.00¼	.00⅛	.50–.75	Jan. 1
(See also butter freezing rates.)					
Buckwheat flour, per bbl	.15	.10½	.10	.50	Oct. 1
Cabbage, per bbl	.25	.20	.20	......	......
Cabbage, per crate	.10	.18	.08	......	......
Calves (per day), each	.10	......	......	......	......
Calves, per lb	.00¾	.00½	.00⅜	......	......
Canned and bottled goods, per lb.	.00¼	.00⅛	.00⅛	......	......
Celery, per case	.15	.10	.10	......	......
Cheese, per lb.	.00⅓	.00¼	.00⅛	.50–.60	Jan. 1
Cherries, per quart	.00½	.00⅓	.00⅓	......	......
Cider, per bbl	.25	.15	.15	......	......
Cigars, per lb	.00⅓	.00¼	.00⅛	......	......
Cranberries, per bbl.	.25	.20	.15	......	......
Cranberries, per case	.10	......	......	......	......
Corn meal, per bbl.	.15	.12½	.10	......	......
Dried and boneless fish, etc., per lb.	.00⅓	.00⅛	.00⅛	.50	Nov. 1
Dried corn, per bbl	.12½	.10	.10	......	......
Dried and evaporated apples, per lb.	.00⅛	.00⅒	......	.50	Nov. 1
Dried fruit, per lb	.00⅙	.00⅛	.00⅛	.40–.50	Nov. 1
Eggs, per case	.15	.12½	.10	.50–.60	Jan. 1
Figs, per lb	.00⅛	.00⅛	.00⅒	......	......
Fish, per bbl	.20	.18	.15	.75	Oct. 1
Fish, per tierce	.15	.13	.12½	.50	Oct. 1
(See also fish freezing rates.)					
Fruits, fresh, per bbl	.25	.20	.20	......	......
Fruits, fresh, per crate	.10	.08	.08	......	......
Furs, undressed, hydraulic pressed, per lb.	.00½	.00¼	.00¼	1.00	Oct. 1
Furs, dressed, per lb	.03	.02½	.02	8.00	Oct. 1
Ginger ale, bottled, per bbl	.20	.15	.15	......	......
Grapes, per lb.	.00½	.00⅓	.00¼	2.00	May 1
Grapes, per basket.	.03	.02	.01	......	......
Grapes, Malaga, etc., per keg	.15	.12½	.12½	......	......
Hops, per lb	.00⅓	.00¼	.00⅛	......	......
Lard, per tierce	.25	.20	.20	1.00	Nov. 1
Lard oil, per cask	.25	.20	.20	1.00	Nov. 1
Lemons, per box	.15	.12½	.10	.50	Nov. 1
Macaroni, per bbl	.20	.15	.12½	......	......
Maple sugar, per lb	.00⅓	.00¼	.00⅛	.40–.50	Nov. 1

TABLE XXIII—Continued
Rates for Cold Storage

Goods and Quantity	First Month	Each Succeeding Month	In Large Quantities, per Month	Season Rate per Bbl. or 100 Lbs.	Season Ends
Maple syrup, per gallon	.01½	.01¼	.01	....	....
Meats, fresh, per lb	.00¾	.00½	.00⅜	....	....
Nuts, of all kinds, per lb	.00¼	.00⅛	.00⅓	.40–.50	Nov. 1
Oatmeal, per bbl	.20	.15	.12½	....	....
Oil, per cask	.25	.20	....	....	....
Oil, per hogshead	1.00	.80	....	....	....
Oleomargarine, per lb	.00⅓	.00	.00⅛	....	....
Onions, per bbl	.15	.12¼	.10	.50–.60	May 1
Onions, per box	.12½	.10½	....	....	....
Oranges, per box	.15	.12½	.10	.50	Nov. 1
Oysters, in tubs, per gal	.05	.04	....	....	....
Oysters, in shell, per bbl	.50	.40	.30	....	....
Peaches, per basket	.10	.08	.07	2.00	Jan. 1
Pears, per box	.20	.15	....	.60	May 1
Pears, per bbl	.40	.30	....	1.20	May 1
Pigs' feet, per lb	.00⅓	.00¼	.00¼	1.00	Nov. 1
Pork, per tierce	.20	.15	.15	....	....
Potatoes, per bbl	.25	.20	.20	....	....
Preserves, jellies, jams, etc., per lb	.00¼	.00⅛	.00	....	....
Provisions, per bbl	.25	.20	.20⅛	....	....
Rice flour, per bbl	.20	.15	.12½	....	....
Sauerkraut, per cask	.25	.20	.15	.60–.75	Nov. 1
Sauerkraut, per ½ bbl	.15	.12½	.10	....	....
Syrup, per bbl	.30	.25	.20	1.00	Oct. 1
Tobacco, per lb	.00⅓	.00¼	.00⅛	....	....
Vegetables, fresh, per bbl	.25	.20	.15	....	....
Vegetables, fresh, per case	.15	.10	.08	....	....
Wine, in wood, per bbl	.25	.25	....	....	....
Wine, in bottles, per case	.10	.10	....	....	....

large storage establishments to have separate rooms and spaces for the different classes of goods stored.

Storage Rates. All storage business is conducted for revenue and it is important that such a schedule of rates be adopted as will allow a reasonable profit for the capital, skill, and experience necessary to run a cold storage establishment. Rates vary with local circumstances, conditions, and with the products stored and the time they are kept in storage, so that no general rule can be laid down. Capacity for storage and the demand therefor as well as the competition to be met must all be considered. As a rough guide, Siebel gives the data presented in Table 23, the figures being averages of rates prevailing in a number of large cold-storage establishments.

REVIEW QUESTIONS.

PRACTICAL TEST QUESTIONS.

In the foregoing sections of this Cyclopedia numerous illustrative examples are worked out in detail in order to show the application of the various methods and principles. Accompanying these are samples for practice which will aid the reader in fixing the principles in mind.

In the following pages are given a large number of test questions and problems which afford a valuable means of testing the reader's knowledge of the subjects treated. They will be found excellent practice for those preparing for Civil Service Examinations. In some cases numerical answers are given as a further aid in this work.

REVIEW QUESTIONS

ON THE SUBJECT OF

COMPRESSED AIR

PART I

---

1. Define *free air; atmospheric air; absolute pressures; absolute temperature.*

2. What is a perfect gas, and why may air be considered a perfect gas?

3. State Boyle's Law; Charles's Law.

4. Define *isothermal compression* and *adiabatic compression*, and explain why neither can be secured in actual air-compression.

5. Explain the effect of heat of compression in actual air-compression.

6. Draw an ideal air-compressor indicator card, and explain what the various lines and points upon it indicate.

7. What is the *mechanical efficiency* of an air-compressor; the *volumetric efficiency;* the *compression efficiency?*

8. How is the mechanical efficiency of an air-compressor found from its indicator cards?

9. How is the volumetric efficiency of an air-compressor found from its indicator cards?

10. Explain the effect of clearance on compressor performance, stating the different elements making up the total clearance.

11. How is the compression efficiency of an air-compressor computed?

12. Explain the effect of high altitudes upon the efficiency and output of an air-compressor.

13. Define the five losses which cannot be avoided in air-compression, and explain why these cannot be avoided.

14. What is the great distinction between steam-engine and air-compressor practice?

15. Why is steam compounding particularly advantageous on air-compressors?

16. Describe the two classes of regulators or governors for steam-driven compressors.

17. Explain the distinction between the methods of regulation for steam-driven and power-driven compressors.

18. Describe the seven classes of governors or regulators for power-driven compressors.

19. Why is a starting unloader desirable on power-driven compressors?

20. Define the straight-line air-compressor, and state its advantages and disadvantages.

21. Explain the pressure relations in the simple straight-line compressor and in the double tandem compound straight-line machine.

22. Describe the duplex compressor, and define its advantages and disadvantages.

23. Explain the pressure relations in the duplex air-compressor.

24. Why are water jackets necessary on air-compressor cylinders?

25. Define the system of compound air-compression, and explain why different degrees of compounding are necessary at different pressures.

26. Describe the essentials of the successful intercooler.

27. Describe the advantages of compound air-compression.

REVIEW QUESTIONS

ON THE SUBJECT OF

COMPRESSED AIR

PART II

---

1. Describe the two classes of air-compressor valves, and explain the advantages and limitations of each.

2. Describe the five essentials of a successful air-inlet valve.

3. Describe the four essentials of a successful air-discharge valve.

4. Discuss the nine points affecting the mechanical efficiency of an air-compressor.

5. Discuss the causes of ignitions and explosions in air-compression, and state how these may be avoided.

6. Describe the two general classes of air-power plants, and state why certain classes of compressors are to be preferred for each.

7. Explain fully why dry air is desirable for transmission purposes.

8. What are the functions of the *antecooler* and the *aftercooler?*

9. State why a primary receiver is essential, and explain its proper installation.

10. What are the functions of the secondary receiver?

11. What are the requirements of a successful air-pipe line?

12. Describe the *loop system* of air transmission.

13. Describe the *dense-air system,* and state its advantages and limitations.

14. What is the theory of air reheating, and under what conditions is it desirable?

15. Describe *isothermal expansion* and *adiabatic expansion.*

16. Draw a diagram showing the isothermal, adiabatic, and practical curves of air compression and expansion, and explain the losses.

REVIEW QUESTIONS

ON THE SUBJECT OF

REFRIGERATION

PART I

---

1. State in what way the refrigerating machine was developed and describe the early forms of machines, giving approximate period when the first refrigerating machine was perfected.

2. Who invented the first practical machine used in commercial work, and of what type was it?

3. About when was the ammonia absorption process first invented and developed?

4. Define the term *refrigeration* and show the difference between a refrigerating machine and a heat engine.

5. What is the nature of heat and how is it measured? Define the two principal heat units in common use.

6. Explain the difference between temperature and the actual amount of heat, as stated in heat units, existing in a given substance.

7. What is a thermometer and how is it made? State the *fixed points* on the two thermometer scales in common use, and show how these are determined.

8. What is *absolute zero*, and why is it impossible to cool a substance to this point?

9. Define the terms *specific heat*, *latent heat*, and also define the unit of *plant capacity*.

10. Name the three ways in which heat transfers may be made.

11. State how *cold* is produced and show the various ways by which temperature reduction may be affected.

12. Define the terms *critical temperature* and *critical pressure*. What bearing do these factors, known as *critical data*, have on the value of a given substance for use in refrigerating work?

REVIEW QUESTIONS

ON THE SUBJECT OF

REFRIGERATION

PART II

---

1. Of what general type is the Triumph compressor, and how is the stuffing box of this compressor constructed and operated?

2. What is the essential feature in the operation of the Linde compressor?

3. Is any water jacket used with a Linde machine? If not, why not?

4. Are vertical machines usually of the single-acting or double-acting type, and why?

5. What is the distinguishing features in the construction of the Great Lakes compressor?

6. What are the main points of difference in the construction of compressor cylinders on machines using *ammonia* and for those using *carbon dioxide?*

7. What should be the length of stroke of a carbon dioxide machine as compared with the diameter of the cylinders?

8. Name the principal uses of a small refrigerating plant and state the advantages in the use of such machines for refrigeration instead of ice.

9. Name the several principal causes of loss in refrigerating compressors.

10. Name the principal types of ammonia condensers in common use and discuss the advantages and disadvantages of each.

11. Name two principal methods of distributing water over condensers. Show how an even distribution of the water is had.

12. Name the two principal types of cooling towers in use. Discuss the principle on which all such towers are operated.

REVIEW QUESTIONS

ON THE SUBJECT OF

REFRIGERATION

PART III

---

1. What is the first essential for an operator of a successful refrigeration plant?

2. How can the practical operating engineer tell whether the valves on his compressor are working properly?

3. Name three ways of detecting ammonia leaks in the different parts of a refrigeration plant.

4. State how you would proceed in case it became necessary to remove gases from the ammonia condenser.

5. How could you tell when the gases collected in the condenser had all been taken out?

6. Assuming the temperature of the supply water in a plant to be 65 degrees, how many heat units would have to be removed from a pound of water to freeze it into ice? Show how the result is obtained.

7. What are the three principal systems of ice manufacture in use at the present time?

8. Describe the *can* system of making ice and define what is meant by the term *tank surface.*

9. Name the different parts of the equipment in a *can ice plant.*

10. Name the parts of a complete distilling system for a can ice plant, and trace the course of steam from the time it leaves the engine until it is ready to go into the cans as distilled water.

11. Name two forms of steam condensers and describe the construction of the Triumph condenser.

12. Why is it necessary to reboil condensed water in a can ice plant?

INDEX

The page numbers of this volume will be found at the bottom of the pages; the numbers at the top refer only to the section.

Note.—For page numbers see foot of pages.

Note.—For page numbers see foot of pages.

Note.—For page numbers see foot of pages.

Note.—For page numbers see foot of pages.

Note.—For page numbers see foot of pages.

Note.—For page numbers see foot of pages.

Printed in the United Kingdom
by Lightning Source UK Ltd.
134840UK00001B/183/A

9 781408 600740